The Challenge of Change

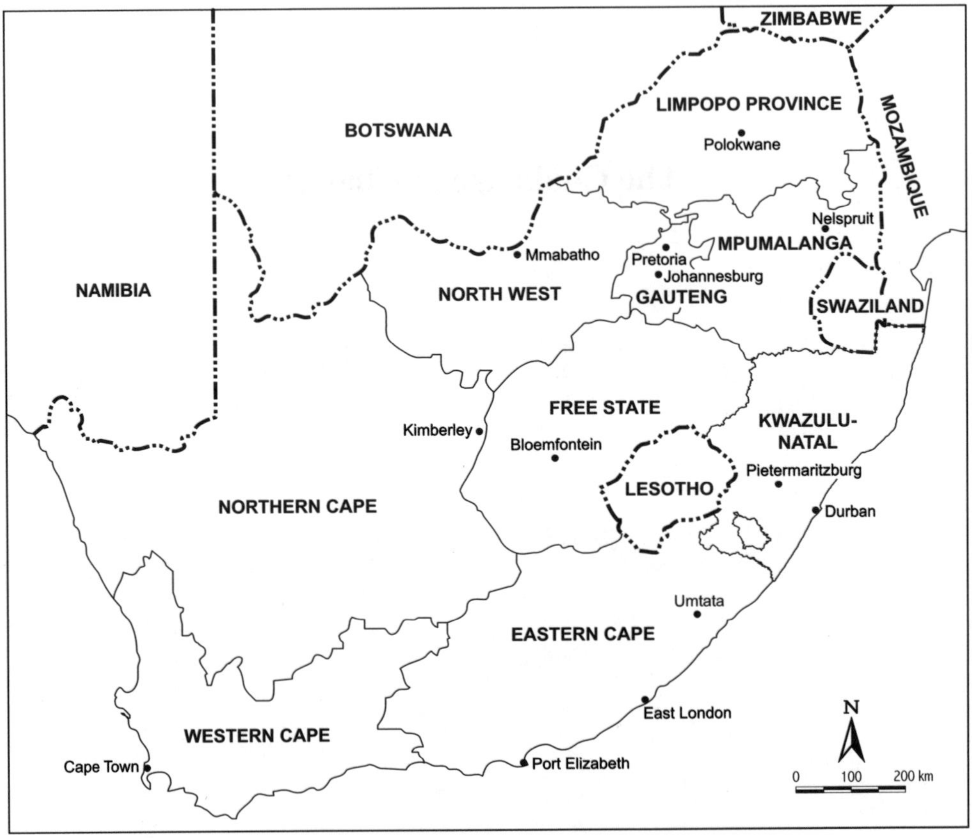

Map of South Africa

The Challenge of Change

Agriculture, Land and the South African Economy

Edited by
Lieb Nieuwoudt and Jan Groenewald

Sponsored by

DEPARTMENT: AGRICULTURE
REPUBLIC OF SOUTH AFRICA

UNIVERSITY OF NATAL PRESS
Pietermaritzburg

Published in 2003 by University of Natal Press
Private Bag X01
Scottsville 3209
South Africa
E-mail: books@nu.ac.za
Website: www.unpress.co.za

ISBN 1-86914-032-X

Cover design by Anthony Cuerden
Front cover photograph by Harry Lock

Typeset by Alpha Typesetters cc, New Germany
Printed and bound by Interpak Books, Pietermaritzburg

Contents

Foreword

Agriculture shaped the history and development of South Africa in the past, and can be expected to play a critical role in its democratisation and growth in the future. Agriculture supports the very existence and productive energy of the nation through food production; contributes to economic growth by generating incomes and providing employment; and presents a social welfare net to the most vulnerable members of our society in the rural areas. The National Department of Agriculture (NDA) is keenly aware of the uniquely strategic importance of the sector to the future of the country and the well-being of its people, and seeks to make others aware of this reality.

Over the years our agriculture has grown into a modern industry capable of feeding the nation and some of our neighbours, while also processing high-value products which earn the country valuable foreign exchange. South African agriculture is therefore deservedly classified among the world's most competitive sectors, and a major role player in international trade. However, as with all aspects of our country's bitter apartheid past, mainstream agriculture excluded the nations' black majority and was founded on dispossession and racial patronage. The Department, therefore, faces the challenge of maintaining the gains made in the past, while also promoting equitable participation and transformation for a better future.

South African agriculture was deregulated through the Marketing of Agricultural Products Act of 1996. The removal of statutory measures improved the technical efficiency of the sector, but also had some unintended negative consequences in so far as promoting black participation. At the same time our newly democratised country joined the World Trade Organisation, and the sector had to relinquish historical trade protection measures. Despite the initial hardships, the industry made a smooth transition into the international trade arena, and is presently regarded as one of the world's most free and influential. While globalisation has borne fruit for the industry, South Africa and others continue to push for the elimination of market access barriers, domestic support and export subsidies by developed countries.

The NDA is responsible for the national growth and development of the sector, and provides leadership to the nine provincial Departments of Agriculture, as well as the national parastatals responsible for research, financing and marketing. It has also drawn a sector plan with the country's two main agricultural unions: the National African Farmers' Union (NAFU) and Agricultural Union of South Africa (AgriSA). Taking the lead from government's service motto 'Batho Pele', which means putting people first, the NDA continues to engage all stakeholders, particularly those previously excluded, to unite a sector painfully splintered by apartheid.

It is our wish that all who read this book do not only become better informed about our industry, but are also motivated to actively contribute to its vital mission.

Bongiwe Njobe

Director-General, National Department of Agriculture

Foreword

This book on South African agriculture is published to coincide with the 25th international conference of agricultural economists hosted by South Africa for the first time in Durban in August 2003. It has always been a dream of the Agricultural Economics fraternity of South Africa to host the International Association of Agricultural Economics (IAAE) conference on South African soil. South Africa's involvement in the IAAE goes back many years, and South Africans have attended all conferences apart from two when visas were refused as a result of isolation during the apartheid years. The enthusiasm with which the South Africans engaged in many of the IAAE meetings led to them gaining four seats (max) on the Council. This was followed by a submission in 1990 to organise an IAAE inter-conference symposium in southern Africa at Swakopmund, Namibia, and succeeded in bringing together more than 80 agricultural economists throughout Africa. In 1994 South Africans supported the IAAE's 22nd conference in Harare in large numbers and contributions. The dawning of a new South African nation also provided a new status to South African agricultural economists. With the full support of the African group, a proposal to the Council at the 1997 conference in Sacramento to host the 2003 conference in South Africa was enthusiastically accepted by the IAAE Council. As a prelude to this event, South Africa hosted an inter-conference symposium in 1998 at Badplaas, Mpumalanga. The year 2003 has finally arrived and the Agricultural Economic Association of South Africa is proud to host this prestigious event. Indeed, dreams do come true!

The book is distributed to all delegates as a memento of the South African IAAE conference, but also serves as an overview of South African agriculture and its economic realities and policy challenges. This is a collective effort by leading researchers in the agricultural economics profession in South Africa under the leadership of Lieb Nieuwoudt and Jan Groenewald. They did a tremendous job in lining up the authors and getting the copy proofread and edited in time for the publication. Jan and Lieb are household names in agricultural economics in South Africa and most of the contributors to the book are their former students. Their effort in putting this project together is applauded and it also demonstrates why they are so respected by the profession in South Africa.

The fact that this book serves as an important benchmark in that it reviews policy and challenges in the agricultural sector, links up with the launch of a strategic plan for South African agriculture in 2002, which intends to remove dualism in the South African agriculture sector and aims to promote a vibrant, equitable, competitive and sustainable sector. The sector is going through challenging times as it deals with the process of black empowerment and land reform, but at the same time has to increase its international competitiveness. We hope this book serves as yardstick against which the sectors' performance can be measured in 10 year's time.

Johann Kirsten
President, Agricultural Economics Association of South Africa, 2001–03

Acknowledgements

We would like to thank Glenn Cowley, publisher at the University of Natal Press, Sally Hines, editor, and the rest of the University Press team who pulled together to make this book happen in a short space of time. We are grateful to Edgar Bruggemann for making useful suggestions to improve the text.

Appreciation is also expressed to Mark Darroch, Gerald Ortmann and Mike Lyne for assisting in reviewing chapters and to Christel Barnard for providing secretarial services.

Fact File

Grazing land (million hectares)	83.93
Potentially arable land (million hectares)	16.74
Nature conservation (million hectares)	11.79
Forestry (million hectares)	1.43
Irrigated land (million hectares)	1.35
Population (2001) (million)	44.56
Gross value (R million 2000–01)	
Field crops	16 796
Horticulture	12 708
Animal products	19 486
Agricultural exports (R million, 2000)	15 820
Agriculture's contribution (%) to GDP (2000)	3.2
Average rainfall (mm)	497
% of country receiving less than 500 mm	65
Large-scale commercial farms (number)	50 000
Small-scale farmers (number)	240 000
Expected inflation rate (%) (June 2003)	7.7
Expected mortgage bond rate (%) (June 2003)	15.5
Farming debt as a % of assets	33.6

US$1=R8 (June 2003)

Abbreviations

ACP	African, Caribbean and Pacific
AE	adult equivalent
AGOA	Africa Growth and Opportunity Act
AgriSA	Agricultural Union of South Africa
AGROCON	agricultural outlook conferences
AMD	Agricultural Markets Division
ANC	African National Congress
ARC	Agricultural Research Council
BASIS CRSP	Broadening Access and Strengthening Input Market Systems Collaborative Research Support Program
BATAT	Broadening Access to Agriculture Thrust
BEE	Black Economic Empowerment
BLNS	Botswana, Lesotho, Namibia and Swaziland
BLS	Botswana, Lesotho and Swaziland
BTT	Board on Tariffs and Trade
CAP	Common Agricultural Policy
CBI	Cross-Border Initiative
CBPWP	Community Based Public Works Programme
CCA	Common Customs Area
CCSP	Cane Contractor Support Programme
CGE	computable general equilibrium
CGIAR	Consultative Group on International Agricultural Research
CMA	Common Monetary Area
CMIP	Consolidated Municipal Infrastructure Programme
COMESA	Common Market for Eastern and Southern Africa
CPI	Consumer Price Index
CSIR	Council for Scientific and Industrial Research
CV	coefficient of variation
DACST	Department of Arts, Culture, Science and Technology
DANIDA	Danish Agency for Development Assistance
DLA	Department of Land Affairs
DPRU	Development Policy Research Unit
DRC	Democratic Republic of the Congo
DRC	domestic resource cost
DTI	Department of Trade and Industry
DWAF	Department of Water Affairs and Forestry
EPA	Economic Partnership Agreement
ESTA	Extension of Security of Tenure Act
EU	European Union

EU-SA TDCA	European Union-South African Trade, Development and Co-operation Agreement
FAO	Food and Agricultural Organisation
FARA	Forum for Agricultural Research in Africa
FTA	Free Trade Agreement
fte	full-time equivalent
FWES	farm-worker equity sharing
GATT	General Agreement on Tariffs and Trade
GDP	Gross Domestic Product
GEIS	General Export Incentive Scheme
GM	genetically modified
GMOs	genetically modified organisms
GNP	Gross National Product
GSP	Generalised System of Preferences
IED	Income Equalisation Deposit
IMF	International Monetary Fund
IMT	irrigation management transfer
ITCA	Inter-departmental Committee on Agriculture
IWMI	International Water Management Institute
LCRF	Land Reform Credit Facility
LDC	Least Developing Countries
LFAs	Local Farmers' Associations
LRAD	Land Redistribution for Agricultural Development
MCC	Mill Cane Committees
MFN	most favoured nation
MGLC	Mill Group Local Committees
MINMEC	Minister Members Executive Council
MOU	memorandum of understanding
NAFTA	North American Free Trade Agreement
NAFU	National African Farmers' Union
NAMC	National Agricultural Marketing Council
NARF	National Agricultural Research Forum
NARS	South African National Agricultural Research System
NDA	National Department of Agriculture
NEDLAC	National Economic Development and Labour Council
NEPAD	New Partnership for Africa's Development
NFI	net farm income
NGO	non-governmental organisation
NRF	National Research Foundation
NWA	National Water Act
NWRS	National Water Resources Strategy
OECD	Organisation for Economic Co-operation and Development
PAMs	Policy Analysis Matrices
PDA	Provincial Department of Agriculture
PDLAs	Provincial Departments of Land Affairs
PIP	Presidential Imperative Programme
PL	poverty line

PPP	purchasing power parity
PRIIF	Poverty Relief and Infrastructure Investment Fund
PSE	Producer Subsidy Equivalent
PSUs	primary sampling units
PTA	Preferential Trade Area
R&D	research and development
RDP	Reconstruction and Development Programme
RTAs	Regional Trade Agreements
RV	recoverable value
SASA	South African Sugar Association
SABS	South African Bureau of Standards
SACCAR	Southern African Centre for Co-operation in Agricultural and Natural Resources Research and Training
SACGA	South African Cane Growers' Association (Canegrowers)
SACU	Southern African Customs Union
SADC	Southern African Development Community
SADCC	Southern African Development Co-ordination Conference
SAFEX	South African Futures Exchange
SAM	Social Accounting Matrix
SASEX	South African Sugar Association's Experiment Station
SGDT	Small Grower Development Trust
SLAG	settlement/land acquisition grant
SLIP	Sugar Logistics Improvement Project
SPS	sanitary and phyto-sanitary
SSA	Sub-Saharan Africa
SSGs	small-scale growers
THRIP	Technology and Human Resources for Industry Programme
TFP	Total Factor Productivity
UAF	Umthombo Agricultural Finance
UNCTAD	United Nations Conference on Trade and Development
USA	United States of America
USAID	United States Agency for International Development
VAT	value added tax
W&S	Wine and Spirits
WRC	Water Research Commission
WTO	World Trade Organisation

Contributors

Gerhard Backeberg: Director, Water Research Commission, Pretoria

Richard Bates: Independent Consultant, Durban, formerly Manager of the Financial Aid Fund (FAF)

Gerhard Coetzee: Director, Finance and Enterprise Group, Ebony Consulting International (Pty) Ltd and Extraordinary Professor, University of Pretoria

Mark Darroch: Senior Lecturer, Agricultural Economics, University of Natal

Tommy Fényes: Professor, Economics, Vista University, Pretoria

Gavin Fraser: Professor, Agricultural Economics, University of Fort Hare

Jan Groenewald: Professor, Agricultural Economics, University of the Free State

Sheryl Hendriks: Director, Food Security Programme, School of Agricultural Sciences and Agribusiness, University of Natal

André Jooste: Senior Lecturer, Department of Agricultural Economics, University of the Free State

Johann Kirsten: Professor, Agricultural Economics, University of Pretoria

Frikkie Liebenberg: Manager, Policy and Planning, Group for Development Impact Analysis, ARC, Pretoria

Mike Lyne: Professor, Agricultural Economics, University of Natal

Charles Machethe: Professor, Agricultural Economics, University of Pretoria

Moraka Makhura: Rural Policy Analyst, Development Bank of Southern Africa

Nico Meyer: Project Economist, Development Bank of Southern Africa

Madime Mokoena: Senior Researcher, Human Sciences Research Council

Nomakhaya Monde: Agricultural Economist, Eastern Cape Department of Agriculture and Land Affairs, Dohne

Simphiwe Ngqangweni: Senior Lecturer, Agricultural Economics, University of Pretoria

Lieb Nieuwoudt: Professor, Agricultural Economics, University of Natal

Gerald Ortmann: Professor, Agricultural Economics, University of Natal

Howard K. Sigwele: Agricultural Economist, Botswana

Patrick Sokhela: Managing Director, Makatini Cotton (Pty) Ltd

Wim van Averbeke: Professor, Pretoria Technikon

Herman van Schalkwyk: Professor, Department of Agricultural Economics, University of the Free State

Nick Vink: Professor, Agricultural Economics, University of Stellenbosch

PART 1

HISTORIAL LEGACY AND PRESENT SITUATION

CHAPTER 1

Agriculture in the National Economy

Nick Vink and Johann Kirsten

1.1 INTRODUCTION

The three most important economic features of South African agriculture are its dualistic structure (commercial and subsistence sectors), the process of deregulation of commercial agriculture that has taken place over the past two decades, and the attempts to 'deracialise' the sector since 1994. These features must be seen against the background of the country's resource endowment. Of the 100.6 million hectares of agricultural land, only some 14% receives enough rainfall for arable farming, while the remainder is used for extensive grazing (83.9 million hectares), forestry and nature conservation. Only 1.35 million hectares of the arable land available is irrigated, and yields at least a third of total agricultural output (Directorate: Agricultural Statistics, 2002).

Some 60 000 commercial farmers occupy almost 87% of the total agricultural land in the country, and produce more than 95% of marketed output. In contrast, African smallholder farmers are found mostly in the former homeland areas, which make up some 13% of the agricultural land.[1] These areas were established under the well-known Land Acts of 1913 and 1936, and are characterised by traditional forms of land tenure, which were regulated by a series of laws and regulations, mostly proclaimed in terms of the Black Administration Act of 1927.

The purpose of this chapter is to describe the performance of this important sector of the South African economy. The chapter first describes the main policies that influence agriculture, and then analyses recent macroeconomic trends in the country. A detailed description of some key facets of agricultural sector performance follows, and a concluding section summarises agriculture's current position in the South African economy.

1.2 AGRICULTURAL AND OTHER POLICIES AFFECTING SOUTH AFRICAN AGRICULTURE IN THE 1990s

The evolution of the South African agricultural sector during the twentieth century is characterised by periods of structural change, starting with the initial steps aimed at the territorial segregation of white and black farmers, and followed by a period of

[1] This despite land reform initiatives over the past eight years, that are discussed in the text.

government support to agriculture and growth in agricultural output up to the early 1980s. This led to the commercialisation of white farming through the adoption of modern mechanical and biological technologies, resulting in consistent growth in output within a policy environment heavily favouring increased production by large-scale owner-operated farms using wage labour. The policy environment of racial discrimination and price distortions could, however, not be sustained and the pressures on agriculture began to mount during the 1980s, forming a prelude to the even more dramatic policy shifts that occurred after the democratic elections in 1994. The most important post-elections policy initiatives included the promulgation of new legislation relating to marketing policy for farm commodities and to water policy, land reform, institutional restructuring in the public sector, and trade policy and labour market reform.

1.2.1 Marketing policy changes

Until early in 1998, the marketing of most agricultural products in South Africa was regulated by statute, largely under the 22 marketing schemes introduced from 1931 and especially from the time of the 1937 Marketing Act. The Marketing of Agricultural Products Act, 47 of 1996, changed the way in which agricultural marketing policy was managed, not least by opening the sector to world market influences in a manner that could hardly have been anticipated a decade earlier. For example, the powers given to the Maize Board to set prices for local maize producers were ended when the Board was abolished. The Act set up the National Agricultural Marketing Council (NAMC), whose task was to dismantle the existing commodity Control Boards, and subsequently to manage and monitor limited state intervention in the sector.

1.2.2 Promulgation of the new Water Act, 36 of 1998

Changes resulting from the implementation of the new Water Act (a process described by Carter, 1996) that are expected to impact most severely on agriculture include: the higher priority afforded to human water use (including preferential access for smallholder farmers) and the environment; the termination of riparian rights; the implementation of an integrated catchment management system; the termination of subsidised water prices; and greater cross-border co-operation between southern African countries. Slow progress in the implementation of the Act has, however, limited the impact to date.

1.2.3 Land reform initiatives

The Department of Land Affairs completed the process of land reform policy design with its White Paper in 1996 (Department of Land Affairs, 1997), while implementation had started in 1994. Land reform was to consist of land restitution, land redistribution and tenure reform programmes. The Department initially spent much time and effort in mobilising communities and assisting them in accessing government settlement/land acquisition grants (SLAGs) to acquire land. However, in most cases, farms financed with these grants and settled by groups of households were too small to support all the beneficiaries as full-time farmers, hence raising concern about the sustainability of the process and the slow pace of land reform. After eight years of state sponsored land reform, less than one million hectares of the available agricultural land was transferred through the above programmes, while market transactions have resulted in twice the area of land being transferred to black owners (Graham & Lyne, 1999; Kirsten & Van Zyl, 1999).

Thus, a new approach to land reform was implemented from 2001 to replace the SLAG programme with the Land Redistribution for Agricultural Development (LRAD) programme (Ministry of Agriculture and Land Affairs, 2001), which provided for an extended scale of grants, dependent on an increasing own equity contribution from emerging commercial farmers. Unlike the SLAG programme, there is no means test for the LRAD programme, implying that aspirant commercial farmers who could not access SLAGs can now access LRAD grants. This approach fits directly with the new vision of the Ministry of Agriculture and Land Affairs to assist in establishing a class of commercial black farmers. Concerns have, however, been raised about how LRAD is being implemented, especially with regard to incentives for agents to broker land transactions (see Chapter 4). The success of LRAD post land redistribution will also depend on farmers having access to support services such as input supplies, markets, research, extension, finance, information and infrastructure, some of which would be provided by the private sector.

1.2.4 Institutional restructuring in the public sector

One of the main features of South African agricultural policy in the 1990s was the extent of the institutional restructuring that took place. There were three main reasons for these processes: Firstly, some institutions (for example, the Development Bank, the Land Bank, the Agricultural Research Council, the Development Corporations in the former homelands, etc.) that were seen as being too closely aligned with apartheid policies aimed at 'development' of the former homeland areas or at favouring commercial farmers (see Callear & Mthethwa, 1996; Development Bank of Southern Africa and Land & Agricultural Policy Centre, 1997) were subjected to restructuring programmes to realign them to a new mandate in support of the development priorities of the new government.

Secondly, the public sector agencies supporting the agricultural sector were subjected to the same processes of 'provincialisation' that came about with the adoption of the Interim Constitution. During this process a change was also effected in the relationship between the national and provincial Departments of Agriculture and farmer lobby groups. Thirdly, agricultural institutions in the public sector were reoriented to fit with new policy directions, in particular relating to the changes to agricultural marketing policy described above.

1.2.5 Labour market policy

Until the 1980s, farm workers in South Africa had little legal protection, inter alia, of their rights to organise and to basic conditions of employment. The Agricultural Labour Act, 147 of 1993, addressed this shortcoming to some extent, but it was only after 1994 that farm worker rights were brought in line with workers elsewhere in the economy. Four major labour laws in South Africa, including the Labour Relations Act (1995), the Basic Conditions of Employment Act (1997), the Skills Development Act (1998) and the Employment Equity Act (1998), now also apply to the agricultural sector. In addition, a minimum wage for farm workers, differentiated by higher-income and lower-income regions, was implemented in March 2003.

1.2.6 Trade policy reforms

Quantitative restrictions, a multitude of tariff lines, a wide dispersion of tariffs, and formula, specific and *ad valorem* duties and surcharges, characterised South Africa's trade regime before 1994. In agriculture, quantitative restrictions, specific duties, and price controls, import and export permits and other regulations were replaced by tariffs after South Africa became a signatory to the Marrakech Agreement in 1994. At present, however, virtually all tariffs are below the bound rates of this Agreement.

In South Africa, the average tariff cascades from a relatively high rate on consumer goods to moderate on intermediate goods and low on capital goods. This pattern, which is typical of protection in many developing countries, implies that less progress has been made in rationalising *effective* protection. While certain manufacturing industries have benefited directly from such support, traditional agricultural export sectors such as wine have been able to base their export strategies on growth in a protected domestic market.

The three most important trade relations in the southern African region include the Southern African Customs Union (SACU), which exhibits the deepest level of integration, the Southern African Development Community (SADC), and the South Africa–Zimbabwe bilateral agreement. Of the extra-regional influences, the Lomé (and now Cotonou) Agreement, the Africa Growth and Opportunity Act (AGOA) of the United States of America (USA), and South Africa's separate bilateral agreement with the European Union (EU) are most influential.

1.2.7 State funding to agriculture

State spending on the farm sector, measured as the budgeted amounts for the national Department of Agriculture plus the agricultural budgets of the nine provinces, amounted to R2.5 billion in 2001. In real terms, this was less than 45% of the budget of the Department of Agriculture plus that of the budgets of the former homeland departments in 1988. The decline in state spending in agriculture is also illustrated by the rapid decline in government funding of agricultural research. Base line funding for agricultural research provided by government through the parliamentary grant system dropped from a high of R337 million in 1997–98 to R262 million in 2001–02 – equivalent to only 55% in real terms of the parliamentary grant it received in 1992.

1.2.8 State support for rural infrastructure

Accurate contemporary data on the impact of government sponsored rural development programmes on the livelihoods of rural people are not readily available. A recent publication (Everatt & Zulu, 2001) shows, for example, that three of the largest rural development programmes in South Africa can hardly be said to have affected the lives of many people:

(a) The Community Based Public Works Programme (CBPWP) has delivered a total of 2 830 items of infrastructure, mostly to rural communities, since 1994 – a rate of only 470 per year, or less than two projects per magisterial district per year.
(b) A total of only R1.3 billion was spent in rural areas under the Consolidated Municipal Infrastructure Programme (CMIP) to September 2000.
(c) The most recent data available show that, of the mere R253 million available to the Poverty Relief and Infrastructure Investment Fund (PRIIF) of the then Department of Welfare in 1998–99, some 72% was spent in rural areas, while some 32% was spent on agricultural projects.

Nevertheless, a range of basic services have been provided in South Africa's rural areas:

(a) A free basic water policy has been introduced, giving four million more South Africans access to a basic amount of water free of charge, bringing the total people reached since 1994 to over seven million.
(b) In the five calendar years to the end of 1998, 2.3 million electricity grid connections were made. In the three years since 1999, 1.2 million new connections were made, bringing the total to 3.48 million.
(c) In 2001, over 20% of the national budget was allocated to education. Combined with improved learning and teaching, one result is that the end-of-school pass rate increased to 61% in 2002.
(d) The number of people benefiting from the Child Support Grant grew from 28 000 about two years ago to more than 1.1 million in July of 2002. About 65 000 children gain new access to the grant every month. Government planned to register three million children for the Grant by 2005, but this target will be met by 2003.

When added to the financing of municipal infrastructure by the Development Bank and the increase in the real value of pensions and other welfare payments, these amount to potentially substantial transfers to rural people.

1.2.9 The strategic plan for agriculture

Agricultural policy in the early part of the twenty-first century will largely be driven by a new strategic plan for South African agriculture released in November 2001. The plan was tabled through a collective effort between commercial farmers, the National African Farmers' Union (NAFU) and the national Department of Agriculture following an instruction from President Thabo Mbeki.

The strategic plan highlights the vision for the agricultural sector as 'a united and prosperous agricultural sector'. This vision implies sustained profitable participation in the South African agricultural economy by all stakeholders, recognising the need to maintain and increase commercial production, to build international competitiveness and to address the historical legacies and biases that resulted in skewed access to land and other resources, and skewed representation in all other aspects of the sector. In addressing the challenges facing the sector and achieving the vision, the strategic plan will consist of three core strategies, namely to:

(a) enhance equitable access and participation in the agricultural sector;
(b) improve competitiveness; and
(c) ensure sustainable resource management.

These core strategies are to be supported by enabling strategies aimed at ensuring:

(a) good governance;
(b) integrated and sustainable rural development;
(c) knowledge and innovation;
(d) international co-operation; and
(e) safety and security.

The need to integrate black farmers into mainstream agriculture is recognised as important, and already a number of initiatives have been launched to achieve this goal. It is also noteworthy to realise that the empowerment process should take place in an industry that remains competitive and sustainable. These are the challenges for South African agricultural policy in the immediate future.

1.3 TRENDS IN MACROECONOMIC PERFORMANCE IN SOUTH AFRICA

The South African economy grew at a relatively high rate (average annual real GDP growth above 3.5%) for the first 35 years after 1945, although the rate of growth had already started to decline after 1970. The growth rate slowed significantly during the 1980s and the first half of the 1990s, but increased to above 2% per year since the democratic elections of 1994. This trend is commonly described as the real cost of apartheid, and the social and political instability that preceded the first democratic elections in the country in 1994. However, it is also clear that the decline in the contribution of minerals to the economy, especially gold after the gold price boom of the late 1970s, was a major contributing factor (Maasdorp, 2002). Real GDP per capita, measured in Rand[2] terms, peaked in 1981 (when it was R16 347), but started to increase again after 1994 to reach R14 321 in 2001. Part of the reason for the recent increase in per capita income is the rapid decline in the population growth rate, which peaked at about 3% per year in the 1960s, increased at about 2.4% per year from 1980–96, and has since declined to less than 1% (Maasdorp, 2002).

This increase in per capita income, together with a decline in the real price of purchasing a typical basket of food on a household level (Ebony Consulting International, 2001), has resulted in an increase in spending on non-durable goods (mostly food) in the economy as a whole. While this has benefited agriculture, it must be noted that the cost of purchasing a scientifically determined, nutritionally balanced basket of food is greater than most South African rural households can afford (ECI, 2001).

Exports have become one of the main components of economic growth in the South African economy. Total exports increased almost fourfold from under R70 billion in 1991 to over R250 billion in 2001 (Department of Trade and Industry, 2002). In addition, the composition of the country's export portfolio has changed rapidly. The contribution of value-added manufactured goods has been responsible for all of the increase in exports, doubling its share from 16% to almost 34% since 1992, while the share of raw products (principally minerals) has declined from almost 58% to less than 42%, and the contribution of semi-manufactured goods has remained constant. At the same time, the composition of the import portfolio remained relatively constant, with the share of manufactured imports declining slightly and the share of raw material imports increasing by about 3%. During this period total imports increased from R46 billion to R216 billion, with the result that the trade surplus increased (DTI, 2002).

This growth performance has been accompanied by a decline in inflation, a weakening of the exchange rate, and an increase in the rate of unemployment. Inflation[3] increased from below 5% per year in the early 1960s to a peak of above 20% in the mid-1980s. The rate declined to below 5% in 2000, and has subsequently increased to above 10%. The recent increase (since December 2001) has partly been the result of an unprecedented increase in food prices, triggered by the collapse in the external value of the currency in the second half of 2001 (Vink & Kirsten, 2002). Nevertheless, the strengthening of the Rand in the second half of 2002 has resulted in a consensus expectation that the rate of inflation will decline to closer to the official target rate of between 3% and 6% during 2003. The external value of the Rand,

[2] South African Rand. The exchange rate in 2002 was approximately US$1 = 1Euro = R10.00, while the currency has appreciated by more than 10% since then. Because of the instability of the exchange rate, that it has substantially weakened over the past decade, and the wide difference between the purchasing power parity (PPP) and the nominal exchange rate, as reported by the World Bank, US$ rates are used only when necessary in this chapter.

[3] These trends represent prices in the metropolitan areas. While prices are known to be higher in non-urban areas, there is no evidence to suggest that the non-urban rate of inflation is higher (Vink & Kirsten, 2002). Metropolitan rates are, therefore, used as a proxy for the overall rate of inflation.

measured in real terms on a trade-weighted basis, declined from an index of above 130 in the mid-1980s, stabilised up to the mid-1990s, and has since lost a further 20% of its value (DTI, 2002). The nominal value of the Rand declined from above parity to the US$ in the early 1970s to its current rate of approximately US$1=R9.00, after reaching a low of approximately US$1=R14.00 at the end of 2001.

These changes in growth and the internal and external value of the Rand have been accompanied by a decline in the overall rate of non-agricultural employment in South Africa (agricultural sector employment is discussed in section 1.4.3). The overall rate of employment grew at less than the rate of population growth until 1990, and then declined in the 1990s by almost 2% per year. This decline was spread throughout the economy: employment in the mining sector declined by 45.6% from 1989–99, in the manufacturing sector by 18.3%, while more than 100 000 jobs were lost in the government sector. The structure of the labour market is illustrated in Table 1.1, where it is also evident that the decline in employment has now stabilised.

Table 1.1: Labour market trends in South Africa using the official definition of unemployment, 2001–02.

		September 2001	*February 2002*
A	Total employed	10 833 000	11 393 000
B	Total unemployed (official definition)	4 525 000	4 738 000
C	Total economically active = A + B	15 358 000	16 130 000
D	Total not economically active	12 006 000	11 543 000
E	Total aged 15–65 years = C + D	27 365 000	27 673 000
F	Official unemployment rate = (B/C) x 100%	29.5%	29.4%
G	Labour market participation rate = (C/E) x 100%	56.1%	58.3%
H	Labour absorption rate = (A/E) x 100%	39.6%	41.2%

Source: Statistics South Africa (2002)

These trends in growth, inflation, the exchange rate and employment are the result of the macroeconomic policy regime followed by the new democratic government from 1994. The major monetary and fiscal policy instruments used have been interest rates and actions to reduce the budget deficit, respectively. While negative real interest rates were maintained in the late 1970s and early 1980s, and again from 1987–89 (refer to Chapter 13, Figure 13.1 for trends in real interest rates), the average lending rate of banks to their best clients (the prime overdraft rate) has been above 10% since about 1974. The deficit before borrowing and debt repayment of the national government as a ratio of GDP amounted to 1.5% in fiscal 2001–02, compared with 2% in the previous fiscal year. The ratio has declined consistently since reaching a peak of 7.3% of GDP in 1993 (Strydom, 2002). These trends have affected agriculture primarily through the exchange rate, the level of interest rates and the inflation rate.

1.4 AGRICULTURE IN THE SOUTH AFRICAN ECONOMY

The annual value of commercial agricultural production in South Africa is now about R45 billion, while the GDP contribution (value added) is around R25 billion. Nominal growth in agricultural production has been about 11.5% per year since

1965. However, the economy as a whole has grown by 14.4% per year over the same period. The result is that agriculture's share of GDP has declined from 9.1% in 1965 to 3.1% in 2001. While this is lower than the average contribution of 9% for middle-income countries (World Bank, 2002), the combined contribution of the primary sectors to GNP in South Africa was 10.5% in 2000, given the 7.3% contribution of the mining sector. The contribution of the latter was 19.7% in 1965, and reached a peak of above 25% in 1980. When upstream and downstream industries are added, the GDP contribution comes to some 13%.

1.4.1 The composition of agricultural output

Given that most of the country is unsuited to cultivation, livestock production accounts for the largest component of output (44% in 2000). Field crop production (31% of total output in 2000) declined from almost half of total output in 1978–79 to less than a third in 2000. During this time animal production increased its share somewhat, while horticultural production increased its share by 9%. The former is due to the increase in intensive livestock production, while the latter is due to the liberalisation of foreign trade and the country's formal re-entry into international markets.

While agricultural exports have grown rapidly, especially from 1990, agricultural imports have grown even faster, partly as a result of the rapid reduction in tariffs to below the bound rates of the Marrakech Agreement. The ratio of exports to imports has, therefore, declined substantially, partly because of the increase in manufactured food products (Vink *et al.* 2002). While there has been rapid growth in agricultural trade with the rest of the world, the country's total exports and imports have been growing even faster.

1.4.2 Investment in agriculture

Following the processes of deregulation described above, real gross domestic fixed investment in agriculture increased by 24% in real terms in 1996 over 1995, while investment figures in 1997 were 9% lower than 1996 but still up on 1995 levels by 13% (Directorate: Agricultural Statistics, 2002). Investment (or gross capital formulation) in agriculture declined between 1997–99 due to a few poor seasons and some elements of rural insecurity. Investment increased again in 2000 and 2001, with the gross capital formation in 2001 back at nominal levels similar to that in 1996.

1.4.3 Employment in agriculture

1.4.3.1 Employment trends

Recent research on the labour market in South Africa shows that there has been a considerable redistribution of income to African, Coloured and Indian households over the past decade, with economic growth identified as the greatest source of income for redistribution. However, the gap between rich and poor households has not narrowed (i.e. most redistributed income accrued to wealthier households), and is not expected to narrow in the near future. As a result, the rural poor, including farm and other rural workers and the rural unemployed, have not (in aggregate) benefited, nor are they expected to benefit in the absence of state intervention. Further, while the wages of less skilled workers rose relatively faster than average wages from the mid-1980s, rural workers were generally excluded from this increase in unskilled wage levels, as they were not unionised. Table 1.2 shows the most recent macro level data on farm employment in South Africa, and indicates that the sector shed about 180 000 regular employees, and about 210 000 casual and seasonal workers, between 1985–96.

Table 1.2: Farm employment in South Africa, 1985–96.

	1985	*1990*	*1991*	*1992*	*1993*	*1994*	*1995*	*1996*
Regular	807 341	728 414	702 323	656 772	647 839	625 244	628 925	625 451
Casual, seasonal	516 411	456 262	413 239	394 425	491 588	302 185	289 810	304 690
Total	1 323 694	1 184 676	1 115 562	1 051 197	1 139 427	927 429	918 735	930 141

Source: Statistics South Africa (2002): Agricultural censuses and surveys

Agricultural employment has declined at a slower pace than employment in the economy in general. A higher economic growth rate over the past two decades may have resulted in a less pronounced downward trend in employment. During the 1980s, seasonal employment increased until 1987, then dropped sharply in 1992, and then showed a sufficiently large increase in 1993 to cause an increase in overall employment in the sector. The category of casual and seasonal employees is notoriously difficult to estimate, so this increase may be no more than a measurement error. On the other hand, the large increase in exports of fruit (the sector that is the largest user of casual and seasonal labour) that occurred during this period may have resulted in an increase in jobs.

Research also shows that the decline in total employment has been accompanied by a shift in the skills composition of the labour force. Table 1.3 shows this shift between 1970–95, and between 1995–99, for agriculture relative to the shift for the whole country.

The following trends are evident from these data:

(a) The total level of employment has increased in South Africa by some one million jobs, or about 12%, since 1995, although the total increase has been smaller than the growth in the economically active population. Agriculture lost 50% of its employment between 1970–95.

(b) During the process of declining total employment in agriculture, there has been a substantial shift in the nature of jobs in favour of more skilled workers. The employment of managers, for example, increased by more than six fold since 1970, and by almost 25 000 people, or 370%, since 1995, while the employment of professionals and technicians increased by 150.4% since 1970.

(c) This shift in the skills profile of farm workers has favoured male workers over female workers, and Coloured workers over African workers. The number of male farm workers, for example, decreased by 37.9% from 1970–95, and of women by 71.9%. Simultaneously, the number of African workers declined by 58.8%, while the number of Coloured workers increased by 88.4% over the 25-year period.

Table 1.3: Shifts in the skills base of agricultural workers versus the total skills base in South Africa, 1975–99.

	Agriculture			*Total*		
	1970–95 (% change)	*1995–99 (% change)*	*Number added (1995–99)*	*1970–95 (% change)*	*1995–99 (% change)*	*Number added (1995–99)*
Managers	633.2	369.65	24 301	272.0	37.81	190 789
Professional	150.4	257.61	1 744	311.9	72.55	234 735
Technicians		28.93	899		0.28	2 969
Clerks	281.7	–8.63	–1 045	45.1	–4.49	–50 648
Sales		109.01	9 414		16.97	180 634
Skilled agriculture*		218.88	216 531		343.51	395 500
Crafts	64.5	97.64	14 259	–4.4	25.15	275 313
Operators		2.69	3 469		1.05	11 434
Elementary*	–58.3	–34.27	–311 897	–54.2	–33.60	–974 438
Unspecified	92.9	377.81	2 928	466.2	68.41	75 381
Total	–50.1	–3.33	–39 397	17.6	12.04	1 131 647

Source: Adapted from Bhorat (2000); Bhorat (2001)

* It is possible that there was a reclassification of 'elementary' workers as 'skilled agricultural workers' between 1995 and 1999. If this were the case, farm employment would still have dropped by 95 366 workers.

1.4.3.2 Trends in wage distribution

The wage distribution by industry in South Africa shown in Table 1.4 indicates that farm workers and domestic workers earn the lowest wages among those formally employed in the country, while the financial services and electricity sector workers are paid the highest (this remains true even when the 20% estimated in kind payment is added to the agricultural wage). The skewed distribution of farm wages is also evident, as the ratio of mean to median income in agriculture of below 50% is not found in any other sector of the economy. The difference in farm wages between areas could be due to payments in kind and a lack of labour mobility.

Table 1.5 shows the relative rate of increases in real wages in the South African economy from 1970–98. The growth rate in agricultural wages was higher than the average for the economy, and higher than all sectors except for mining. Real hourly wages in agriculture grew by 46.14% from 1970–98, at an average rate of 1.6% per year. Despite this increase in the real wage, the unit cost of labour, measured as the

Table 1.4: Mean and median hourly wages by industry in South Africa, 1997.

Industry	*Mean (Rand)*	*Median (Rand)*	*Median as % mean*	*Agriculture as % of sector*
Agriculture	3.57	1.68	47	100.00
Mining	11.95	7.67	64	29.87
Manufacturing	13.07	8.73	67	27.31
Electricity	16.08	11.11	69	22.21
Construction	9.83	6.39	65	36.32
Trade	10.77	7.07	66	33.15
Transport	14.37	10.16	71	24.84
Finance	18.26	11.46	63	19.55
Services	17.87	13.85	78	19.97
Domestic	4.10	2.60	63	87.10

Source: Statistics South Africa (2002)

ratio of the total cost of labour to the total value of output, has remained relatively stable over time. In 1970, 16 cents was spent on labour for every R1 of output produced. This decreased to 13 cents in 1980, increased to 19 cents in 1994 and decreased to 17 cents in 1998.

Table 1.5: Growth rate in real wages per sector in South Africa, 1970–98 (%).

Sector	*Total increase*	*Average annual real growth (%)*
Agriculture	46.14	1.65
Total mining	105.20	3.76
Total manufacturing	5.17	0.18
Total services	27.22	0.97
Total economy	40.55	1.45

Source: Based on wage data from Statistics South Africa (2002)

1.5 THE LOCATION OF AGRICULTURAL PRODUCTION

Agriculture (including nature conservation and forestry) uses 82.3% of the total land in South Africa, ranging from as low as 44% in Gauteng (the metropolis centred on Johannesburg) to over 70% in seven of the other provinces. More importantly, grazing land (natural pastures in desert and semi-desert areas of the country) makes up almost 70% of all agricultural land, and more than 70% of four of the provinces. The areas with the most arable land include the Western Cape, the Free State, Mpumalanga, Gauteng and North West provinces. However, high potential arable land is concentrated on the eastern seaboard (KwaZulu-Natal and Mpumalanga), while the Western Cape, Eastern Cape and the Free State have the most medium potential land (McKenzie *et al.*, 1989).

These data show that most of the southern and the western interior of the country is semi-arid.[4] The southwestern Cape has a winter rainfall, and the southern Cape has year-round rainfall. The rest of the country (the eastern seaboard and the northern and eastern interior) has summer rains. The southern and western interior is suitable for extensive livestock (wool and mutton sheep and beef) production only. Livestock farming (dairying, intensive poultry and pig farming, etc.) is also practised in the arable areas of the country, generally nearer to the major metropolitan markets or on the coast where access to imported animal feeds is easier. The country is a net importer of meat, mainly lower quality grade meat, sourcing most of its imports from Botswana and Namibia.

South Africa's horticultural production includes deciduous, citrus and subtropical fruit, vegetables and flowers. Deciduous fruits comprise table grapes (under irrigation along the Orange River in the Northern Cape, and in parts of the Western Cape), wine grapes (in select winter rainfall regions of the Western Cape), apples (under irrigation mostly in the Ceres and Elgin/Grabouw valleys in the winter rainfall regions of the Western Cape and the Langkloof in the Eastern Cape) and stone fruit (in the winter rainfall regions of the Western Cape). Some deciduous fruit is grown in the interior (for example, in the eastern parts of the Free State). Citrus fruit is grown in the Olifants River valley along the west coast of the Western Cape, and in the Lowveld areas of Limpopo and Mpumalanga, while subtropical fruit is grown in those areas and in KwaZulu-Natal. Most fruit produced in the country is exported. Vegetables are grown in the Lowveld, in the southern Cape, and near to all the main metropolitan areas. Less than 5% of vegetable output is exported.

The most important field crops grown in South Africa are maize, sugar and wheat. The country is a net exporter of maize over the longer term[5] and sugar, and a net importer of wheat. Maize is produced in the highveld regions of the interior (Free State, North West and Mpumalanga), sugar in the KwaZulu-Natal coastal areas and in the Lowveld areas of Mpumalanga, while summer wheat is produced in the Free State and winter wheat in the Western Cape.

1.6 TRENDS IN AGRICULTURAL PRODUCTION IN SOUTH AFRICA[6]

When the prices of farm inputs change, the profitability of the agricultural sector also changes. In the longer term, farmers adapt to such changes by either decreasing their level of input use, by increasing output from a constant level of input use, or by a combination of these. In each case, productivity has been increased. This section analyses trends in factor productivity, profitability, flexibility in input substitution, and aggregate support in South African agriculture.

---

[4] The per capita availability of fresh water resources is 1 187.5 cubic metres, compared to a global average of 8 240.40 m^3.
[5] Production varies according to the rainfall levels, and so this is not true in every year.
[6] This section has been adapted from Van Zyl *et al.* (2001).

1.6.1 Growth in Total Factor Productivity

Table 1.6 shows trends in the TFP (Total Factor Productivity) ratio – the most comprehensive measure of productivity growth – for commercial agriculture in South Africa since 1947.

Table 1.6: Trends in TFP in South Africa, 1947–96.

Period	*Terms of trade*	*TFP*	*Net farm income*
1960–1980	–0.18	2.05	4.03
1980–1990	–2.58	0.96	–3.73
1980–1996	–1.80	1.19	–2.23
1990–1996	–0.91	1.56	0.32
1960–1996	–1.01	1.66	1.20

Source: Van Zyl *et al.* (2001)

A number of important conclusions can be drawn from the data in Table 1.6:

(a) The domestic terms of trade for intermediate and capital goods for commercial farmers were negative between 1960–96, implying that input prices they paid were rising faster than the output prices they received throughout the period.
(b) The rate at which the domestic terms of trade turned against commercial farmers worsened in the period around 1980 (i.e. during the first phase of deregulation – see below) and improved subsequently, but was still far higher than during the period 1960–80.
(c) TFP growth slowed during the first phase of deregulation, then increased again thereafter;
(d) From 1980–90, when inflation rates in South Africa had reached their peak and TFP growth was at its weakest, Net Farm Income (NFI) growth was negative. However, by 1990, TFP growth had recovered enough to cause a positive annual growth in NFI up to 1996.

1.6.2 Profitability

In agriculture, one standard measure of profitability is NFI, which is gross revenues minus ordinary costs of production, including depreciation, salaries and wages, interest paid and rent paid. However, because agriculture is so dependent on the climate, NFI fluctuates annually, and is thus less useful as a macro-level indicator of trends in the sector. An alternative proxy for profitability is to analyse trends in land prices over time, since expected returns to land are a key driver of land values. For example, over the period 1995–99, farmland prices rose markedly in the Western Cape, probably indicating strong export performance by deciduous fruit farmers, while farmland prices fell in KwaZulu-Natal, in part reflecting lower world sugar prices. Land prices fluctuated in the Northern and Eastern Cape, but also fell markedly in the Free State as maize profitability declined over this period.

1.6.3 Trends in the elasticity of input substitution

The TFP results estimate the extent to which farmers have reacted to the cost-price squeeze, and clearly, one of the main solutions was to change not only the volume of inputs used, but also the particular input mix. Past research shows that farmers' ability to substitute inputs was severely constrained by state intervention in the sector, but that this ability improved as a result of the first stages of deregulation in the 1980s (Van Zyl & Groenewald, 1988; Sartorius von Bach & Van Zyl, 1991). More recent research (see Vink, 2000a) confirms that these trends have continued.

1.6.4 The aggregate level of protection given to South African farmers

The Producer Support Estimates were calculated for 17 selected commodities, which together make up 69% of the total gross value of agricultural production in South Africa (Kirsten *et al.*, 2000), as shown in Table 1.7.

Table 1.7: Percentage Producer Support Estimate per commodity in South Africa, 1996–98.

Commodity	*1996*	*1997*	*1998*
Wheat	–21.70	16.09	19.44
Maize	–9.68	–1.55	–18.25
Barley	–41.14	–13.25	–12.34
Oats	–290.41	–75.58	–140.28
Grain sorghum	–38.73	–16.48	–15.56
Total other grains	–123.43	–35.10	–56.06
Sunflower seed	–41.18	–49.04	–10.79
Groundnuts	–79.71	–92.77	–92.24
Soya beans	2.04	10.31	–38.91
Total oil seeds	–152.34	–90.74	–137.11
Sugar	39.60	42.62	39.55
Dairy	2.85	16.39	20.92
Beef and veal	10.77	13.64	21.20
Pork	–27.43	–11.99	–0.03
Poultry	3.38	6.23	–15.78
Sheep meat	47.66	40.36	49.28
Wool	0.42	0.34	14.97
Eggs	–9.02	9.38	–32.41
Livestock products	4.09	10.62	8.30
Total Producer Support Estimate for SA (%)	1.78	10.89	4.18

Source: Kirsten *et al.* (2000)

A negative number shows a transfer from producers to society, and a positive number reflects a transfer to producers. While the support estimate varies widely between different commodities, and across the years for each commodity, these data show that the Producer Support Estimate is lower than most countries in the world, with the exception of Australia and New Zealand (Kirsten *et al.*, 2000).

1.7 CONCLUSION

The South African economy has grown, albeit modestly, since the democratic elections of 1994. This growth ended almost three decades of declining per capita incomes, the result of the country's increasing isolation and poor economic policies. South Africa's agricultural sector has continued to account for a smaller share of a growing national GDP, and now contributes less than 4% of GDP. Nevertheless, the performance of the sector during this period was heavily influenced by government agricultural and macroeconomic policies. The result is that the sector is now, at the macro level, increasing productivity and receiving (in aggregate) less support from the state. Some major future challenges will be how to empower new commercial farmers in an industry that must remain competitive and sustainable, and how to manage recent changes in new labour, trade and water policy.

REFERENCES

BHORAT, H. (2000). The impact of trade and structural changes on sectoral employment in South Africa. *Development Southern Africa* 17(3): 437–466.

BHORAT, H. (2001). Labour market challenges in the post-apartheid South Africa: a country profile. Unpublished mimeo. Development Policy Research Unit, University of Cape Town.

CALLEAR, D. & MTHETHWA, T. (1996). South African agricultural development corporations: where to from here? Unpublished paper. Land and Agricultural Policy Centre (LAPC), Pretoria.

CARTER, R.C. (1996). Policy development in the water sector: the South African experience. In Abrams, L.J. & Howsam, P. (eds). *Water Policy: Allocation and Management in Practice.* Proceedings of an International Conference. UK: Cranfield University.

DEPARTMENT OF LAND AFFAIRS. (1997). *White Paper on South African Land Policy.* Pretoria: Government Printer.

DEPARTMENT OF TRADE AND INDUSTRY (DTI). (2002). http://www.dti.gov.za (Accessed 5/11/02).

DEVELOPMENT BANK OF SOUTHERN AFRICA & LAND AND AGRICULTURAL POLICY CENTRE. (1997). The reorganisation of state-run and parastatal-run agricultural projects. Development Bank of Southern Africa and Land and Agricultural Policy Centre, Midrand.

DIRECTORATE: AGRICULTURAL STATISTICS. (2002). *Abstract of Agricultural Statistics.* Pretoria: National Department of Agriculture.

EBONY CONSULTING INTERNATIONAL (ECI). (2002). The food security effects of the deregulation of agricultural marketing in South Africa. A report to the National Agricultural Marketing Council, Pretoria.

EVERATT, D. & ZULU, S. (2001). Analysing rural development programmes in South Africa, 1994–2000. *Development Update* (Quarterly Journal of the South African National NGO Coalition and INTERFUND) 3(4): 1–38.

GRAHAM, A. & LYNE, M. (1999). Land redistribution in KwaZulu-Natal: analysis and comparison of farmland transactions in 1997 and 1998. *Agrekon* 38(4): 516–525.

KIRSTEN, J.F. & VAN ZYL, J. (1999). Approaches and progress with land reform in South Africa. *Agrekon* 38 (Special issue): 326–342.

KIRSTEN, J.F., GOUSE, M., TREGURTHA, N., VINK, N. & TSWAI, J. (2000). Producer subsidy equivalents (PSE) for South African agriculture for 1996, 1997, 1998. Report to the National Department of Agriculture, 6 March, Pretoria.

MAASDORP, G. (2002). Economic survey, 1970–2000. In Jones, S. (ed.). *The Decline of the South African Economy*. Cheltenham, UK: Edward Elgar: 7–30.

MCKENZIE, C., WEINER, D. & VINK, N. (1989). Land use, agricultural productivity and farming systems in southern Africa. Unpublished research report. Development Bank of Southern Africa, Midrand.

MINISTRY OF AGRICULTURE AND LAND AFFAIRS. (2001). *Land Redistribution for Agricultural Development*. Pretoria: Ministry of Agriculture and Land Affairs.

SARTORIUS VON BACH, H.J. & VAN ZYL, J. (1991). Have recent structural changes caused agriculture to become less rigid? *Development Southern Africa* 8(3): 399–404.

STATISTICS SOUTH AFRICA. (2002). http://www.statssa.gov.za (Accessed 5/11/02).

STRYDOM, P.D.F. (2002). Macroeconomic policy, 1970–2000. In Jones, S. (ed.). *The Decline of the South African Economy*. Cheltenham, UK: Edward Elgar: 31–50.

VAN ZYL, J. & GROENEWALD, J.A. (1988). Flexibility in inputs substitution: a case study of South African agriculture. *Development Southern Africa* 5(1): 2–13.

VAN ZYL, J., VINK, N., KIRSTEN, J. & POONYTH, D. (2001). South African agriculture in transition: the 1990s. *Journal of International Development* 13(6): 725–739.

VINK, N. (2000a). Farm profitability and the cost of production inputs in South African agriculture. A report to the National Department of Agriculture, Pretoria.

VINK, N. (2000b). Agricultural policy research in South Africa: challenges for the future. *Agrekon* 39(4): 432–470.

VINK, N. & KIRSTEN, J. (2002). Pricing behaviour in the South African food and agricultural sector. A report to the National Treasury, Pretoria.

VINK, N., TREGURTHA, N. & KIRSTEN, J.F. (2002). South Africa's changing agricultural, food and beverage imports: implications for SADC suppliers. A report to the World Bank, University of Stellenbosch.

WORLD BANK. (2002). *World Development Report 2002*. Washington, DC: IBRD.

CHAPTER 2

Structure and Production in South African Agriculture

Tommy Fényes and Nico Meyer

2.1 INTRODUCTION

This chapter describes the agricultural sector within its environment with specific reference to structure and production in order to highlight the agricultural potential at a national and provincial level as determined by the natural resource base. Land-use patterns in the developing (formerly homeland) farming areas also receive attention. In these areas the land tenure system and land-use patterns differ remarkably.

An analysis of the demographic profile by province is provided. The agrarian structures are noted as constantly changing as the land and agrarian reform programmes are gaining momentum. Agriculture's contribution to the economy is highlighted and the most important field crops, horticultural crops and livestock are analysed. The chapter concludes with the current state of the farming industry where the analysis of farm income by main branch shows important structural changes in production patterns.

2.2 NATURAL RESOURCE ENDOWMENT

The development potential of South Africa is determined by its natural and human resources. The nature and distribution of the natural agricultural resources (climate, soils and terrain, vegetation and water) determine the agricultural potential of the country.

2.2.1 Climate

South Africa, with a mean annual rainfall of 497 mm (considerably less than the world mean of 860 mm), may be described as mainly a semi-arid region with acute water shortages. There are three main rainfall areas: the winter rainfall area in the Western Cape; an all-year-round rainfall area along the southern Cape coast; and a summer rainfall area that covers the remainder of the country.

There are also great variations in both rainfall and runoff. Annual rainfall varies from 100 mm in the west to more than 2 000 mm in the KwaZulu-Natal and Mpumalanga provinces in the east (DBSA, 2000). Over 65% of the country receives less than 500 mm of rain per year and about one fifth of the country obtains less than 200 mm. Only 3% of South Africa receives rain throughout the year, while 86% of the country only gets summer rain.

Temperatures in South Africa are strongly affected by altitude, latitude and ocean currents. Temperature is also an important factor influencing agricultural potential and crop adaptability. Except for small high-lying areas and the eastern escarpment, most of South Africa has hot summers and a long growing season. High temperatures also influence evaporation which varies from 1 100 mm to over 3 000 mm per year. Few areas are frost-free in the winter months.

2.2.2 Soils and terrain

The surface area of South Africa falls naturally into two major physiographic regions: the interior plateau, comprising extensive level surfaces with a mean altitude of 1 200 m; and the marginal lands which lie between the plateau and the coast. A third major element is the Great Escarpment, which forms the boundary between the plateau and the marginal lands (DBSA, 2000).

About 31% of South Africa's surface area is level to gently undulating, while 19% is steep, rocky and mountainous. The Drakensberg, a part of the Great Escarpment, forms the boundary between the interior plateau and marginal lands towards the coast, and has a marked effect on the climate and vegetation of the country.

2.2.3 Water resources

2.2.3.1 Water availability in South Africa

The availability of water in South Africa correlates with the physical location of the sub-continent and the prevailing high-pressure systems associated with dry conditions and low rainfall. Rainfall distribution also correlates strongly with the occurrence of the warm Agulhas current in the Indian Ocean flowing southwards along the eastern seaboard and the cold Benguella current in the Atlantic Ocean flowing northwards along the west coast. Topographical features such as the Drakensberg mountain range and escarpment extending from north to south parallel to the east coast have a major impact on the distribution of rainfall.

Water in South Africa is a scarce resource. Annual rainfall varies from as much as 2 000 mm/yr in the wet east along the escarpment to less than 100 mm/yr in the dry west coast. This is compounded by the fact that South Africa is periodically afflicted by severe and prolonged droughts which often end in severe floods.

2.2.3.2 Water demand

During the first half of the 1900s, water development centred around the provision of water for agricultural purposes. During the past 40 years, urban and industrial water demand have both increased at a significantly higher rate and the total demand for water will reach 26 billion m³ by the year 2010 (see Table 2.1). Available supplies in some regions will start to fall short before the end of 2100. Irrigation demand comprised almost 51% of total water in 1990 and consumed more water than any other sector. Irrigation demand is, however, expected to decrease to about 46% by the year 2010 (WRC, 1997).

Table 2.1: Water use in South Africa, 1990 and 2010.

Water demand	*1990 million m³ per annum*	*Percentage*	*2010 million m³ per annum*	*Percentage*
Direct use:				
Municipal/domestic	2 281	12.0	4 477	17.3
Industrial	1 448	7.6	2 961	11.4
Mining	511	2.7	649	2.5
Power generation	444	2.3	900	3.5
Irrigation	9 695	50.9	11 885	45.9
Stock watering	288	1.5	358	1.4
Nature conservation	182	1.0	191	0.7
Indirect use:				
Forestry runoff reduction	1 427	7.5	1 700	6.6
Ecological use/estuaries and lakes	2 767	14.5	2 767	10.7
Total	19 043	100	25 888	100

Source: Water Research Commission, 1997

Agriculture is the largest user of water, consuming as much as 50% of total demand. In 1981, the latest year for which data were available at the district level, 824 000 hectares were irrigated, or 5.9% of the 14.3 million hectares with crop potential.[1] In 1988, 230 000 irrigated hectares were in government water schemes, 300 000 hectares were controlled by irrigation boards, and 70 000 hectares were in the homelands. According to the Department of Agriculture, the total area under irrigation in South Africa today is 1.2 million hectares.

2.3 AGRICULTURAL POTENTIAL

For a given crop and level of management, agricultural potential is largely determined by the interaction of natural resources such as soil, climate and terrain. Primary production is a quantitative expression of vegetative matter (i.e. harvested yield) which the natural environment at any specific location, per unit area or for a given period of time, can produce. Primary production also gives an indication of the sustainable production one can expect. According to estimates of primary production by province, the areas of the highest natural environmental potential are

[1] This compares to 33% in Asia.

in the east (see Figure 2.1). Primary production changes most rapidly in the southern and southwestern coastal areas and along the eastern seaboard, mainly in conjunction with rainfall gradients. Figure 2.1 shows that KwaZulu-Natal and Mpumalanga have the highest average of primary production of plant material, with 20% of Mpumalanga producing more than 9.5 tons of plant material per hectare in a growing season and 80% producing more than 6.6 tons.

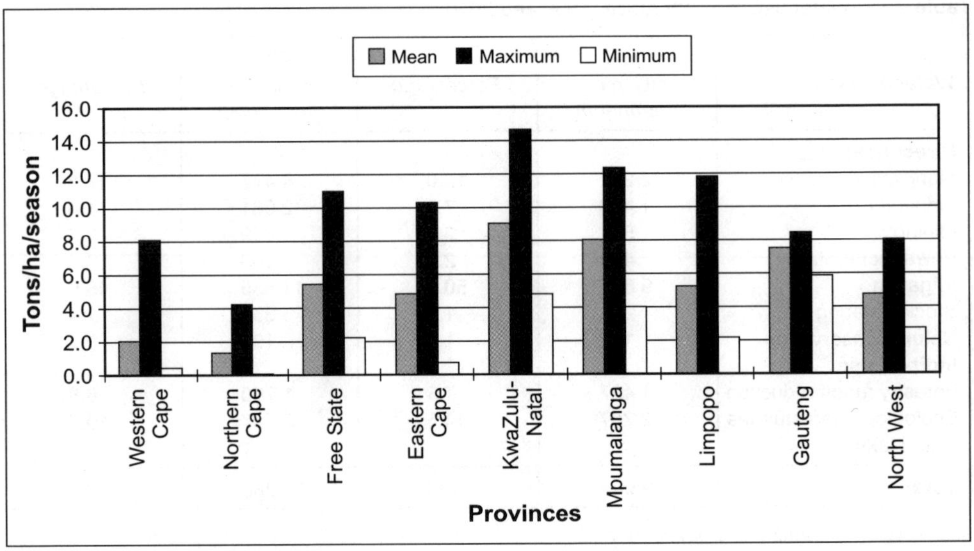

Figure 2.1: Mean annual primary production (tons/ha/season) by province.

2.4 LAND-USE PATTERNS

The total land area of South Africa comprises 122.3 million hectares of which farmland occupies approximately 82.3% (Abstract, 2002). Commercial and developing agriculture occupy surface areas of 86.2 million hectares and 14.5 million hectares respectively. The total area under commercial agriculture is about six times that of developing subsistence agriculture. Total potential arable land in South Africa comprises only 14.2 million hectares compared to grazing land which comprises 72 million hectares. The remaining area is comprised of nature conservation land (11 million hectares), forestry (1 million hectares) and other land (6.8 million hectares).

2.4.1 Land-use patterns in commercial agriculture

The Free State, KwaZulu-Natal and Mpumalanga provinces have almost 45% of the arable land in South Africa, covering a total surface area of 4.2 million hectares, 1.2 million hectares and 1.7 million hectares respectively. Approximately 22% of the arable land in commercial farming areas is classed as high potential, 90% of which is found in KwaZulu-Natal and Mpumalanga (Meyer, 1998). These regions are situated in the moist eastern half of South Africa. Medium potential arable land in commercial areas occupies almost 29% of total available arable land, 75% of which is found in the Western Cape, Free State, Limpopo and North West. Almost 57% of medium and low potential arable land is found in the Free State (See Figure 2.2).

Only 1.1% of the total surface area of South Africa can be irrigated. About 178 000 hectares of dryland could be irrigated, but the utilisation of this land is unlikely due to the present water demand and supply situation (Meyer, 1998). Demand for water keeps increasing and some of the faster growing industrial and metropolitan areas are situated in dryish parts which are far from the largest dams and rivers. Almost three-quarters of the potential irrigation area is found in KwaZulu-Natal and the Eastern Cape, and consists of deep valleys, which are very uneconomic to develop. A shift to planted pastures is most evident in the marginal commercial cropping regions.

Natural grazing provides a basis for extensive stock farming, and the most important stock grazing regions are the Northern Cape, Mpumalanga, KwaZulu-Natal and Limpopo. The carrying capacity of the veld varies considerably between these provinces and an overestimation of the carrying capacity of the veld often results in overstocking.

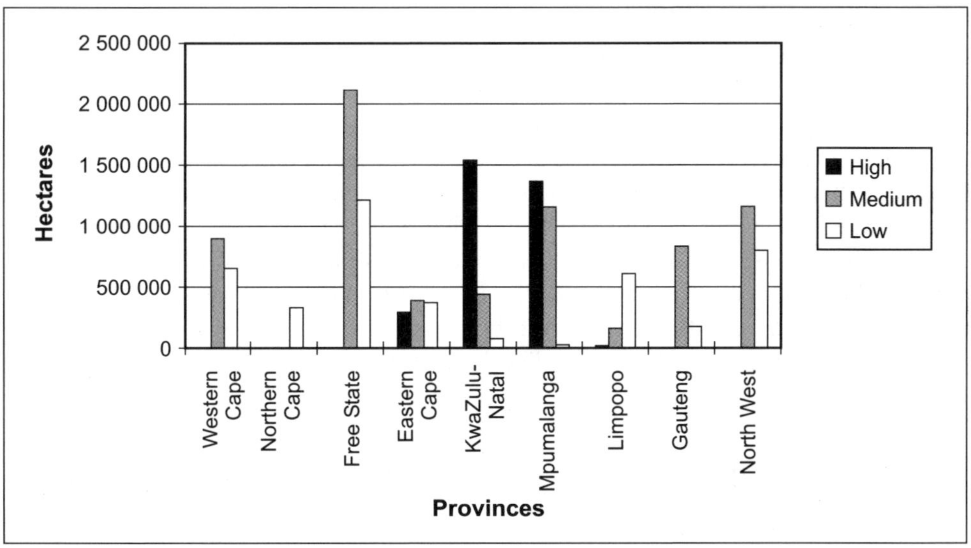

Figure 2.2: Potential arable land in the commercial farming regions of South Africa.

Source: Development Information Business Unit (1997)

2.4.2 Land-use patterns in developing farming areas

Developing farming areas occupy 17.1 million hectares or 13.9% of the total surface area of South Africa (DBSA, 2000). While the percentage of land allocated for agriculture is relatively high in these areas, a large portion is utilised for settlement outside proclaimed towns. The majority of inhabitants of the rural areas are women, children and aged people who occupy the land for purposes of social security and not necessarily for agricultural production. It is estimated that potential arable land comprises 11.1–16.6% of the total area. Only 40–80% of this land is cultivated in any given year. The total irrigation potential of developing areas is estimated at 200 000 hectares (the sub-classification of this land is currently not available). It appears that although some of the developing areas are situated in the moist eastern part of South Africa, steep terrain limits the amount of available arable land and increases the soil erosion potential (Meyer, 1998).

Veld grazing in these areas has a relatively higher potential in terms of possible carrying capacity than in commercial areas. The present stocking rate, however, exceeds the carrying capacity in virtually all the developing farming areas. Overgrazing has affected the quality of arable land to such an extent that it is no longer suitable for crop production in many places.

Forestry land in the developing areas occupies nearly 256 268 hectares of the total estimated potential of 845 877 hectares. A significant potential for expansion of forestry land exists in KwaZulu-Natal and the Transkei portion of the Eastern Cape (DBSA, 2000).

2.4.3 Nature conservation

Nature conservation areas, both public and private, roughly equal the area of the former homelands. Total area under conservation is estimated at around 16 million hectares, or 13% of the total land area of South Africa. About 11 million hectares of these areas are privately owned and used for eco-tourism and game farming. South Africa's 229 nationally protected areas (state forests, parks, and reserves) occupied around five million hectares or 4% of the total land area in 1991. The relative size of the nature conservation area under public ownership is not uncommon when compared to other countries in the world. However, in terms of the absolute number of protected areas, South Africa's number is much higher than for other middle-income economies (DBSA, 2000). The latter may be attributed to the extremely diverse flora in the country.

2.5 DEMOGRAPHIC PROFILE OF SOUTH AFRICA

2.5.1 Population size and growth

Table 2.2 shows that the total population of South Africa grew from 42 million in 1996 at an average annual growth rate of 1.56% to 44.7 million in the year 2000. The composition of population groups in 2000 were as follows: Africans represented 77.4%, whites 11.7%, Coloureds 8.4% and Indians 2.5%.

The average population growth rate in South Africa is expected to continue to decline over the coming decades. The growth rate of the African population has decreased from 2.45% in 1991 to an estimated 1.65% in 2001 while that of the total population has dropped from 2.13–1.43% over the corresponding period.

Table 2.2: Population profile of South Africa, 1996 and 2000.

Population groups	*1996*	*%*	*2000*	*%*	*% Growth rate 1996–2000*
Africans	32 245 180	76.74	34 609 550	77.42	1.78
Indians	1 055 840	2.51	1 102 690	2.47	1.09
Coloureds	3 546 170	8.44	3 759 770	8.41	1.47
Whites	5 173 620	12.31	5 232 600	11.70	0.28
Total	42 020 810	100.00	44 704 610	100.00	1.56
Male	20 771 720	49.43	22 010 880	49.24	1.17
Female	21 249 090	50.57	22 693 730	50.76	1.32
Urban	23 350 964	55.57	24 006 376	53.70	0.70
Rural	18 669 846	44.43	20 698 234	46.30	2.70

Source: DBSA (2000)

Population growth has important implications for the provision of infrastructure – for example meeting basic needs such as housing, health, transport and education, as well as provision of basic food types. Although lower population growth rates will put less pressure on the provision of infrastructure and basic needs, our declining economic performance and growth rate have increased unemployment and poverty significantly. In recent decades, population growth exceeded GDP growth which had a depressing effect on per capita income.

2.5.2 Spatial characteristics

There are significant differences in population numbers and densities between the provinces, i.e. the eastern seaboard is more densely populated than the western interior. The highest concentration of the South African population is found in KwaZulu-Natal (19.8%), followed by Gauteng (17.4%), Eastern Cape (15.1%), Limpopo (11.4%) and Western Cape (9.4%) (see Table 2.3).

Processing and distribution plants for basic food types, among other things, should take this population distribution into account, seeing that the distribution of the production volumes of products is not necessarily synchronised with this population distribution. Population densities vary widely between provinces: from as high as 390 persons per km² in Gauteng to two persons per km² in the Northern Cape.

Table 2.4 summarises the population density details per province. Increases in population density, especially in former homeland areas, have impacted on access to land, especially land suitable for agricultural purposes, and has increased the rate of landlessness among the rural population. It is not clear whether the present high rates of urban migration will release rural land pressure for South Africa as a whole.

Table 2.3: Population distribution by province, 1996.

Province	*Urban* %	*Rural* %	% *Distribution*
Western Cape	88.85	11.15	9.40
Northern Cape	71.27	28.73	2.00
Free State	69.75	30.25	6.28
Eastern Cape	36.45	63.55	15.06
KwaZulu-Natal	42.27	57.73	19.83
Mpumalanga	61.78	38.22	10.68
Limpopo	10.42	89.58	11.36
Gauteng	96.42	3.58	17.43
North West	37.09	62.91	7.96

Source: Statistics South Africa (1996)

Table 2.4: Population density by province, 2000.

Province	*1996*	*2000*	*Area* km^2	*Population per* km^2
Western Cape	3 974 542	4 205 555	129 386	32.50
Northern Cape	837 828	874 818	363 389	2.41
Free State	2 740 788	2 902 176	129 437	22.42
Eastern Cape	6 475 780	6 892 058	170 616	40.40
KwaZulu-Natal	8 774 924	9 349 905	91 481	102.21
Mpumalanga	2 906 308	3 108 835	82 333	37.76
Limpopo	5 089 220	5 456 963	116 824	46.71
Gauteng	7 748 572	8 217 992	21 025	390.87
North West	3 475 848	3 696 308	118 710	31.14
Total	42 023 810	44 704 610	1 223 201	36.55

2.5.3 Demographic profile of rural areas

Within the rural areas there is a marked difference in the nature and level of economic activity and levels of welfare. These inequalities, coupled with the uneven distribution of income in agriculture, have, according to Brand (1990), contributed to a disproportionate prevalence of poverty in rural compared to urban areas, and some questions, such as the distribution of access to land between the different groups, will become increasingly important issues with respect to these rural areas.

Continuous and fluctuating movement of people, according to Eckert (1994), provides one of the main dynamic forces intimately linking the rural and urban economies of southern Africa. Three accelerating factors that will enhance this movement are:

(a) the disappearance of former homeland boundaries;
(b) a possible reduction in civil service employment and a collapse of the former homeland economies; and
(c) a reduction in the labour demand on commercial farms.

2.5.4 Commercial farming areas

In general, the features of commercial farming areas are that of a high migration of economically active population from poorer to economically stronger areas, farm evictions, retrenchments, unbalanced economic structure and an uneven distribution of economic and natural resources. The densification of commercial farming areas is increasing as migration from the former homeland areas to the rural towns escalates. Thus, significant shifts in the spatial distribution of people are possible and migration within rural areas is expected to be significant. Further, this could be complicated by a substantial influx of people from countries to the north who are presently competing effectively for service and informal openings in the larger cities.

2.5.5 Former homelands

The main feature that characterises the demographic profile in the former homelands is high population growth coupled with a migration from former commercial farming areas, very unequal age/sex distribution in the productive age groups (20–50) and a high dependency ratio (DBSA, 2000). Women outnumber men because men work as migrant workers.

2.6 AGRARIAN STRUCTURE

One of the objectives of the agrarian reform programmes in South Africa is to change the ownership and control patterns of land. An increasing flexibility of the farming sector is manifested in changing land-use patterns and farm size and the workings of an active and robust land market. This flexibility will be enhanced by the repeal of the Sub-division of Agricultural Land Act of 1970 which allowed sub-division of a holding only through ministerial approval.

2.6.1 Number of farms

The number of farms in the mainly commercial farming sector increased from 81 432 in 1921 to 119 556 in 1952, before declining to 59 960 by 1983. From 1983–88 the total number of farms increased slightly to 67 010. In 1996, farming units in the commercial farming sector numbered 60 938 (Table 2.5).

2.6.2 Total area of farms

The amount of land held (owned or leased) by the commercial sector has risen and fallen in line with the trend in the number of farming units, but at a more moderate rate – from 82.2 million hectares in 1930 to 91.4 million hectares in 1964, gradually declining thereafter to 85.8 million hectares by 1988 and 82.2 million in 1996 (World Bank, 1996).

2.6.3 Farm size

Data on changes over time in farm size in South Africa are difficult to obtain. Official data show an unambiguous trend towards larger farms and a smaller number of farmers in the commercial sector up to the beginning of the 1980s. As the reduction in the number of farming units outpaced the decline in the size of the commercial sector, the size of the farming units has tended to increase over the past 30 years. Nationwide, farm sizes increased from 738 hectares per farm in 1953 to 1 280 hectares per farm in 1988 and 1 349 hectares per farm in 1996 (World Bank, 1996).

Table 2.5: Commercial farming units: number and size, 1996.

Provinces	*Number*	*Hectares*	*Average farm size (ha)*
Western Cape	9 759	9 766 969	1 001
Northern Cape	6 730	29 734 978	4 418
Free State	11 272	11 342 502	1 006
Eastern Cape	2 342	756 946	323
KwaZulu-Natal	6 338	10 327 660	1 629
Mpumalanga	5 037	4 068 401	808
Limpopo	4 675	4 544 012	972
Gauteng	7 273	5 488 613	755
North West	7 512	6 179 490	823
Total	60 938	82 209 571	1 349

Source: Statistics South Africa (1996)

2.7 FARMING AND THE ECONOMY

Agriculture's contribution to the economy can be highlighted by making use of a range of measures. Probably the most important role of agriculture is providing sufficient food to the nation at affordable prices. Not only is this important in a macro and regional context, but especially at a household level. The importance of the sector is also highlighted by the relative contribution of agriculture to the GDP, its share of the labour force and its contribution as an earner of foreign exchange.

2.7.1 Contribution towards GDP

Agriculture's contribution to GDP in South Africa (at current factor cost) increased from R190 million in 1946 to R25 375 million in 2000 (DBSA, 2000). The real economy had grown at an average annual growth rate of 3.8% between 1946–95. It has slowed down and stagnated since the 1980s. The value added by the agricultural sector had grown by only 2.4% during this period, but being heavily dependent upon climatic circumstances, its growth was more sporadic than that of the overall economy (see Table 2.6).

Table 2.6: Average annual growth rate of South African GDP, 1946–95.

	Agriculture	*Total*
1946–1950	4.6	4.5
1950–1955	4.6	4.7
1955–1960	0.8	4.0
1960–1965	0.0	6.4
1965–1970	3.0	5.5
1970–1975	3.7	4.1
1975–1980	4.1	2.8
1980–1985	–2.7	1.1
1985–1990	4.4	1.9
1990–1995	–0.8	0.7

The growth in the GDP was accompanied by a high degree of diversification of the economy. The lower growth rate of the agricultural sector relative to that of the overall economy resulted in its share in the GDP steadily declining. This has been part of a broader transformation of the economy over the past century from one dependant on the primary sector (agriculture and mining) to a broadly diversified manufacturing and services economy. Table 2.7 shows the marked changes in the structure of South Africa's economy in recent years.

There is little question of the view that modernising industrialisation is the engine of sustained growth. However, there is dissent in theory as well as in practice about how best to industrialise. Is it to be industry-led or agriculture-led industrialisation, through import-substitution or a trade-led strategy? Also, the need for increased agricultural production and food supply to sustain the industrialisation process is recognised (Meyer, 1998).

South Africa's agriculture contributes substantially to economic development and prosperity. The real contribution of the sector is, however, underestimated and obscured when directly measured. For example, the low contribution of agriculture to GDP of 4.4% generated in 1999 (see Figure 2.3), may create the impression that agriculture is not important to the economy as a whole, but the true value lies in its backward and forward linkages (DBSA, 2000).

Table 2.7: Gross value added by kind of economic activity, R million.

Year	*Agriculture, forestry and fishing*	*% of total*	*Primary sector*	*% of total*	*Secondary sector*	*% of total*	*Tertiary sector*	*% of total*	*Gross value added at basic prices*
1950	405	16.0	727	28.8	581	23.0	1 220	48.3	2 528
1960	559	11.2	1 177	23.6	1 270	25.5	2 541	50.9	4 988
1970	861	7.2	1 914	15.9	3 536	29.4	6 569	54.7	12 020
1980	3 654	6.2	15 788	26.8	16 382	27.8	26 790	45.4	58 972
1990	12 184	4.6	36 291	13.8	81 408	30.9	145 451	55.3	263 151
2000	25 375	3.2	76 938	9.7	194 022	24.4	523 033	65.9	793 993

Source: DBSA (2000)

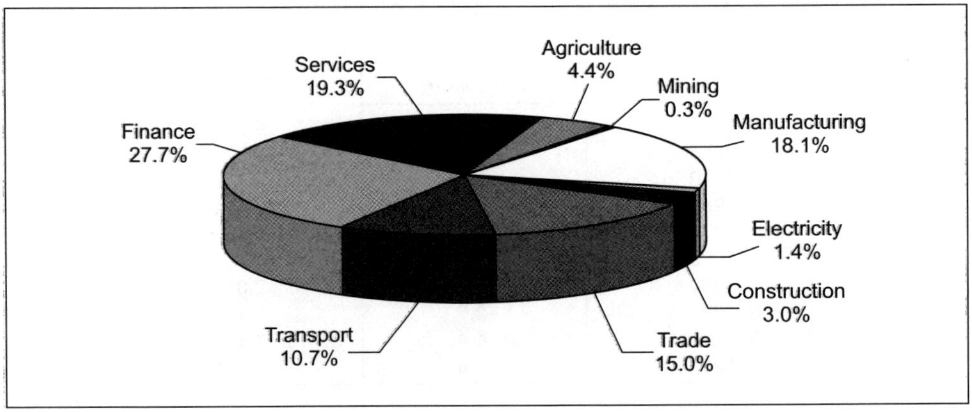

Figure 2.3: GDP at factor cost: per cent contribution by sector, 1999.
Source: WEFA Consultant (2001)

Available evidence lends strong support to the close relationship between agricultural development and overall development. Economists generally agree that the sector's contribution could be as high as 20–30% if one considers the linkages with other sectors of the economy. The relatively small contribution made by agriculture to the country's GDP tends to conceal the sector's most important contribution to the economy in terms of food supply, earner of foreign exchange, economic interregional linkages and employment opportunities. Figure 2.3 depicts the sector contribution of agriculture for 1999 (Meyer, 1998).

Table 2.8 summarises the economic growth rate of South Africa since the 1960s. Average annual real GDP growth was 5.7% during the 1960s, 3.3% during the 1970s, 2.2% during the 1980s and only 1.1% during 1990–97. The impact of agriculture's performance on the economic growth rate can be linked to the negative influence in years when agricultural production was adversely affected by natural factors such as droughts, while favourable periods impact positively on the economic growth rate.

2.7.2 Contribution to employment

Agriculture creates one of every seven job opportunities in South Africa and involves four million people or 10% of the population. Although agriculture, as a rule, is not the leading sector in employment-creating economic development, it must be borne in mind that in absolute terms the agricultural sector was and will always be an important source of employment. It remains important because of the large number of benefiting dependants per farm worker. On the basis of the 1996 Census results, the number of employed people in the labour force in the ten major sectors of the economy was 9.1 million, of whom 814 000 (10%) worked in the agriculture, hunting, forestry and fishing sector (Table 2.9). More than four million people are sustained in this sector.

Table 2.8: Sectoral composition of South Africa's GDP (at factor cost), 1997.

	1960–69		*1970–79*		*1980–89*		*1990–97*	
	Real % change	*% of total*	*Real % change*	*% of total*	*Real % change*	*% of total*	*Real % change*	*% of total*
Primary sector	3.7	22.7	0.1	19.5	0.8	20.3	0.1	13.3
Agriculture	3.2	11.1	3.2	8.0	4.3	6.1	2.4	4.7
Mining	3.9	11.1	–0.6	11.5	–0.6	14.2	–0.6	8.6
Secondary sector	8.3	28.7	5.0	30.6	1.7	31.8	0.6	31.5
Manufacturing	8.5	22.6	5.1	22.6	1.5	24.1	0.4	24.2
Electricity	6.3	2.6	7.1	3.0	5.4	4.0	3.1	4.1
Construction	8.3	3.5	3.4	4.7	–0.1	3.6	–1.3	3.2
Tertiary sector	5.2	49.1	3.9	49.9	3.1	47.9	1.4	55.2
Commerce	7.1	14.5	3.4	13.9	3.4	12.9	1.4	15.8
Transport	5.5	10.0	6.1	9.7	2.2	8.7	1.4	7.7
Finance	6.1	11.9	3.5	13.6	3.3	13.0	2.1	16.3
Community services	3.5	1.9	3.8	1.7	4.3	1.7	1.9	1.9
Less: Service charges	8.1	1.6	3.5	1.8	3.8	2.4	1.6	3.7
General government	4.4	8.8	4.0	9.9	3.2	11.7	2.9	15.0
Other	1.3	3.6	3.1	2.9	3.1	2.3	1.0	2.2
Total	5.6	100.0	3.4	100.0	2.2	100.0	1.1	100.0

Source: SARB (1997)

Within the broad sector of agriculture, hunting, forestry and fishing, the agriculture and hunting sub-sectors together accounted for 92% of all employment opportunities. In terms of the rural areas of the former homelands, the rural survey of 1997 suggests that a total of 2.2 million people are employed (Statistics South Africa, 2002). Real gross capital formation in agriculture increased at a relatively high rate from 1950–97.

Notwithstanding the high rate of capital formation and the decline in agriculture's share in the provision of employment, agriculture is still relatively labour intensive. According to Fényes *et al.* (1987), on average, the agricultural sector uses more

Table 2.9: Employment in urban and non-urban areas, 1996.

	Urban	*Non-urban*	*Total*
Number employed:			
Agriculture and hunting	94 234	655 403	749 637
Forestry and logging	7 921	44 131	52 052
Fishing operation of fish farms	11 209	1 452	12 661
Total	113 364	700 986	814 350
Per cent in each sub-sector:			
Agriculture and hunting	83.1	93.5	92.1
Forestry and logging	7.0	6.3	6.4
Fishing operation of fish farms	9.9	0.2	1.6
Total	100	100	100

Source: Census (1996)

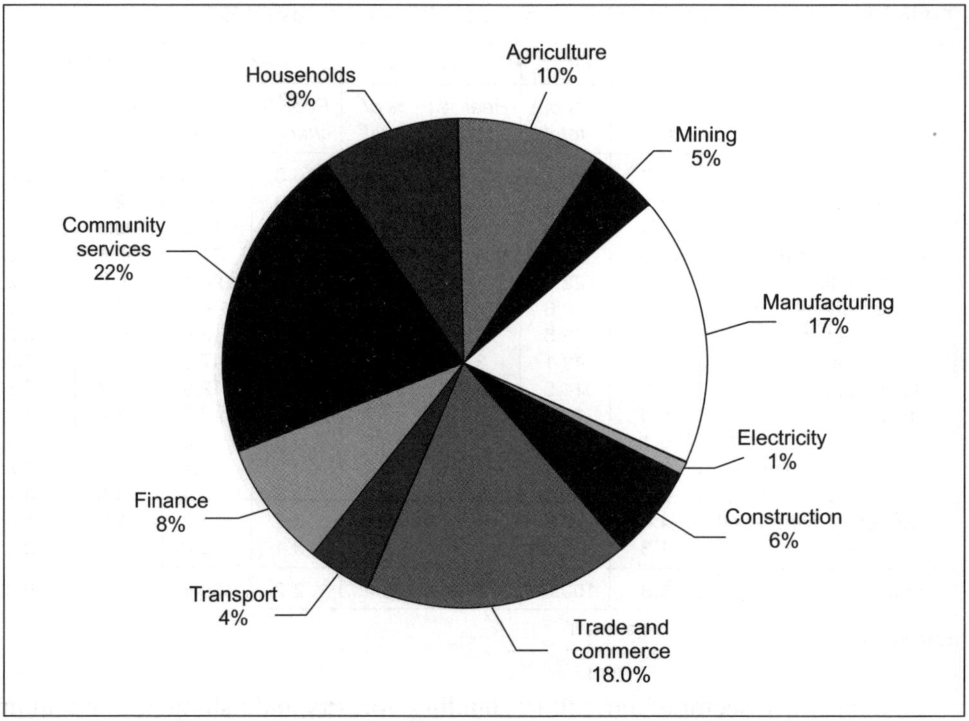

Figure 2.4: Formal and informal employment by sector, 1999.
Source: WEFA Consultants (2001)

workers for every R1 million value added than any other sector in the formal sector of the South African economy. As a factor of production, labour should not be seen in isolation. A closer examination of trends exposes important structural trends.

Apart from economic factors, recent labour unrest and the adoption of land reform and labour legislation pertaining to agriculture have contributed to the eviction of farm workers and unemployment. These trends identified above have definite policy implications. The national government and the provincial authorities should modify those policy aspects which impair job creation opportunities in agriculture. Unless job creation capabilities are enlarged, the problem of large-scale rural unemployment, poverty and social deterioration is unavoidable and will remain unresolved.

2.7.3 Agriculture as earner of foreign exchange

Agriculture is a net earner of foreign exchange. Export earnings by the sector have increased at a faster rate than import expenditure. Table 2.10 shows that export earnings by the sector have increased substantially over the last number of years, and the value of agricultural exports as a percentage of the total exports remained fairly constant at approximately 7–8%.

Table 2.10: Total agricultural exports.

Years	R million		Percentage
	Agricultural exports	*Total SA exports*	
1985	2 382.3	36 410.4	6.5
1986	3 024.6	41 327.8	7.3
1987	3 192.7	42 762.5	7.5
1988	3 759.9	49 360.0	7.6
1989	5 707.4	58 728.4	9.7
1990	4 625.1	60 770.0	8.7
1991	4 769.4	64 356.3	8.9
1992	4 860.2	68 035.0	7.1
1993	5 287.1	80 671.3	6.6
1994	8 055.8	90 021.5	8.9
1995	8 145.6	102 417.5	8.0
1996	11 819.0	125 725.1	9.4
1997	12 508.9	143 413.7	8.7
1998	13 699.4	147 546.7	9.3
1999	14 773.5	165 142.6	8.9
2000	15 819.7	210 022.1	7.5

Source: National Department of Agriculture (2000)

Despite relatively poor production conditions in agriculture (during the 1980s and early 1990s) and persistent recessionary conditions in the economy as a whole in the early 1990s, agriculture has made an enormous contribution to foreign exchange earnings which, in turn, has helped the country to meet its foreign debt obligations under extremely difficult circumstances. The value of agricultural exports has also shown the biggest increase in comparison with other sectors in the economy since 1993. This is mainly as a result of South Africa's deciduous fruit, citrus, wool, mohair, groundnuts, cut flowers and bulbs, subtropical fruit and wine being highly in demand in foreign countries.

Trends in the structure of exports show that during the 1960s wool, mohair and skins constituted about 50% of the total value of agricultural exports. This situation has since changed in that intensive industries – such as deciduous fruit, citrus and wine – now contribute more to the value of exports than wool, mohair, hides and skins and other extensive industries (Table 2.11). This highlights the importance of the optimum utilisation of high potential land, especially land under irrigation in the Western Cape, Mpumalanga and KwaZulu-Natal provinces, and the continued support of the sector through infrastructure development and agricultural research.

2.8 CURRENT STATE OF FARMING INDUSTRY

The relative position of horticultural crops and animal products has strengthened over the past decade at the expense of field crops. Shifts of crop production to the higher potential areas and livestock farming to the drier areas can be expected to continue where livestock production will replace crop production in marginal production areas.

Table 2.11: Export value of agricultural products (f.o.b.), (R1 000).

Product	*1990*	*1995*	*2000*
Sugar	711 840	545 687	1 861 841
Wine	55 669	332 895	1 689 557
Citrus fruit	335 959	722 183	1 582 225
Grapes	212 884	490 554	1 254 665
Preserved fruit and nuts	324 956	611 677	864 625
Apples, pears and quinces	334 124	791 142	702 986
Fruit and vegetable juices	115 403	247 414	657 292
Cigars, cheroots, cigarillos and cigarettes	8 279	107 413	548 624
Maize	593 732	566 202	498 675
Ethyl alcohol	28 048	219 927	381 413
Wool	536 494	298 514	353 035
Food preparations	19 940	237 393	253 627
Apricots, cherries, peaches, plums and sloes, fresh	40 576	115 169	206 298
Tobacco	42 982	51 791	194 903
Dates, pineapples, avocados, figs, guavas, mangoes	52 011	106 149	183 008
Raw skins of sheep and lambs	109 510	229 983	180 763
Undenatured ethyl alcohol	35 002	63 046	167 745
Sugar confectionary	28 082	57 509	165 851
Other meat and edible meat offal	10 260	21 907	158 361
Raw hides and skins of bovine	122 900	101 850	143 121
Other	906 464	2 224 181	3 771 035
Grand Total	4 625 118	8 142 585	15 819 651

Source: Customs & Excise (2001)

2.8.1 Gross value of production

Table 2.12 shows the gross value of production for 1990, 1995 and 2000. The gross value of agricultural production expanded annually by 10.7% between 1990–95 and 6% between 1995–2000, well below the inflation rate.

2.8.1.1 Field crops[2]

South Africa's top field crops by gross value in 2000 were maize, sugar cane, wheat, hay, sunflower seed, tobacco and groundnuts (Figure 2.5).

Production trend analysis of area cultivated and production for the period 1990–2000 reveals the following trends (Table 2.13):

2.8.1.2 Maize

2.8.1.2.1 Maize consumption

The local commercial consumption requirements for white maize for 2000–01 were approximately 4.2 million tons, of which 81% was expected to be used for human

[2] Due to editorial requirements this section was shortened. Detailed information on other products is available from the authors.

Table 2.12: Gross value of production by main branch.

Main branch	*1990*	*1995*	*2000*
Field crops	7 616 718	13 307 172	16 798 599
Horticulture	4 816 993	8 502 983	12 708 041
Animal products	9 520 747	14 708 949	19 485 827
Total	21 954 458	36 519 104	48 992 467

Source: Abstract of Agricultural Statistics (2000)

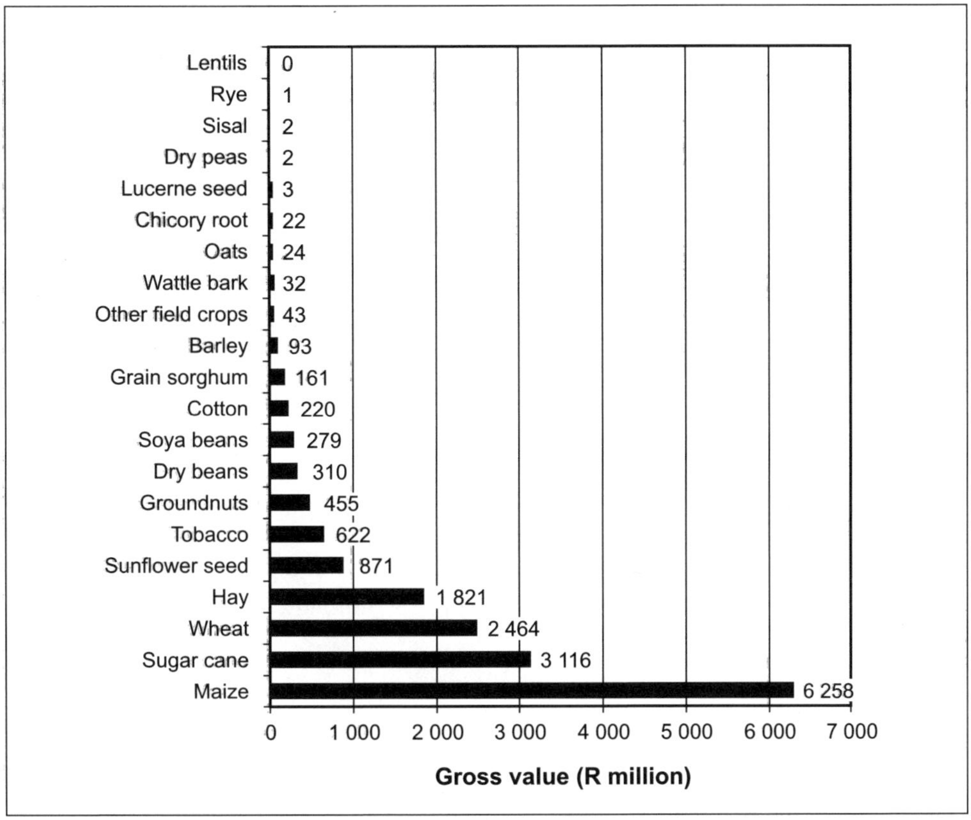

Figure 2.5: Ranking and value of the leading field crops, 2000.

Source: Abstract of Agricultural Statistics (2000) as adapted

consumption; while that for yellow maize was approximately 3.3 million tons. Yellow maize is sold mainly (80%) for animal consumption and for use in the manufacturing of animal feeds.

The maize industry is an important earner of foreign exchange for South Africa through the export of maize and maize products. The international maize market, especially the US market, has a dominant influence on local maize prices.

Table 2.13: Trends in area planted and production, 1990–2000.

Product	Period	Area (1 000 ha)	Annual growth rate (%)	Annual production (1 000 t)	Growth rate (%)	Yield (t/ha)
Maize	1990–2000	3 863	–2.0	8 679	3.3	2.24
Wheat	1990–2001	1 119	–3.6	2 003	2.5	1.84
Sorghum	1990–2000	160	–7.0	362	0.8	2.26
Sunflowers	1990–2000	528	0.6	577	8.9	1.09
Groundnuts	1990–2000	129	–5.3	118	6.2	0.91
Soya beans	1990–2000	89	7.3	124	11.6	1.39
Sugar cane	1990–2001	405	1.4	18 962	4.2	46.86

Source: Calculated from Abstract (2000)

2.8.1.2.2 Maize production

Trend analysis of time series production data shows that the area under maize production decreased by 2% while production increased by 3.3% in the period 1990–2000. During 2000–01, an estimated 2 707 505 hectares were planted to commercial maize – a decrease of 16% compared to the 3 230 440 hectares planted in 1999–2000. Commercial white and yellow maize comprised 1 595 605 hectares and 1 111 900 hectares respectively of the total hectares planted to maize. This represents a decrease of 20% in the case of white maize and 9% for yellow maize. The decrease in plantings can mainly be ascribed to the following: In many cases producers planted sunflower seed, groundnuts and soya beans instead of maize. Following the bumper crop of maize in 1999–2000, which caused a substantial drop in maize prices, and the grading problems experienced in certain areas (Mpumalanga and Free State Province), producers scaled down on their plantings. Furthermore, the deteriorating input/output price ratio and market risks also played a role.

2.8.1.3 Wheat

2.8.1.3.1 Wheat consumption

The composition and the various needs of the population has a major impact on the consumption of the product. A large section of the South African population is poor and is urbanising at a fast rate. Urbanisation results in consumers requiring more ready-to-eat foods. Demand for convenience type foods is growing. Bread is such a product, which is a substitute for maize meal.

In South Africa, wheat is mainly used for human consumption. During the 2000–01 marketing year, approximately 2 444 000 tons of wheat were used for human consumption, 2 000 tons for animal feed and 24 000 tons for seed. An additional 56 000 tons were either used on-farm or the use was not defined, and 84 000 tons were exported in the 2000–01 marketing season. The total demand for wheat for the 2000–01 season was therefore 2 610 000 tons.

2.8.1.3.2 Wheat production

Wheat is the second most important field crop in South Africa. It is planted mainly between mid-April and mid-June in the winter rainfall area and between mid-May and the end of July in the summer rainfall area. Most of the wheat produced in South Africa is bread wheat, with a little durum wheat being produced in certain areas.

Trend analysis of time series production data shows that the area under wheat production decreased by 3.6%, while production increased by 2.5% in the period 1990–2000. The estimated area planted to wheat for the 2001–02 season was 959 250 hectares – an increase of 2.7% in plantings from the 2000–01 season, of which 450 000 hectares (47%) were in the Free State and 374 000 hectares (39%) in the Western Cape. Approximately 15% of the total area planted to wheat is cultivated under irrigation and 85% under dryland conditions. The expected commercial wheat crop for 2001–02 was 2.403 million tons, of which 945 000 tons (39%) were in the Free State, 785 400 tons (33%) in the Western Cape, and 272 500 tons (11%) in the Northern Cape. The expected average yield for commercial wheat is 2.50 t/ha.

2.8.1.4 Horticultural crops

South Africa's top horticultural crops by gross value are deciduous fruit, vegetables, potatoes, citrus fruit, viticulture and subtropical fruit, which represent almost 90% of the gross value of horticultural production (Figure 2.6).

The most important fruit products in the horticultural sector by gross value of production in 2000 were table grapes, followed by apples, pears, peaches and plums. On the basis of volume produced by main branch, apples are the most prominent crop followed by pears and table grapes. The area planted by fruit type is shown in Figure 2.7. Table 2.14 presents the growth rates and sales in the deciduous fruit sector.

2.8.1.5 Deciduous fruit industry

The deciduous fruit industry is the most important branch by gross value of the horticultural industry contributing more than 28.7% of the total value of this branch in 2000–01. It has become a world player in the export market, generating over R4.6 billion a year in export earnings. The number of exporters has grown significantly.

Apples are the most important export fruit while the domestic market and fruit juice industry plays an important role. There are nearly 10 million trees occupying 17 000 hectares mainly in the Elgin/Grabouw, Ceres and Langkloof production areas. Apple production increased from 515 074 tons in 1990–91 to 561 889 tons in 2000–01.

Pear exports totalled 89 454 tons in 2000–01 while 109 610 tons were processed. There are approximately five million trees on 10 000 hectares mainly in the Elgin,

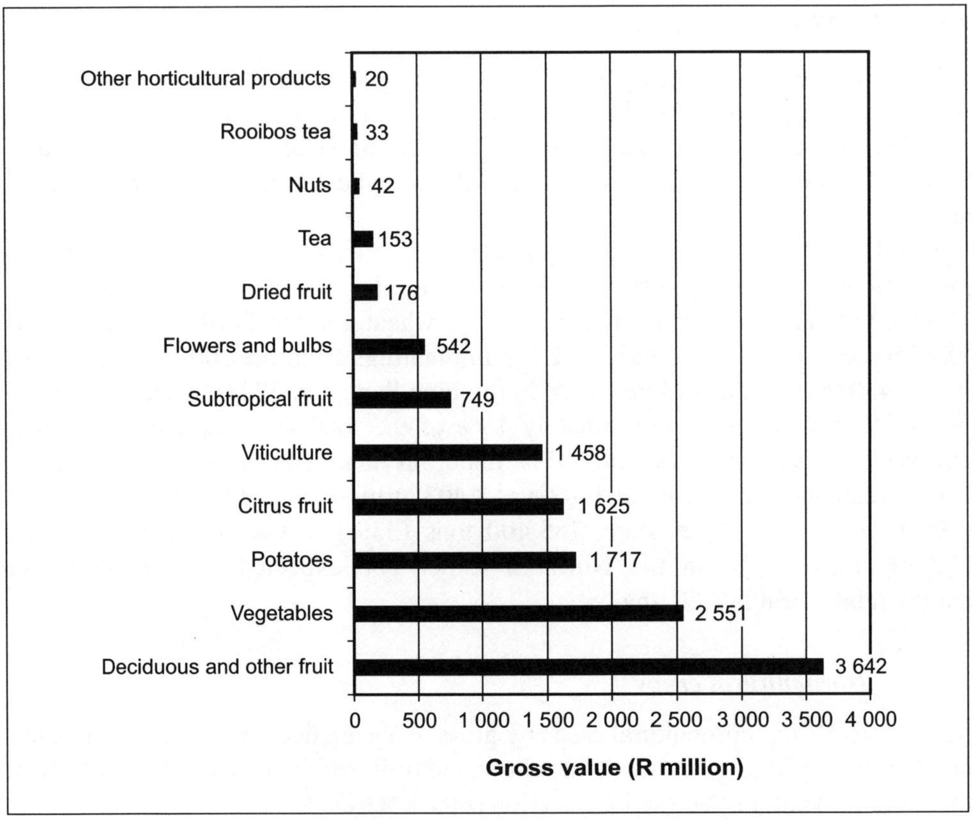

Figure 2.6: Ranking and value of the leading horticultural crops, 2000.
Source: Abstract of Agricultural Statistics (2000) as adapted

Ceres, Stellenbosch and Langkloof production areas. Pear production increased significantly from 208 900 tons in 1990–91 to 267 340 tons in 2000–01 at an average annual rate of 2.5%.

About 6 000 hectares have been planted with table grapes in the Western Cape, 3 600 hectares of which is found in the Hex River Valley in the Worcester district. More than 52% of table grape exports originate in this valley. Other important production areas include the Paarl/Wellington districts in the Western Cape and the sultana production area in the Lower Orange River of the Northern Cape. Table grape production increased from 1.31 million tons in 1990–91 to 1.33 million tons in 2000–01.

Intake figures show that approximately 62% of the total plum exports originate in the Franschoek and Stellenbosch districts of the Western Cape. Nearly 1.6 million trees have been planted on 1 800 hectares. Plum production increased from 19 376 tons in 1990–91 to 39 077 tons in 2000–01.

Peaches are mainly grown in the Franschoek, Paarl and Wolseley production areas. Peach production declined from 1981–91 but increased from 157 100 tons in 1990–91 to 173 548 in 2000–01.

Table 2.14: Deciduous fruit: growth rates in production and sales.

Product		*Period*	*Average volume (tons)*	*Annual growth rate (%)*
Apples	Production	1990–2000	557 799	0.75
	Local sales	1990–2000	123 349	4.63
	Export	1990–2000	206 473	–3.25
	Processing	1990–2000	197 898	1.64
Apricots	Production	1990–2000	63 029	2.78
	Local sales	1990–2000	3 333	–4.58
	Export	1990–2000	2 842	8.99
	Processing	1990–2000	47 499	3.83
Grapes	Production	1990–2000	1 377 707	0.87
	Local sales	1990–2000	22 499	0.68
	Export	1990–2000	114 308	10.87
	Processing	1990–2000	3 820	–0.35
Pears	Production	1990–2000	251 514	3.45
	Local sales	1990–2000	41 350	8.80
	Export	1990–2000	102 804	0.53
	Processing	1990–2000	9 530	4.34
Peaches	Production	1990–2000	190 731	3.21
	Local sales	1990–2000	32 316	–0.78
	Export	1990–2000	3 879	9.22
	Processing	1990–2000	126 522	5.23
Plums	Production	1990–2000	30 868	8.97
	Local sales	1990–2000	6 168	6.25
	Export	1990–2000	23 102	11.00
	Processing	1990–2000	959	–34.49
Wine (litres)	Good wine	1990–2001	487 113 718	4.03
	Distilling wine	1990–2001	187 021 842	–9.40
	Juice	1990–2001	94 863 054	1.03
	Rebate wine	1990–2001	76 765 084	–11.35
	Total	1990–2001	845 763 698	–0.41

Source: Abstract of Agricultural Statistics (2002) as adapted

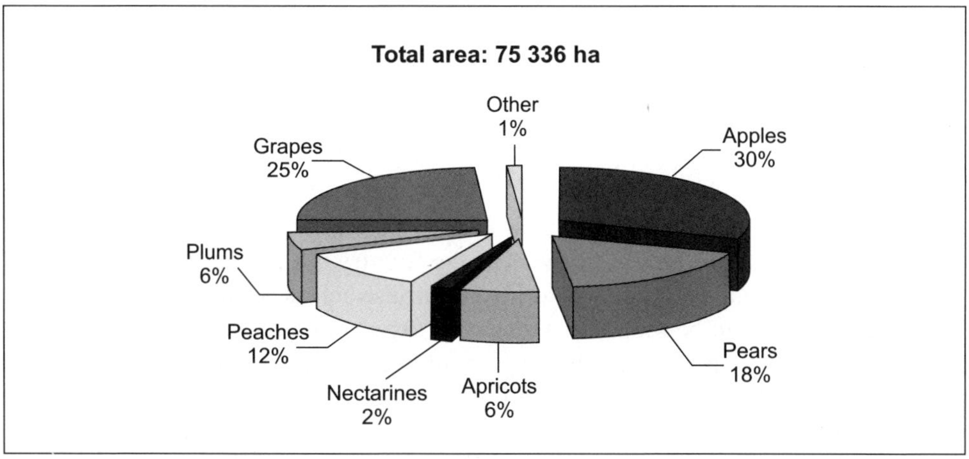

Figure 2.7: Percentage of area planted to diciduous fruit, 2000.

Approximately 65% of apricot trees are found in the Little Karoo production area of Ladismith and the Langkloof in the Western Cape. The crops are largely dried and canned. Production trends show an increase of 48 469 tons in 1990–91 to 66 294 tons in 2000–01 at an average annual growth rate of 3.2%.

2.8.1.6 Animal products

South Africa's top animal products by gross value in 2000 were fowls slaughtered, cattle and calves slaughtered, fresh milk, eggs and sheep and goats slaughtered, which together present 81% of the gross value of livestock and animal products.

2.8.1.6.1 Natural grazing and livestock numbers

The general condition of natural veld in South Africa is poor and is deteriorating at an alarming rate. According to Meyer (1998), the main reason for this condition is an overestimation of the carrying capacity of the veld by farmers which results in an overstocking of the veld. In recent years, the growth of the game farming industry has been threatening conservation farming, mainly as a result of the following:

(a) the conservation of game is practised at the expense of the conservation of vegetation;
(b) the fencing of farms and the increase in game numbers;
(c) insufficient information of game farming and veld management; and
(d) the exclusion of game in determining the carrying capacity of the veld.

According to statistics held by the National Department of Agriculture there were 13.5 million cattle, 25.3 million sheep, 2.4 million goats and 1.5 million pigs in the country in 2001.

The distribution of cattle in the nine provinces was as follows: KwaZulu-Natal (20.2%), Eastern Cape (22.5%), Free State (16.5%), North West (12.7%), Mpumalanga (9.8%), Limpopo (8.9%), Western and Northern Cape (7%) and Gauteng (2%). The low percentage of the Limpopo Province can be ascribed to the increase in game farming.

2.8.1.6.2 Cattle and calves slaughtered

Statistics on beef production in South Africa shows that production is directed almost entirely towards the local market as the country is a net importer of meat and meat products, mainly from neighbouring countries such as Namibia and Botswana. While there appears to be scope for substantial expansion in South African red meat production, for example, by feedlotting should grain/meat price relations make it profitable, there is also scope for considerable local market expansion, particularly if income levels rise. The income elasticity of demand for red meat is fairly high, implying that increased real incomes will lead to substantial increases in South African demand for red meat.

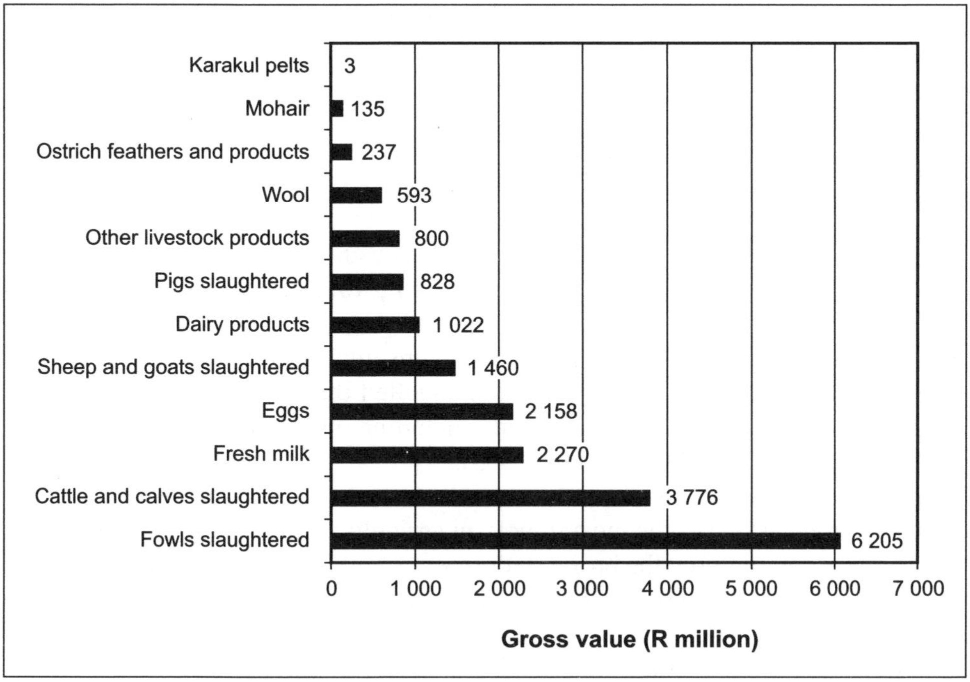

Figure 2.8: Ranking and value of the leading animal products, 2000.
Source: Abstract of Agricultural Statistics (2000) as adapted

Beef and veal (regarded as one commodity) form the main product of the red meat industry. Although economic and technological factors have an influence on production cycles and trends in the meat industry, variation in production (which is mainly extensive) is, in the long run, mainly determined by climatic conditions, especially the rainfall cycle. The build-up phase of stock numbers, i.e. when fewer female animals are slaughtered, is directly linked to long-term weather patterns.

Cattle numbers reached a low of 11.9 million in 1984–85, after which a build-up phase followed. As a result, slaughterings reached a high of 2.4 million carcasses in 1984 and a low of 1.9 million in 1988. Cattle slaughterings show a decrease from 2.5 million in 1990–91 to 2.1 million in 2000–01. Beef prices are directly affected by these cycles, i.e. prices rise during the build-up phase, reach a saturation point and then fall as more cattle are slaughtered. A growth phase in the economy and a concomitant general rise in the demand for red meat usually leads to meat price increases. The part of the livestock cycle associated with increasing numbers also often coincides with a period of decreasing spending power, which makes price variations even more profound.

Low beef prices have major repercussions on the grain industry as the profitability of feedlotting – now using about one million tons of maize, wheat and sorghum annually – will be marginal. Beef prices also have an indirect effect on the dairy industry, i.e. when beef prices are favourable, farmers refrain from marketing calves in anticipation of higher prices. The opposite is also true: when prices are low, farmers market calves at a young age, because it becomes unprofitable to rear them for a longer period.

2.8.1.6.3 Fowls slaughtered

The broiler industry has shown continued growth for over 20 years. According to Rainbow Chickens, the estimated per capita consumption of chicken in South Africa was 13.8 kg in 1990–91. This compares with 15.6 kg beef, 4.1 kg mutton and lamb, and 3.4 kg pork. Unlike red meat that shows cycles in production and consumption, a constant growth in the consumption of chicken has been experienced. The consumption of chicken equalled that of red meat by 1992. By 2002, per capita poultry consumption, including chickens and other poultry products, reached 22.65 kg/yr (NDA, 2003).

An analysis of the historic trends in production shows that the broiler industry can be regarded as one of the most important agricultural industries. On the basis of gross value of production, the broiler industry currently lies in second position after the maize industry. The gross value of poultry meat increased from R2 237 million in 1990–91 to R6 204 million in 2000–01, while its share in the total value of agricultural production rose from 6.7% in 1980–81 to 12.6% in 2000–01. If current growth trends prevail, the broiler industry will become the single most important agricultural industry by the end of the century. Currently an estimated 7.5 million broilers are slaughtered per week, while 1.5 million live broilers are sold to the informal sector (SAPA, 2003).

2.9 CONCLUSION

The structure of South African agriculture is characterised by the limits set by natural resources, climate, population profile, composition, government policies and traditional practices. The number of farms and the total area of farms has risen and fallen in line with the trend in the number of farming units over several decades. The size of commercial farming units has tended to increase over the past 30 years.

The importance of agriculture in the South African economy is highlighted by the delivery of food to the nation at affordable prices, the relative contribution to the GDP, its share of the labour force and its contribution as an earner of foreign exchange. The gross value of agricultural production annually expanded by 10.7% between 1990 and 1995, and 6% between 1995 and 2000, well below the inflation rate.

The top field crops by gross value were maize, sugar cane, wheat, hay, sunflower seed, tobacco and groundnuts. The most important horticultural crops by gross value were deciduous fruit, viticulture and subtropical fruit, which presented almost 90% of the gross value of horticultural production in 2000. South Africa's top animal products by gross value in 2000 were fowls slaughtered, cattle and calves slaughtered, fresh milk, eggs and sheep and goats slaughtered, which together present 80% of the gross value of animal products.

In conclusion, this study shows that the relative position of horticultural crops and animal products has strengthened at the expense of field crops, especially since the 1980s. Shifts of crop production to the higher potential areas and livestock farming to the semi-arid areas can be expected to continue where livestock production will replace crop production in marginal areas.

REFERENCES

ABSTRACT. (2000). *Abstract of Agricultural Statistics*. Pretoria: National Department of Agriculture.

ABSTRACT. (2002). *Abstract of Agricultural Statistics*. Pretoria: National Department of Agriculture.

BRAND, S.S. (1969). The contributions of agriculture to the economic development of South Africa since 1910. PhD thesis, University of Pretoria.

BRAND, S.S. (1986). Agricultural outlook in the TBVC and self-governing states. Agricultural Outlook Conference, Pretoria.

BRAND, S.S. (1990). Political, economic and sociological objectives of irrigation development: present and future. Paper presented at a workshop conducted by the Co-ordinating Committee for Irrigation Research, Pretoria.

COBBETT, M.J. (1987). The land question in Southern Africa: a preliminary assessment. *SAJE* 55(1): 63–77.

CUSTOMS & EXCISE. (2001). *Trade information data base*. Pretoria: Government Printer.

DEVELOPMENT BANK OF SOUTHERN AFRICA (DBSA). (2000). *Development Information Base*. Midrand.

DEVELOPMENT INFORMATION BUSINESS UNIT. (1997). Operational information. Development Bank of Southern Africa, Midrand.

DEVELOPMENT INFORMATION BUSINESS UNIT. (2000). Operational information. Development Bank of Southern Africa, Midrand.

ECKERT, J.B. (1994). The land question in South Africa: present and future perspectives. *Agrifutura Bulletin* 1(2), University of Stellenbosch.

FÉNYES, T.I., VAN ZYL, J. & VINK, N. (1987). Structural imbalances in South African agriculture. *South African Journal of Economics* 56(2/3): 181–194.

KINDLEBERGER, C.P. (1962). *International Economics*. Homewood, Illinois: Richard D. Irwin, Inc.

LINDERT, P.H. & KINDLEBERGER, C.P. (1982). *International Economics*. Homewood, Illinois: Richard D. Irwin, Inc.

MCKENZIE, C.C. & LOUW, A.H. (1990). Completion report: land use, agricultural productivity and farming systems. Unpublished report, DBSA.

MEYER, N.G. (1998). The agricultural potential of South Africa: a provincial perspective on food security and land reform. PhD thesis, University of Pretoria.

NATIONAL DEPARTMENT OF AGRICULTURE (NDA). (2000). Operational Information. Pretoria.

NDA. (2001). *Agricultural Economic Trends*. Pretoria.

SOUTH AFRICAN POULTRY ASSOCIATION (SAPA). (2003). Operational Information, 2000. Johannesburg.

SOUTH AFRICAN RESERVE BANK (SARB). (1997). *Quarterly Bulletin*. Several issues.

STATISTICS SOUTH AFRICA. (1993). *Census of Agriculture, 1993*. Report No 06-01-20. Pretoria: Government Printer.

STATISTICS SOUTH AFRICA. (1996). *Population Census, 1996.* Pretoria: Government Printer.
STATISTICS SOUTH AFRICA. (2002). *South African Statistics.* Pretoria: Government Printer.
VAN AARDT, C.J., VAN TONDER, J.L. & SADIE, J.L. (2000). A projection of the South African population, 1996–2021. Bureau for Market Research, UNISA, Pretoria.
WATER RESEARCH COMMISSION (WRC). (1997). Operational information. Pretoria.
WORLD BANK. (1994). South African agriculture: structure, performance and options for the future. Discussion Paper 6. The World Bank Southern African Department, Washington.
WEFA CONSULTANTS. (2001). Operational information. Pretoria.

CHAPTER 3

Problems and Opportunities in South African Agriculture

Gerald Ortmann and Charles Machethe

3.1 INTRODUCTION

South Africa comprises 122.3 million hectares of which 100.7 million hectares (82.3%) consist of farmland, including potential arable land and grazing land (Directorate: Agricultural Statistics, 2002: 5). The area of potential arable land is only 16.7 million hectares or 13.7% of the total area of the country. Nature conservation comprises 11.8 million hectares (9.6%) and forestry 1.4 million hectares (1.2%). South Africa's total land area represents 13.4% of the area of the United States of America (USDA, 2003a) and five times that of the United Kingdom.

Primary agriculture forms an important part of the South African economy despite its relatively small direct share of less than 4% of the total GDP (Directorate: Agricultural Statistics, 2002: 78). At present there are about 50 000 large-scale commercial farmers employing approximately one million workers, or about 11% of the total formal employment of South Africa. In 2000, these farmers exported about R16 billion (US$2.3 billion) worth of products, or nearly 10% of South Africa's total exports (NDA, 2001).

The agricultural sector also has about 240 000 small-scale farmers, providing a livelihood for more than a million of their family members, and an estimated three million farmers, mostly in the communal areas of the former homelands, who produce food primarily to meet their families' subsistence needs (NDA, 2001).

This chapter introduces some of the important challenges and opportunities facing South African agriculture. Section 3.2 presents some critical problems in South African agriculture, including the dualistic nature of the sector, the challenges of land redistribution, uncertain property rights, the relatively risky nature of crop production, high debt levels of many commercial farms, restrictive labour policies, HIV/AIDS, low productivity in smallholder agriculture, the likely introduction of a rural land tax, under-investment in the rural economy, and competitive pressures on agricultural producers and agribusiness. On the positive side, there are also many opportunities, including market-led land reform, greater access to foreign markets, smallholder participation in value chains, greater commitment from government to agricultural research, institutional change to promote competitiveness, and greater co-operation among major role players in agriculture.

3.2 PROBLEMS IN SOUTH AFRICAN AGRICULTURE

3.2.1 Dualistic nature of agriculture and land redistribution

South African agriculture has a dual nature, with a well-developed commercial sector comprising about 50 000 commercial farmers occupying 86% of agricultural land (Directorate: Agricultural Statistics, 2002; NDA, 2001). Small-scale and subsistence (communal) farmers occupy the remaining 14% of farmland. Past government policies, which restricted blacks to certain regions (homelands) and therefore excluded them from entering mainstream agriculture, was a factor in promoting this dualistic nature of agriculture. The redistribution of farmland in South Africa is high on the government's agenda, and it is generally accepted that it is necessary for political stability and hence economic growth.

The problem of a skewed distribution of land in South Africa still exists despite genuine efforts by the government to redress the imbalances in land ownership. The government intends to transfer ownership of 30% of agricultural land over 15 years (Agriculture and Land Affairs Portfolio Committee, 2000). The government's land redistribution programme aims to 'provide the wider majority of South Africans with access to land for residential and productive use in order to improve their livelihoods, with particular emphasis on the poor, labour tenants, farm workers, women and emergent farmers' (Department of Land Affairs, 1997). To achieve this objective, the government initially introduced a means-tested settlement and land acquisition grant of R15 000 per beneficiary household for the purchase of land and to make other capital investments. This grant was generally not successful in establishing commercial farmers and it was replaced with the Land Redistribution for Agricultural Development (LRAD) programme in 2001, under which beneficiaries can access a range of grants (R20 000 to R100 000) depending on the amount of their own contribution in kind, labour and/or cash. Beneficiaries must provide an own contribution worth at least R5 000. This shift in government policy is in favour of creditworthy land reform projects, and makes larger cash grants available to emerging farmers and farm workers to finance land or equity (shares) in established commercial farming operations.

The limited progress that has so far been achieved in the redistribution of land in South Africa has caused some tensions and frustrations within the disadvantaged communities and some isolated land invasions have been reported in parts of the country.

3.2.2 Uncertain property rights

Land claims (backed by legislation to secure the land rights of labour tenants on commercial farms and to protect farm workers against eviction), land invasions and the measures taken by the Zimbabwean government to redistribute land have created considerable uncertainty among many commercial farmers in South Africa with regard to property rights. In addition, crimes committed on commercial farms, such as the murder of farmers and theft of livestock and other property, have made farming less attractive. The government has tried to remove the uncertainty by declaring that land transfers will occur within the legal framework and is implementing measures to curb crime on farms.

In smallholder agriculture, insufficient security of land tenure and free rider problems associated with communal land ownership are widely considered to be obstacles to agricultural development. Land tenure security is generally regarded as a necessary but insufficient condition for economic (agricultural) development (Roth & Haase, 1998). Strengthening land tenure security in the developing areas of South Africa would improve the performance of smallholder agriculture by strengthening incentives to invest and promoting allocative efficiency through rental transactions in farmland (Lyne & Graham, 2001).

3.2.3 Risk in crop production

The relatively small percentage of land that is potentially arable (13.7%) is an indication that most of South Africa receives a low rainfall that makes crop production limited or impossible. In fact, 32% of the country receives an average rainfall of less than 300 mm per year and 57% receives a rainfall of less than 500 mm per year (Lynch, 2002). South Africa's rainfall is also very erratic; for example at the town of Bothaville, which lies in the main maize (corn) growing area of South Africa, the mean annual rainfall over 117 years was 534 mm with a coefficient of variation (CV) of 28.6%. This area has a maize grain yield potential of 2–3 tons/ha. At the town of Delmas, further east, the mean annual rainfall over 106 years was 681 mm and the CV 25%. The maize grain yield potential in this region is 3–4 tons/ha (Lynch, 2002). The variability in rainfall is reflected in highly variable yields of dryland crops. For example, Figure 3.1 shows the variability in the yield of maize grain, the most important field crop in South Africa with a gross value of R6.258 billion (US$0.71 billion) in 2000–01, or 37% by value of all field crops in that year (Directorate: Agricultural Statistics, 2002: 80). The mean yield over the period 1970–71 to 2000–01 was 2.027 tons/ha with a CV of 30.4%, which is higher than the CV of rainfall at Bothaville and Delmas. The relatively low yield and high CV suggest that other, less suitable areas were also used for maize production.

To put the maize yield and yield variability into perspective, the USA achieved a mean maize grain yield of 6.8 tons/ha over the period 1970–2001 with a CV of 18.3%. The states of Illinois, Iowa and Indiana averaged maize grain yields of 7.5, 7.4 and 7.2 tons/ha, respectively, over the same period with CVs of 18.9, 19.2 and 19.1% (USDA, 2003b). Clearly, the risk involved in maize grain production in South Africa is much higher than in the USA.

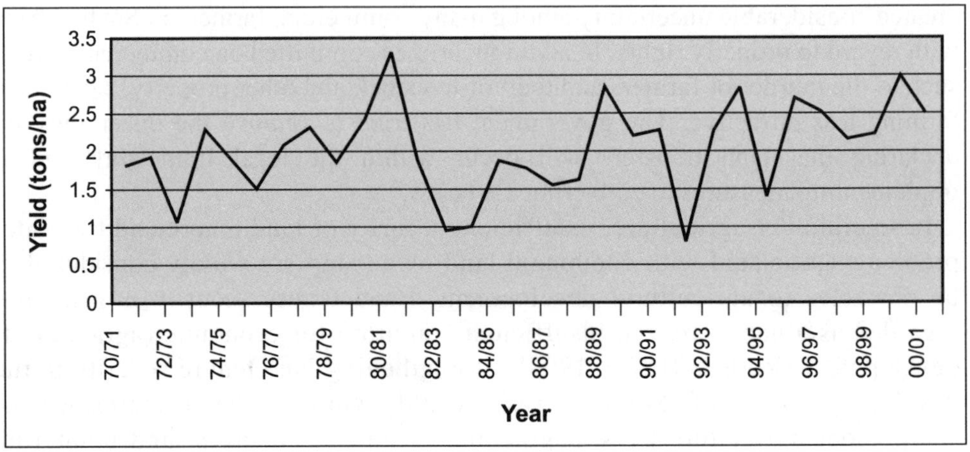

Figure 3.1: Maize grain yield variability in South Africa, 1970–71 to 2000–01 (mean yield=2.027 tons/ha; coefficient of variation=30.4%).

Source: Directorate: Agricultural Statistics (2002)

3.2.4 High debt levels

Farm debt in South Africa was roughly R30 billion (US$4.32 billion) in 2000, or 33.6% of the value of capital assets on commercial farms (Directorate: Agricultural Statistics, 2002: 84–85). This compares to a debt/asset ratio of 13.9% in 1980. The variability in the debt/asset ratio on South African commercial farms over 20 years is presented in Figure 3.2. The mean debt/asset ratio over this period was 24.8% with a coefficient of variation of 19.3%. As a comparison, the debt/asset ratio for farms in the USA in 2000 was 15.5% (USDA, 2003a).

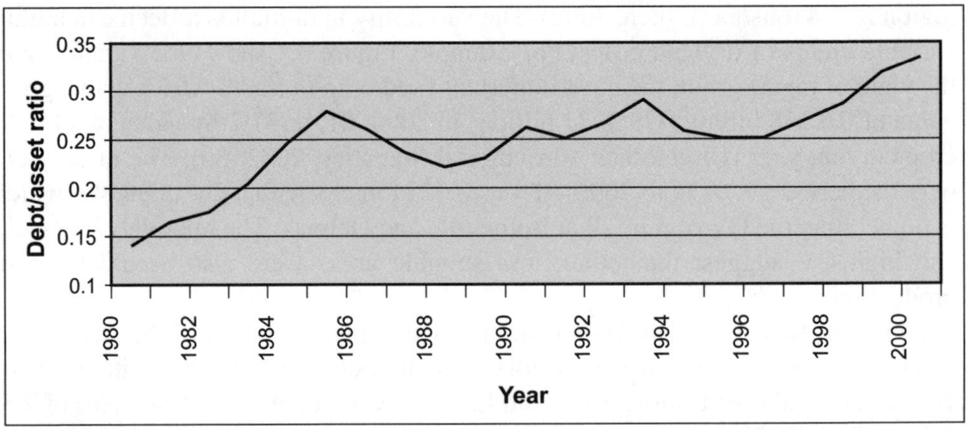

Figure 3.2: Debt/asset ratio on South African commercial farms, 1980–2000 (mean ratio=24.8%; coefficient of variation=19.3%).

Source: Directorate: Agricultural Statistics (2002)

The relative debt position of commercial farms in South Africa has increased sharply since 1980. Following severe droughts in the early 1980s and rising nominal and real interest rates, the debt/asset ratio increased to about 28% in 1985. The ratio then decreased following more favourable weather conditions and declining nominal interest rates, but severe drought in 1991–92 again led to a rising debt/asset ratio. Although the ratio decreased to about 25% in the mid-1990s following good production seasons, the ratio then increased sharply to the highest level ever, namely 33.6% in 2000. In addition to the effects of weather and interest rates, the real value of capital assets (i.e. land and fixed improvements, machinery and livestock) on commercial farms has fallen steadily during the last two decades. For example, the real value in 1980 was R185 billion, R104 billion in 1990 and R64 billion in 2000 (1995=100). The value of land and fixed improvements as a proportion of the total value of capital assets fell from 78% in 1980 to 66% in 1990 and to 58% in 2000 (Directorate: Agricultural Statistics, 2002: 84). Clearly, declining real land values have also contributed to the rising debt/asset ratio.

3.2.5 Restrictive labour policies

A number of labour laws that affect the agricultural sector have been in place since the 1990s: the Occupational Health and Safety Act of 1993; the Public Holidays Act of 1994; the Labour Relations Act of 1995; the Basic Conditions of Employment Act of 1997; and the Employment Equity Act of 1998. The Labour Relations Act governs how employers and employees should deal with each other in the workplace; the Basic Conditions of Employment Act lays down the minimum standards of employment below which employers may not contract; and the Employment Equity Act seeks to promote equity in the workplace and fair treatment in employment by eliminating unfair discrimination and implementing affirmative action measures. The Labour Relations Act and the Basic Conditions of Employment Act are sometimes referred to as 'job security laws' (Siebert, 2001) because they provide protection for employees from unfair dismissal and ensure that employees are properly compensated if they lose their jobs as a result of economic restructuring.

Although these laws have some benefits, they result in higher transaction and wage costs for employers in the agricultural sector by, for example, (a) raising the cost of dismissing and/or downsizing the workforce, and (b) increasing the cost of labour by requiring employers to pay for overtime and work performed on public holidays and Sundays. Some economists hold the view that labour laws increase the cost of labour and thus lead to an increase in unemployment. The reduction in employment attributed to job security laws is estimated at 4% for South Africa (Siebert, 2001). Higher labour costs in the agricultural sector due to job security laws may cause farmers to reduce their labour force, invest in machinery, adopt new technologies (for example, new crop varieties, which are less labour intensive), and employ machinery or labour contractors (Goedecke & Ortmann, 1993; Newman & Ortmann, 1996). This may exacerbate the problem of high unemployment in South Africa, which in 2002 was close to 30% (South African Reserve Bank, 2003).

Minimum wage laws have been introduced in various countries to ensure that low-paid workers receive a decent wage and thus reduce the level of poverty (Ehrenberg & Smith, 1991). Hertz (2002) concludes that a national minimum wage is likely to reduce poverty in South Africa. The government has introduced minimum wage provisions for domestic workers, and from 2003 for farm workers.

Whether minimum wage legislation reduces employment partly depends on the level at which the minimum wage is set relative to prevailing wages (Ehrenberg & Smith, 1991). The standard model of a unified and perfectly competitive labour market predicts that minimum wage provisions result in a decrease in employment if the minimum wage level is set above the market equilibrium (Hertz, 2002). Some analysts believe that the introduction of minimum wage provisions in the agricultural sector will increase the cost of labour relative to productivity, and farmers will respond by cutting down on labour. Different minimum wages have been set for high- and low-income areas. This is expected to impact on the regional movement of labour, while it may influence the regional location of labour-intensive crops. The setting of minimum wages in Zimbabwe in the early 1980s is said to have resulted in the dismissal of thousands of farm employees (Ehrenberg & Smith, 1991).

3.2.6 HIV/AIDS

The socio-economic impact of HIV/AIDS on the agricultural sector and the economy of South Africa is highly significant. HIV/AIDS affects agriculture and the economy negatively by destroying the social capital, weakening institutions and exacerbating the problem of poverty. The United Nations Development Programme has estimated that the Human Development Index of South Africa will be 15% lower in 2010 due to HIV/AIDS (Joint United Nations Programme on HIV/AIDS & United Nations Development Programme, 2000).

Agricultural productivity, labour turnover rates, and production costs on commercial farms are adversely affected by the deaths and ill-health of workers with HIV/AIDS. The capacity of small-scale farming households is limited as HIV/AIDS prevents them from utilising their land effectively – infected members are too weak to perform farming tasks and members with farming skills die from the disease. On the demand side, the death of humans and poor health may result in reduced domestic demand for agricultural products. The impacts of HIV/AIDS come at a considerable cost to a business, whose competitive advantage may be adversely affected due to lower productivity and higher labour costs.

3.2.7 Low productivity in smallholder agriculture

Smallholder agricultural productivity in South Africa is generally regarded as being very low. Improving productivity is a crucial though not sufficient condition for eradicating rural poverty. Increasing smallholder agricultural productivity requires tenure security and improving access to reliable and effective farmer support services such as extension, finance and marketing. Agricultural service institutions in the past were designed along racial lines with the result that smallholder farmers' needs were not adequately addressed. Lyne (1996) notes that smallholder farmers have limited access to factors of production, credit and information. In addition to tenure security, improving smallholder agriculture requires that support services be accessible to the majority of smallholder farmers (Machethe & Mollel, 2000). International experience indicates that with adequate access to farmer support services, smallholder farmers can increase productivity and production significantly. For example, smallholder farmers in Zimbabwe doubled maize and cotton production in the 1980s when extension, marketing and credit services were provided (Rukuni & Eicher, 1994).

3.2.8 Rural land tax

A tax on rural land (including farms) in South Africa is likely to be introduced in terms of the proposed Local Government Property Rates Bill. Considerable debate has been raging on the likely costs and benefits of such a tax, and what an appropriate tax rate should be. Commercial farmers are concerned that policymakers do not understand their cash flow position, as the current annual returns (rents) to farmland are only around 5% of land values. A land tax of 2%, for example, would reduce these returns by 40%, which may cripple many farmers financially. Much higher rates have been mentioned in the press, creating considerable uncertainty for commercial farmers. A land tax will reduce the value of land and therefore the tax base. This means that at a rate of more than 5% all rents will be taxed away. This is tantamount to nationalising the income stream of the land (Nieuwoudt, 1995).

A widely held view is that a land tax may promote a more intensive use of land and the cultivation of unutilised land. Nieuwoudt (1995) has argued convincingly that this view is false. An active land rental market creates an opportunity cost that penalises the non-use of land if it could earn a profit. Nieuwoudt (1995) maintains that the presence of idle land may possibly be attributed to development costs exceeding the expected benefit from that land, and that a land tax will aggravate rather than improve this situation because 'a tax on agricultural land will, in the long run, fall on new investment and as such will be a disincentive to future investment in land improvements' (p. 85). Thus, a land tax would detrimentally affect the productivity of the land.

Clearly, policymakers need to understand the impacts of a land tax on commercial farming in South Africa. They need to consult widely and study the benefits and costs of land taxes in other countries in order to make the 'right' decisions regarding the administration and rate of land tax, and thus reduce the uncertainty for farmers. In the USA, for example, property taxes on farm real estate are generally less than 1% of the real estate value (USDA, 2003c). Commercial farmers in South Africa do not receive the same level of government support (for example, subsidies for inputs or crop insurance) that is received by farmers in the developed countries with whom South Africa trades. This raises the question of whether higher land tax rates in South Africa relative to these countries are justified.

3.2.9 Under-investment in the rural economy

Investment in rural areas in physical and institutional infrastructure is inadequate. For example, deteriorating roads in the rural areas have increased the cost of transport. Crime prevention and control is also inadequate. Many farmers and family members have been robbed or murdered and theft of livestock is a major problem and has financially crippled many farmers. There is also a drain of human capital from agriculture with educated youth seeking vocations in urban areas or overseas. Inadequate investment and the flow of capital out of rural areas are problems that need to be urgently addressed by government.

The wastage and under-utilisation of scarce resources also constitute a considerable problem in rural areas. For example, the absence of well-defined, secure and tradable water rights results in substantial wastage of scarce water resources (Conradie, 2002; Nieuwoudt, 2000). In addition, the absence of secure and tradable property rights in communal areas has led to the degradation of land in those areas. Clearly, this creates an opportunity for induced institutional change that could benefit agriculture and the economy.

3.2.10 Competitive pressures on agricultural producers and agribusiness

Agricultural producers and agribusiness firms in South Africa face tough competition in domestic and international markets. The successful conclusion to the Uruguay Round of the General Agreement on Tariffs and Trade (GATT) in 1994 and the subsequent World Trade Organisation (WTO) have promoted the liberalisation of international trade in agricultural and food products. Local agricultural markets have also been deregulated since the mid-1990s. This has been complemented by rapid advances in technology, particularly of information and communication technology, which has reduced costs and accelerated the availability of relevant information. Customers have also become more demanding with regard to product and service quality. Agricultural producers and agribusiness managers are, therefore, under increasing pressure to improve product and service quality, enhance productivity, and reduce production and transaction costs.

Presently, South Africa's competitive position in the world economy is not highly ranked. For example, the country was ranked 25th in the micro-economic factor section of the 2000 Global Competitiveness Report (Bennett, 2000), and seventh among African countries. According to Porter, as cited by Bennett (2000), South Africa's productive capacity is much greater than that reflected in its GDP per capita, which Porter believes is due to the country's low productivity levels. There is also evidence that some sectors of agriculture and value-adding activities are not competitive in local and international markets, which has been attributed to high input costs combined with low productivity, poor business strategies and inefficiencies, and 'unfair' trade practices by the country's competitors (NDA, 2001). Clearly, there is a need to investigate strategies and consider institutions that will promote South Africa's global competitiveness. A major concern for South African producers is the high levels of subsidies and protection that farmers in the European Union and the USA enjoy. The developed countries have frequently dumped their surplus products in developing countries, including South Africa, which has resulted in depressed product prices for producers. The complaint that 'the playing fields are not level' has often been heard. In response, the government has increased tariffs in order to protect local industries from the effects of dumping.

Symptoms of the major problems facing South African agriculture are manifesting themselves in various ways, including the increased number of bankruptcies of commercial farms and low investor confidence in agriculture.

3.2.11 Bankruptcy of commercial farms

The number of commercial farms annually sequestrated has risen sharply over the last three decades. For example, in the period 1970–2001 the lowest number of annual sequestrations was 41 (in 1970 and 1981) and the highest 575 (in 2001) (Grain SA, 2002). Figure 3.3 shows that there was a sharp rise in sequestrations in the mid-1980s, following severe droughts in the early 1980s and rising nominal interest rates that peaked at over 25% (and real rates at over 10%) in the mid-1980s. Severe droughts were again experienced in the early 1990s, which led to rising sequestrations, followed by a period of favourable weather conditions and lower sequestrations. Since 1996 the number of sequestrated farms has again increased, reaching a record 575 in 2001. This may reflect the combined effects of market deregulation and elimination of government support for commercial agriculture since the mid-1990s (thus increasing risk), continuing relatively high nominal interest rates and the difficulties that many farmers had in adapting to this situation.

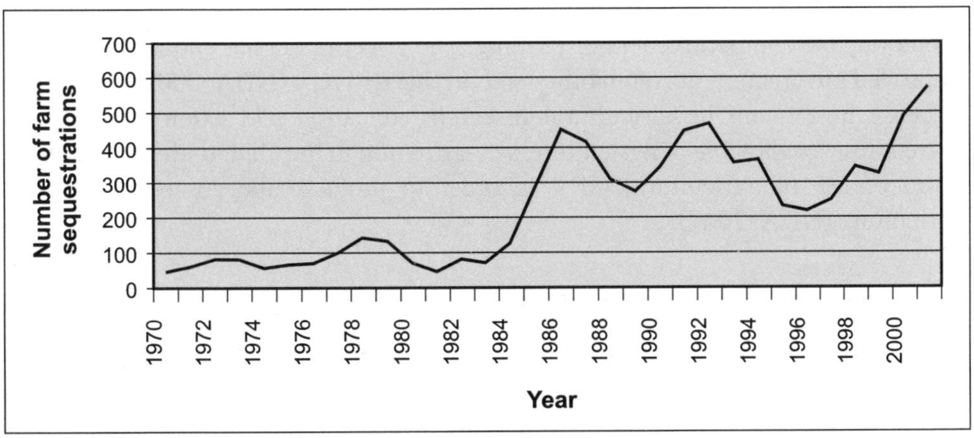

Figure 3.3: Number of commercial farm sequestrations in South Africa, 1970–2001.

Source: Grain SA (2002)

The rate of sequestrations relative to the number of commercial farms has increased even more sharply as the number of commercial farms has decreased from about 91 000 in 1970 to about 50 000 in 2001. This is of concern as extensive farm bankruptcies impose major adjustment costs on the agricultural economy and give rise to demands for government assistance to alleviate financial stress for farmers.

A study by Swanepoel *et al.* (1998) confirmed *a priori* expectations that maize farm and extensive beef farm bankruptcies were negatively related to annual rainfall (business risk factor), but positively related to the lagged aggregate farm debt/asset ratio and lagged real interest rates (financial risk factors). Maize farm bankruptcies also increased as lagged real maize and beef producer prices fell (business risk factors). Beef farm bankruptcies rose with lower lagged real beef producer prices and higher lagged real stockfeed subsidies and transport rebates (business risk factors). Part of the rise in maize and extensive beef farm failures in the study period (1970–94) can therefore be ascribed to changes in agricultural price and macroeconomic policies.

3.2.12 Low investor confidence in agriculture

Recent farm invasions in Zimbabwe, uncertainty regarding land claims and how land redistribution will be managed in future, illegal land occupations, the likely introduction of a rural land tax, ongoing farm robberies and murders, the introduction of minimum wages for farm workers and greater labour union activity on farms are major issues causing low investor confidence in agriculture. In addition, current annual returns (rents) to farmland are only around 5% of land values.

Other issues that cause poor investor sentiment are inadequate, ineffective and inefficient support and service delivery from government agencies. Government has a crucial role to play in creating an enabling environment for investors and promoting the competitiveness of farming. The government has committed itself to improved governance, accountability and service delivery (NDA, 2001). It will also increase investment in agricultural research, education and extension from the current low level of 1% of agriculture's contribution to the GDP to the international norm of 3% of agricultural GDP in order to promote the competitiveness of agriculture (NDA, 2001).

3.3 OPPORTUNITIES IN SOUTH AFRICAN AGRICULTURE

Despite the problems faced by agriculture in South Africa, there are also various opportunities for this sector. These include market-led land reform, greater access to foreign markets, smallholder participation in value chains, increased funding for agricultural research, institutional change, and greater co-operation among major role players in agriculture.

3.3.1 Market-led land reform

Although little progress has been made in redressing the imbalances in land ownership in South Africa, the potential exists for a peaceful resolution of the land issue. The South African government is committed to resolving the land problem and thus avoiding some of the problems experienced in neighbouring countries. The principle of market-based land redistribution has been established, and a major challenge for government and other institutions is how to finance land redistribution on a willing buyer–willing seller basis. South Africa has an active market for commercial farmland and a sophisticated banking sector that finances land transactions. Studies in KwaZulu-Natal have shown that private transactions are redistributing more land wealth than government-assisted transactions (Graham & Lyne, 1999).

The peaceful resolution of the land problem would remove uncertainty about property rights in the agricultural sector, which would have a positive impact on this sector and the economy as a whole. However, the efficient use of commercial farmland and other agricultural resources should not be compromised in the long term.

3.3.2 Deregulation of agricultural marketing and access to foreign markets

Agricultural markets in South Africa have been deregulated since the mid-1990s in tandem with the deregulation of world markets following the GATT agreements in 1994. Although deregulation of markets leads to uncertainty for producers, it also creates opportunities for them, particularly for 'alert entrepreneurs' (Pasour, 1981), who see profit opportunities in such markets. Trade opportunities for South African managers have proliferated and international pressure is being exerted on the country to liberalise its trading system even further (Van der Merwe & Otto, 1997).

South Africa has established a number of preferential trade relationships with countries in and outside the Southern African Development Community (SADC) region. Examples include the agreement on trade, development and co-operation with the European Union, preferential trade arrangements with the USA, the provision of market access through the Generalised System of Preferences by countries such as Norway, Switzerland, Hungary, Canada and the Czech Republic, the trade agreement to strengthen trade and investment with SADC countries, and the Southern African Customs Union free trade agreement. Through these trade agreements and arrangements, farmers have access to foreign markets and earn the foreign exchange needed to develop the country's economy. Smallholder farmers in particular could benefit significantly from these agreements provided appropriate measures are taken to ensure that they produce export quality products.

South African commercial farmers have generally adapted well to deregulated markets, many switching to producing commodities that are exported (for example, fruit, flowers and other niche products). The depreciating Rand has boosted earnings from agricultural exports (including the more 'traditional' products, such as maize grain and sugar). Farmers have been very innovative in finding markets and adapting their marketing strategies to suit changing circumstances. For example, maize farmers are increasingly making use of forward contracts and futures markets to better manage price risks (Bown *et al.*, 1999). In an uncertain world farmers will attempt to obtain more information on marketing opportunities if the expected benefits exceed the costs. There is no doubt that the Internet will play an increasingly important role in future as a useful information source for farmers, particularly with regard to marketing opportunities.

3.3.3 Smallholder participation in value-added chains

Most smallholder farmers in South Africa have not had the opportunity to participate in the production of high-value products due to limited resources and institutional constraints. Efforts of government and other institutions to remove these constraints are likely to open up opportunities for smallholder farmers to participate in value-added chains. Participation in value chains will be further enhanced by the strengthening of linkages between smallholders and commercial farmers on the one hand, and farm/non-farm linkages on the other (Machethe *et al.*, 1997). Successful contractual arrangements between small-scale farmers and processors have been established in the sugar, forestry and cotton industries, while some equity-sharing arrangements in the wine and fruit industries (where some commercial farmers restructured their operations as companies with their farm workers as shareholders) have been successful at redistributing land and wealth and promoting participation of employees in business decision making (Lyne & Graham, 2001; Ortmann, 2001).

3.3.4 Agricultural research

Various research studies have indicated that agricultural research is a high payoff investment (see for example, Khatri *et al.*, 1996; Thirtle *et al.*, 1998; Townsend *et al.*, 1997). It disproportionately benefits the poor because the proportion of their budget spent on food purchases is relatively high, and it can promote the competitiveness of South African agriculture and agribusiness on export markets (by reducing costs and improving product and service quality). Effective agricultural research can therefore promote economic growth. The objective of government to increase investment in agricultural research, education and extension from the present 1% of agricultural GDP to about 3% will promote resource productivity in the agricultural sector.

3.3.5 Institutional change

There is a need for private and public institutions to help promote the competitiveness of commercial farms and agribusiness firms, and enhance the productivity of communal farmers and the competitiveness of emerging farmers in

South Africa. Commercial agriculture and agribusiness are creating institutions (such as food safety standards and strategic partnerships), adopting existing private and public institutions (for example, the international quality assurance standard ISO 9000, and HACCP [Hazard Analysis and Critical Control Point] to improve food safety) or restructuring to add value to products and services, reduce costs and gain access to export markets (Ortmann, 2000). Government should focus its relatively scarce resources on providing physical and legal infrastructure (such as secure property rights and contract enforcement) to reduce transaction costs, including risk, so that markets work efficiently, promote agricultural research, and develop appropriate institutions to promote the efficient use of natural resources (for example, markets for land and water).

3.3.6 Co-operation among major role players

The Strategic Plan for South African Agriculture was published in November 2001. It is a critically important document because it is a product of government and the agricultural industry represented by members of AgriSA (commercial farmers and agribusiness) and the National African Farmers' Union (NAFU). The objective of this co-operation was to compile a common agricultural perspective 'to which government and industry would commit their efforts and resources in its implementation. The implications of this partnership are that government and industry now share a common perspective on the sector's strategic issues' (Didiza, 2001). The three strategic goals of access and participation, competitiveness and profitability, and sustainable resource management support the vision of a united, non-racial and prosperous agricultural sector.

3.4 CONCLUSION

There are a number of problems facing South African agriculture. One is its dualistic nature, comprising a well-developed commercial sector occupying 86% of farmland and a smallholder sector that grows products mainly for subsistence needs. The problem of a skewed distribution of farmland persists despite concerted efforts by government to redress the imbalances in land ownership. A more equitable distribution of farmland is regarded as necessary for political stability and economic growth. There has been a shift in government policy towards creditworthy land reform projects aimed at establishing commercial farmers. Uncertainty about property rights among commercial farmers has also emerged following numerous land claims, land invasions and crime. For smallholders, insecure land tenure and free-rider problems associated with communal land are considered to be obstacles to agricultural development.

The relatively risky nature of crop production in South Africa is seen as another problem area. Nearly 60% of the country receives an annual rainfall of less than 500 mm. The variability of production is high. For example, the variability of maize grain yield is reflected in a relatively high coefficient of variation (CV) of 30.4%

(mean yield of 2 tons/ha), compared to a CV of around 19% in the main corn producing states of the USA (mean yields of over 7 tons/ha). South African farmers are also faced with relatively high debt/asset ratios (mean ratio of 33.6% in 2000) compared to farmers in the USA (mean ratio 15.5% in 2000). The relative indebtedness of commercial farmers has increased sharply over the last two decades due to droughts, high nominal and real interest rates, deregulation of markets and declining real land values. The rate of farm sequestrations has also increased with rising debt/asset ratios. A tax on agricultural land is likely to be implemented, which will reduce farmers' current annual returns (rents) and land values.

Restrictive new labour laws applied to agriculture have also increased wage and transaction costs of employing labour. Many commercial farmers have responded by reducing their labour force, investing in machinery and labour-saving technologies, and employing machinery and labour contractors. The implementation of minimum wage legislation from 2003 is expected to exacerbate the already high unemployment problem in South Africa. The high incidence of HIV/AIDS among farm workers adversely affects labour turnover rates, farm productivity, production costs and business competitiveness.

Investment in the rural economy in terms of physical and institutional infrastructure (for example, roads, crime prevention and control) is inadequate and needs to be urgently addressed by government. There is also a drain of human capital from rural areas with educated youth seeking vocations elsewhere. Institutional change is also urgently needed to address the wastage and under-utilisation of scarce resources due to the absence of well-defined, secure and tradable property rights (for example, for land and water). All of these issues lead to low investor confidence in South African agriculture.

There are, however, also many opportunities for South African agriculture. The potential exists for a peaceful solution to the land redistribution issue, to which the government has committed itself. The principle of market-based land redistribution has been established, and a peaceful resolution would remove uncertainty about property rights in agriculture, which would have a positive impact (improved investor sentiment) on this sector and the economy as a whole. The government has also committed itself to increased investment in agricultural research, extension and education from the present 1% of agricultural GDP to 3%. This can promote the competitiveness of South African agriculture and effect economic growth.

The deregulation of local and international markets and the establishment of preferential trade agreements with a number of countries and trading blocks have created many trade opportunities for South African farmers. Small-scale farmers could also benefit from these agreements through their participation in value-added chains, particularly for high value commodities such as horticultural and animal products. However, problems that constrain access to resources and markets will have to be addressed for such participation to be successful. Government has a responsibility to provide and maintain physical infrastructure and the institutions needed to promote efficient resource and product markets.

REFERENCES

AGRICULTURE AND LAND AFFAIRS PORTFOLIO COMMITTEE. (2000). Minutes of the Agriculture and Land Affairs Portfolio Committee, Pretoria.

BENNETT, J. (2000). SA's unfulfilled potential. *Sunday Times Business Times*, 9 July: 6.

BOWN, A.N., ORTMANN, G.F. & DARROCH, M.A.G. (1999). Use of maize marketing alternatives and price risk management tools by commercial maize farmers in South Africa. *Agrekon*, 38(3): 275–301.

CONRADIE, B. (2002). The value of water in the Fish-Sundays Scheme of the Eastern Cape. Report No. 987/1/02, Water Research Commission, Pretoria.

DEPARTMENT OF LAND AFFAIRS. (1997). *White Paper on South African Land Policy*. Pretoria.

DIDIZA, T. (2001). Foreword to *The Strategic Plan for South African Agriculture*. Pretoria: National Department of Agriculture.

DIRECTORATE: AGRICULTURAL STATISTICS. (2002). *Abstract of Agricultural Statistics*. Pretoria: National Department of Agriculture.

EHRENBERG, R.G. & SMITH, R.S. (1991). *Modern Labour Economics: Theory and Public Policy*. New York: Harper Collins.

GOEDECKE, E.J. & ORTMANN, G.F. (1993). Transaction costs and labour contracting in the South African forestry industry. *South African Journal of Economics* 61(1): 67–83.

GRAHAM, A.W. & LYNE, M.C. (1999). Land redistribution in KwaZulu-Natal: an analysis and comparison of farmland transactions in 1997 and 1998. *Agrekon* 38(4): 516–525.

GRAIN SA. (2002). Personal communication. Bothaville.

HERTZ, T. (2002). Forecasting the effects of pending minimum wage legislation on poverty in South Africa. Paper prepared for the Global Labor Standards and Living Wages Working Group Meeting, Political Economy Research Institute, University of Massachusetts.

JOINT UNITED NATIONS PROGRAMME ON HIV/AIDS & UNITED NATIONS DEVELOPMENT PROGRAMME. (2000). *HIV/AIDS and Human Development South Africa*. New York: United Nations.

KHATRI, Y., THIRTLE, C. & VAN ZYL, J. (1996). Public research and development as a source of productivity change in South African agriculture. *South African Journal of Science* 92: 143–150.

LYNCH, S.D. (2002). Personal communication. Senior Research Fellow, School of Bioresources Engineering and Environmental Hydrology, University of Natal, Pietermaritzburg.

LYNE, M.C. (1996). Transforming developing agriculture: establishing a basis for growth. *Agrekon* 35(4): 188–192.

LYNE, M.C. & GRAHAM, D.H. (2001). The impact of land redistribution on tenure security and agricultural performance in KwaZulu-Natal. *Agrekon* 40(4): 656–668.

MACHETHE, C.L. & MOLLEL, N.M. (2000). Extension and support services for smallholder agricultural development in South Africa: who is the smallholder farmer? In Cousins, B. (ed.). *At the Crossroads: Land and Agrarian Reform in South Africa into the 21st Century*. Cape Town and Braamfontein: Programme for Land and Agrarian Studies and National Land Committee.

MACHETHE, C.L., REARDON, T. & MEAD, D. (1997). Promoting farm/nonfarm linkages for employment of the poor in South Africa: a research agenda focused on small-scale farms and agroindustry. *Development Southern Africa* 14(3): 377–394.

NATIONAL DEPARTMENT OF AGRICULTURE (NDA). (2001). *The Strategic Plan for South African Agriculture*. Pretoria.

NEWMAN, R.A. & ORTMANN, G.F. (1996). Machinery and labour contracting among commercial farmers in KwaZulu-Natal. *South African Journal of Economic and Management Sciences* 20(Summer): 85–97.

NIEUWOUDT, W.L. (1995). The impact of a land tax on future investments: a note. *South African Journal of Economics* 63(1): 85–90.

NIEUWOUDT, W.L (2000). Water market institutions: lessons from Colorado. *Agrekon* 39(1): 58–67.

ORTMANN, G.F. (2000). Promoting competitiveness in South African agriculture and agribusiness: the role of institutions. *Agrekon* 39(4): 367–400.

ORTMANN, G.F. (2001). Industrialisation of agriculture and the role of supply chains in promoting competitiveness. *Agrekon* 40(4): 459–489.

PASOUR, E.C. (1981). A further note on the measurement of efficiency and economies of farm size. *Journal of Agricultural Economics* 32(2): 135–146.

ROTH, M. & HAASE, D. (1998). Land tenure security and agricultural performance in southern Africa. Broadening Access and Strengthening Input Market Systems Paper, University of Wisconsin, Madison.

RUKUNI, M. & EICHER, C.K. (eds). (1994). *Zimbabwe's Agricultural Revolution.* Harare: University of Zimbabwe.

SIEBERT, W.S. (2001). Investment, employment and South African labour laws: an international comparison. FMF Monograph No. 30, The Free Market Foundation, Cape Town and Johannesburg.

SOUTH AFRICAN RESERVE BANK. (2003). http://www.reservebank.co.za (Accessed 5/2/2003).

SWANEPOEL, D.S., ORTMANN, G.F. & DARROCH, M.A.G. (1998). Causes of bankruptcy amongst maize and extensive beef farmers in South Africa: 1970–1994. *South African Journal of Economic and Management Sciences* NS 1(1): 52–72.

THIRTLE, C., TOWNSEND, R.F., AMADI, J., LUSIGI, A. & VAN ZYL, J. (1998). The rate of return on expenditures of the South African Agricultural Research Council (ARC). *Agrekon* 37(4): 621–631.

TOWNSEND, R., VAN ZYL, J. & THIRTLE, C. (1997). Assessing the benefits of research expenditures on maize production in South Africa. *Agrekon* 36(4): 585–597.

UNITED STATES DEPARTMENT OF AGRICULTURE (USDA). (2003a). United States state fact sheet. Economic Research Service, US Department of Agriculture. http://ers.usda.gov/StateFacts/US.HTM (Accessed 8/1/2003).

USDA. (2003b). State level data for field crops: grains. National Agricultural Statistics Service, US Department of Agriculture. http://www.nass.usda.gov:81/ipedb/grains.htm (Accessed 9/1/2003).

USDA. (2003c). Farm real estate values, rents, and taxes. Economic Research Service, US Department of Agriculture. http://www.ers.usda.gov/publications/ah712/AH712-4.PDF (Accessed 5/2/2003).

VAN DER MERWE, A. & OTTO, R.J. (1997). International marketing developments and the effects on South African agriculture. *Agrekon* 36(4): 434–452.

PART 2

PLANNING AND ACTION FOR THE FUTURE

CHAPTER 4

Land Redistribution in South Africa

Past Performance and Future Policy

Mike Lyne and Mark Darroch

4.1 INTRODUCTION

South Africa's legacy of racially biased land ownership was formalised in 1913 by the Natives Land Act. This legislation restricted African land ownership to native reserves where the principal mode of tenure was 'customary' and administered by traditional leaders. By 1991, these former homelands covered 17 million hectares, or roughly 13.9% of the national area (NDA, 2000: 5). In addition to these laws that prevented Africans from owning land in white farming areas, other legislation prevented white farmers from leasing their land to black tenants and sharecroppers. In some cases African farmers resorted to exchanging their labour for the use of land owned by white farmers – the infamous labour tenancy arrangements that have persisted in parts of KwaZulu-Natal and Mpumalanga provinces.

This discouragement or prohibition of black farmers was one of the major forces that drove Africans out of the commercial farming areas into the former homelands. Between 1960 and 1980 the population of the former homelands increased from 4.5 to 11 million people (Turner & Ibsen, 2000: 2). Evictions from white farms accelerated in the early 1990s, partly in response to commercial farmers' concerns about legislation intended to improve the security and working conditions of their labour. Some moved to the overcrowded homelands, while many erected shacks in urban shanty settlements. It is now estimated that almost 13 million of South Africa's 40 million residents live in the former homelands, and that over 80% of rural people in South Africa in 1993 were living in poverty (Turner & Ibsen, 2000: 2; South African Data Profile, 2002).

This was the situation awaiting South Africa's first democratic government as it took power in 1994 and began to deal with land and agrarian reform. Faced with the need to balance strong demands from the dispossessed with the need to preserve the commercial farming sector and a fragile political compromise, the African National Congress (ANC)-led government opted for a three-pronged land reform policy:

(a) land restitution – a legal process whereby people who can prove that they were dispossessed of their land after 1913 can regain their land or receive due financial compensation for it;
(b) land tenure reform – which aims to address insecure tenure in the former homelands; and
(c) land redistribution – which aims to redress the racial imbalances in rural land ownership.

As the current political and economic crisis in Zimbabwe clearly illustrates, land redistribution is not a sufficient condition for political stability and economic growth. To achieve these goals, the commercial farming sector must be restructured to narrow wealth and income gaps between farmers of different races, and to promote the efficient use of land and other agricultural resources in the long term.

The South African government's land restitution, tenure reform and redistribution strategies, first publicised in the 1994 Reconstruction and Development Programme, and the principles underlying them, were entrenched in the country's new constitution approved by parliament in 1996. The most significant principle is contained in the 'property clause' which insists on the payment of fair, market-related compensation for land taken in the public interest, including via land restitution and land redistribution. Land redistribution through voluntary market transactions was to be the leading edge of land reform, transferring 30% of white-owned farmland to over 800 000 previously disadvantaged households over five years. The ANC expected that this would result in a highly efficient small-scale farm sector that would generate economic growth. Note that the terms 'previously disadvantaged' or 'historically disadvantaged' used in this chapter refer to people in South Africa who were previously excluded from land markets on the basis of racial segregation.

This chapter analyses the past performance of land redistribution policies, in particular comparing how government programmes have performed relative to private transactions in redistributing farmland. Unfortunately, such comparative data exist only for the province of KwaZulu-Natal where census surveys of farmland transactions have been conducted during 1997–2001 as part of a Broadening Access and Strengthening Input Market Systems Collaborative Research Support Program (BASIS CRSP) commissioned by the United States Agency for International Development (USAID). Nevertheless, these data provide the detail necessary to examine the performance of key land redistribution instruments, and to comment on recent policy initiatives intended to achieve the government's new and somewhat less ambitious goal of transferring 30% of white-owned farmland to emerging farmers over a period of 15 years (Ministry of Agriculture and Land Affairs, 2001: 5). This new timeframe and target group reflects, in part, poor performance of the government grant programme first used to redistribute land (see section 4.2.3.1).

The chapter first gives a brief overview of the three main strategies in South Africa's land reform programme – land restitution, tenure reform and land redistribution – with special emphasis on the instruments used to achieve (or impede) land redistribution. Section 4.3 then compares the relative effectiveness of public and private efforts to redistribute farmland in KwaZulu-Natal, in terms of the total area redistributed, the number of transactions, and the quality and total value of farmland redistributed. Section 4.4 concludes the chapter with some policy suggestions to help the government meet its goal of redistributing commercial farmland in order to promote rural stability and economic empowerment through improved access to land markets and productive use of farmland (Department of Agriculture, 2001: 8–9).

4.2 LAND REFORM INSTRUMENTS IN SOUTH AFRICA

4.2.1 Land restitution

Restitution was never meant to be an open-ended process. Following the promulgation of the Restitution of Land Rights Act, 22 of 1994, members of the public were invited to submit land claims before a deadline that was extended to 31 December 1998. A total of 68 878 restitution claims were registered with the Land Claims Commission, which investigates the validity of each claim, and recommends a resolution to the Land Claims Court for approval or adjudication in the event that a claim is contested.

This legal process started very slowly. By September 2000, 12 623 households had received a total of 268 306 hectares (Turner & Ibsen, 2000: 26) – or less than 1% of the land available for redistribution. However, following an instruction to the Commission by President Thabo Mbeki to finalise all land claims by the end of 2005, the pace accelerated remarkably. By October 2002, more than half of the claims (35 173) involving 84 928 households had been settled (Natal Witness, 2002). Most (70%) of these settled claims related to urban rather than rural land, and only one-third of the R1.8 billion awarded to beneficiaries was used to purchase land (437 021 hectares). Turner & Ibsen (2000) raise a concern that these land transfers could create new rural 'dumping grounds' as they are not linked to a development process that offers livelihoods to beneficiaries.

4.2.2 Land tenure reform

Officially, land in the former homelands belongs to the state. In the past, traditional leaders, magistrates and the Department of Agriculture each played a role in administering this land in a predictable way. In recent years these governance systems have become less predictable and tenure has become increasingly insecure in the former homelands. In 1996 an interim law (Interim Protection of Informal Land Rights Act, 31 of 1996) was passed to protect homeland residents against abuses such as the sale of their land by corrupt traditional leaders, while a new Land Rights Bill was being developed with the aim of transferring ownership from the state to people living on the land.

The logistics of assigning exclusive rights to individuals accustomed to more flexible and overlapping customary rights were, however, far too complex, and the drafters of the Bill opted for a system of 'protected rights' that would be determined through a local process of negotiation facilitated by Land Rights Boards. These rights could be assigned to individuals or to groups constituted as legal entities with democratically elected management committees. The Bill was ready to go to parliament before the second round of democratic elections in May 1999, but was withdrawn for political reasons as it challenged the power base of traditional leaders.

On taking office in June 1999, the new Minister of Agriculture and Land Affairs, the Honourable Thoko Didiza, shelved the Bill, stating that new legislation would be drafted to enable the transfer of communal land to 'tribes' as well to individuals and communities. August 2002 saw the release of Draft 8 of the Communal Land Rights Bill. While earlier versions of this Bill were criticised (Cousins, 2002) for sanctioning the transfer of title to tribal authorities and entrenching the arbitrary and sometimes corrupt administration of land by traditional leaders, Draft 8 comes closer to the original Land Rights Bill and, predictably, has been denounced by the Inkatha Freedom Party as an attempt to dis-empower traditional leaders.

In the meantime, the Department of Land Affairs has focused its attention on enforcing the provisions of the Land Reform (Labour Tenants) Act, 3 of 1996 (which aims to secure the land rights of labour tenants on commercial farms), and ESTA, the Extension of Security of Tenure Act, 62 of 1997 (which aims to protect the farm workers against unlawful eviction). Anecdotal evidence suggests that these Acts have damaged labour relations on many farms and, despite their good intentions, have encouraged commercial farmers to substitute machinery for labour, thereby accelerating lawful evictions of redundant farm workers and their dependants.

4.2.3 Land redistribution

Acceptance of the property clause in the South African constitution indicates that land reformers in South Africa had generally accepted the notion of market-based land redistribution prior to 1994. In the months leading up to the first democratic election, debates on this issue focused on what the future government should do to help disadvantaged people access the land market. Should they offer cash grants to prospective buyers, or should public funds be used in a way that would give disadvantaged people better access to mortgage loans from commercial banks?

4.2.3.1 Cash grants

In their proposals for land reform in South Africa, the World Bank recommended the use of cash grants to help historically disadvantaged farmers to finance land purchases (World Bank, 1993: iii). This advice reflects the view that poor people are unable to finance land with mortgage loans because the market value of land exceeds what proponents of this view call its 'productive value' (Binswanger *et al.*, 1993). According to Deininger & Binswanger (1992: 9–12), the market value of farmland in South Africa exceeded its productive value owing to a long history of input and product subsidies granted to white commercial farmers. From 1994–99, the South African government vigorously pursued a policy of market liberalisation in commercial agriculture and, at the same time, implemented the settlement/land acquisition grant (SLAG) of R16 000 (initially R15 000) per beneficiary household.

In terms of the SLAG programme, historically disadvantaged South Africans who were landless and poor could apply for a cash grant to purchase and develop farmland. In practice, beneficiary households had to pool their meagre grants in order to buy a (whole) farm from a willing seller. The group established a legal entity (usually a community land trust or communal property association) that was formally registered as the owner of the property. In most cases, farms financed with land grants and settled by groups (of up to 500 households) were much too small to support all of the beneficiaries as full-time farmers. The Department of Land Affairs (DLA) anticipated that emerging farmers would use the grant to leverage loan finance for additional land. However, most creditworthy farmers did not qualify for a land grant as the means test which applied to potential beneficiaries precluded individuals with a monthly household income greater than R1 500.

By the end of 2000, the Ministry of Agriculture and Land Affairs had approved 484 projects under the SLAG programme, transferring a total of 780 407 hectares to 55 383 households of which some 14% were headed by women (Turner & Ibsen, 2000: 12). Taken together, land restitution and land redistribution had transferred roughly one million hectares, or less than 1.2% of the 86 million hectares of white-owned farmland, to disadvantaged South Africans over a period of six years. Unimpressed with the performance of the SLAG programme, Minister Didiza imposed a moratorium on further SLAG projects in July 1999 while the Departments of Agriculture and Land Affairs redesigned the grant programme.

The Minister released her proposals for a new programme, Land Redistribution for Agricultural Development (LRAD), late in 2000. The LRAD programme was implemented in August 2001 after several revisions. It differs from SLAG in one major respect: beneficiaries do not have to be poor to qualify for a minimum grant of R20 000 – and those who have more savings and who can raise bigger loans to finance their farms qualify for successively larger grants. A beneficiary must inject equity and debt capital totalling at least R400 000 to qualify for a maximum grant of R100 000 (Ministry of Agriculture and Land Affairs, 2001: 8). This marks a distinct shift in the South African government's land redistribution policy away from poverty alleviation and group settlement, in favour of settling prospective farmers on their own farms. In its first year, LRAD redistributed approximately one million hectares of farmland in South Africa (Shabane, 2002). Criticism of the LRAD programme has focused largely on its likely exclusion of the poor rural majority seeking land primarily for residential rather than agricultural purposes.

4.2.3.2 Diminishing, finite interest subsidies

The explanation put forward by Binswanger *et al.* of why poor people are unable to buy land with mortgage loans was disputed by Nieuwoudt & Vink (1995), who countered that the removal of subsidies would not make it any easier for prospective buyers to finance land purchases in South Africa because the underlying problem was relatively high inflation. Inflation raises immediate costs (higher nominal interest charges) while deferring returns (future earnings increased by inflation). The result is inadequate cash flow to service debt repayments during the critical early years after land purchase.

An investment in land is similar to an investment in the stock market in that current returns (dividends) are usually low relative to capital growth. Empirical evidence from South Africa shows that the average annual current return to agricultural land seldom exceeds 5% of its market value. During periods of inflation when nominal interest rates are high (say 17% per year) relative to the current return on agricultural land, conventional mortgage loans with constant repayment schedules create a formidable but temporary liquidity problem for borrowers who are unable to make a substantial down-payment on the purchase price of a farm. This liquidity problem diminishes over time if expected inflation raises future earnings relative to the constant flow of loan repayments. One method of addressing the liquidity problem is to graduate the loan repayments by subsidising interest charges at a decreasing rate over a finite period of time.

Nieuwoudt & Vink (1995: 509–517) showed that cash grants and diminishing interest rate subsidies impact differently on a beneficiary's future expected cash flow. In particular, an interest rate subsidy that diminishes at the expected rate of inflation can completely eliminate cash flow problems in the first few critical years after entry, whereas a cash grant (drawing the same level of public funding) does not. In their case study they show that the interest rate subsidy will phase out after 11 years if the annual inflation rate is 12% and the beneficiary pays (an affordable) 5% interest rate on the full purchase price in the first year. A variant of this graduated repayment concept has been used by private sector sugar millers to sell farmland to emerging commercial farmers in KwaZulu-Natal since 1995.

When Illovo Sugar Limited invited applications for 20 medium-scale sugar cane farms (ranging from 55–105 hectares in area) in 1995, none of the more than 100 previously disadvantaged applicants could afford an equity contribution large enough to reduce the size of a conventional mortgage loan down to a level that could be serviced from farm income. To mitigate this problem, Illovo agreed to sell the farms to successful applicants at market-related prices and to invest 18% of the purchase price with Ithala Bank. This capital amount, plus accrued interest, would then fund a finite interest rate subsidy for mortgage loans that Ithala granted to each farmer. In effect, Illovo discounted the price of its land by 18%, and Ithala used these funds to reduce annual interest payments by a diminishing amount over time. The current mortgage loan rate was reduced from 16.5% down to 10% in the first year. The subsidy declines to zero at the end of year six, in line with expected increases in nominal income associated with an annual expected inflation rate of about 10%. The buyer then pays the full annual interest rate of 16.5% for the remaining 14 years of the 20-year mortgage loan.

To put the size of these transactions into perspective, the average market price of a medium-scale sugar cane farm is about R900 000 of which 18% or R162 000 is invested by Ithala Bank to finance a finite, diminishing interest rate subsidy on its mortgage loan. The buyer pays the full purchase price and is expected to make a down-payment of at least 10% (R90 000). Ithala provides a mortgage loan for the balance (R810 000) and the seller receives a net amount of R738 000 (R900 000 less R162 000) for the land. Clearly, the programme is elitist as it benefits emerging farmers who are relatively wealthy and creditworthy. Nevertheless, it attracted support from other sugar cane estate owners and by March 2001 had financed 106 medium-scale farms with a total market value of almost R100 million – mainly during 1997–98. Early indications are favourable in the sense that the medium-scale farmers are meeting their loan obligations (Simms, 1997). Growing concern about levels of reinvestment on these farms has, however, prompted a detailed study of their performance and ways to improve the scheme (Mashatola, 2001).

One criticism of the above graduated repayment scheme is that it focused on medium-scale farms and excluded smaller, more affordable farms that were creditworthy. The scheme emphasised medium-scale farms because (a) the sugar millers providing the interest subsidy insisted that land should be sold to full-time farmers, and (b) subdivision and transaction costs are largely fixed costs that raise the unit price of land as farm size diminishes. Had part-time farmers (i.e. applicants capable of financing family drawings and debt repayments partly from non-farm income) been considered, a larger number of smaller, creditworthy farms could have been sold, exposing buyers to lower levels of leverage and less financial risk. Smaller farms would also be more feasible if transaction, subdivision and survey costs were lower, or if the borrower could augment his or her down-payment with a public cash grant. Medium-scale sugar cane farmers were effectively precluded from SLAG by its means test, but now qualify for LRAD grants. Eleven of the initial 54 LRAD grants made in KwaZulu-Natal during 2002 were awarded to medium-scale sugar cane farmers who will participate in the next round of the graduated repayment scheme.

In summary, a finite and diminishing interest rate subsidy on mortgage loans used to finance the purchase of agricultural land improves the risk exposure of both the borrower and the lender when inflation and nominal interest rates are relatively high. Finite interest rate subsidies create fewer distortions in capital markets because they do not become entrenched. Moreover, the present value of a phased interest rate subsidy is significantly lower than that of a comparable cash grant. In other words, the state would have fewer liquidity problems of its own if it sponsored a diminishing interest rate subsidy rather than a cash grant of equivalent value to a beneficiary.

4.2.3.3 Subdivision of farmland

In South Africa, the cash flow problem associated with land purchase has been compounded by another impediment to private land transactions – the Subdivision of Agricultural Land Act, 70 of 1970. In terms of this Act, farm owners must get permission from the government to subdivide their land. This introduces uncertainty and delays that add to the costs of registering, surveying and transferring affordable parcels of land to aspiring farmers. Although Act 70 has been rescinded, President Mbeki has not yet signed the repeal into law. The delay has been attributed to the absence of national zoning legislation regulating the conversion of agricultural land into residential or industrial uses (Graham, 2000: 19). Repeal of Act 70 will make it easier for the many poor and part-time farmers who will be rationed out of the proposed LRAD programme to finance smaller, more affordable farms.

4.2.3.4 Equity-sharing projects

Costs, delays and uncertainty associated with the formal transfer of small subdivisions of land in South Africa have contributed to the growing popularity of farm-worker equity-sharing (FWES) projects as a means of redistributing wealth and incomes while maintaining or improving agricultural performance (Eckert *et al.*, 1996: 693–712.). In 1998 it was estimated that about 50 FWES projects had been initiated in South Africa, mostly in the Western Cape (Department of Land Affairs, 1998: 2) and it is clear that this number has increased substantially in recent years. For example, in December 2001, the Land Reform Credit Facility (LRCF) – discussed in section 4.2.3.5 – approved loans for a further 11 projects (LRCF, 2001a: 3). Today, FWES projects are spread across all nine of South Africa's provinces and involve wine, fruit, vegetables, olives, poultry, cut flowers, dairy and eco-tourism enterprises (Knight *et al.*, 2002).

In general, these projects are company operations in which financial equity is owned by workers, former owners, managers and other investors (Ngqangweni & Van Rooyen, 1995: 211–214) in the form of tradable shares that define their individual rights to vote for directors and to benefit from the profits and capital gains generated by the company. This is quite distinct from a co-operative or collective farming enterprise where voting and benefit rights are egalitarian and non-marketable, resulting in free- and forced-rider problems that undermine incentives to invest time and money in the enterprise (Sykuta & Cook, 2001). In addition, FWES projects benefit from experienced management (frequently the former white owner) and have been able to attract finance from commercial banks and venture capitalists (Knight & Lyne, 2002). These private financiers have a strong incentive to help their clients build sound business organisations and to train worker-shareholders for their active participation in a successful company.

While workers do recognise the future potential for dividends and capital gains, field interviews show that workers' immediate interest in joining equity-sharing projects centres more on the opportunity to influence managerial decisions affecting wages, working conditions, housing and tenure security for their families (Department of Land Affairs, 1998: 5). Concerns about the lack of worker empowerment in early FWES projects (Surplus Peoples Project, 1999) appear to have been addressed in more recent projects (Knight & Lyne, 2002). Case studies analysed by Knight *et al.* (2002) show that both skills transfer and gender representation are positively related to the workers' share of equity in the enterprise. In essence, empowerment requires that workers buy a significant shareholding. Initially, farm workers had to finance their equity in the company with loans, creating the usual cash flow problems. This situation changed in 1996 when the DLA allowed farm workers to finance equity with SLAG grants, and more recently (May 1999) when it piloted the LRCF. The LRAD programme explicitly supports equity-sharing projects (Ministry of Agriculture and Land Affairs, 2001: 5), as does the strategic plan for South African agriculture (Department of Agriculture, 2001: 9).

4.2.3.5 The Land Reform Credit Facility

The LRCF was established with the intention of drawing private sector finance and human capital into commercially viable land reform projects. As an alternative, but partial, solution to the liquidity problem associated with conventional mortgage loans, the LRCF offers unsubsidised loans with deferred or graduated repayment schedules to commercial banks who finance, on similar terms, equity-sharing projects and land purchased by aspiring farmers (Department of Land Affairs, 1998: 21–26). In essence, the LRCF inherits the project's (temporary) cash flow problem. Private lenders and investors who apply for loans from the Facility are expected to conduct their financial evaluation and screening of projects thoroughly, adhering to sound business criteria, as they are putting their own resources at risk. The LRCF is administered by just one full-time manager, as there is no need to conduct or review these financial analyses. Rather, the manager's principal task is to approve loan applications submitted by accredited financial intermediaries according to criteria (for eligibility, outreach, empowerment and land reform) established by the DLA. In the case of equity-sharing projects, the land reform criterion requires that at least 10% of the LRCF loan must finance the purchase of land or a long-term lease to secure worker-shareholders' rights to residential property.

The LRCF was initially capitalised at a level of R63 million via a R32 million grant made by the Department of Land Affairs that was matched by grants from the European Union (R29.4 million) and The Danish Agency for Development Assistance (DANIDA) (R1.7 million) (LRCF, 2000: i). Simulation exercises based on anticipated loans with deferment periods ranging from one to three years (longer deferment periods render most projects unprofitable) indicated that approximately R15 million could be disbursed annually without reducing the real value of the LRCF fund to a level where it would not recover in the longer term. Although the

DLA's moratorium on land grants prevented commercial banks from financing all but highly elitist projects (whose beneficiaries did not require grants), the Facility approved loans worth R32 million between May 1999 and April 2001, with applications for another R34 million pending its re-capitalisation.

Of the R32 million approved by the LRCF during its pilot phase, R4.8 million financed mortgage loans made to individual farmers and R27.2 million financed long-term loans made to equity-sharing projects. In May 2001, the DLA and the European Union accepted proposals to re-capitalise the LRCF. The DLA approved a grant of R60 million during the 2001–02 financial year, and the European Union is likely to provide a further grant of R10 million (LRCF, 2001b). ABSA Bank, the Facility's largest client, intends to make much greater use of the LRCF in future, especially if the DLA accepts a recommendation to allocate a portion of the proposed LRAD grants to the LRCF and to fast-track these grants to the beneficiaries of projects financed by the Facility (LRCF, 2001c: 53).

4.3 PUBLIC AND PRIVATE LAND REDISTRIBUTION IN KWAZULU-NATAL

While it is possible to acquire information about the value and area of farmland redistributed through the LRCF and Ithala Bank, very little is known about other private land transactions. In 1997, USAID commissioned researchers from the School of Agricultural Sciences and Agribusiness at the University of Natal to monitor all farmland transactions in KwaZulu-Natal as part of its BASIS CRSP. This section reports the results of annual census surveys of the Deeds of Transfer recorded in KwaZulu-Natal over the period 1997–2001, paying particular attention to the rate of land redistribution and the relative effectiveness of private market transactions and those financed with SLAG grants from government. The data are drawn from Lyne & Darroch (2001) and the most recent annual census survey of farmland transfers.

KwaZulu-Natal is the province with the largest population (about nine million people) but it is relatively small in area (about nine million hectares). At the time of South Africa's political democratisation in 1994, 50.04% of the farmland (including provincial parks) area of 8.19 million hectares in KwaZulu-Natal was controlled by a small minority of white owners. It is estimated that 6 755 large-scale commercial farms accounted for 4.13 million hectares (Lyne & Ortmann, 1996: 70). Of the remaining 4.06 million hectares, 2.84 million hectares was occupied by about three million black South Africans under communal tenure, 0.04 million hectares was privately owned by 'non-whites', 0.42 million hectares was farmed by the state itself (including forests) and the remaining 0.76 million hectares was used for provincial parks.

The census surveys required that all land transactions recorded each year in KwaZulu-Natal had to first be stratified into urban and rural transactions. Within the rural group, transactions listed separately for each parcel, or subdivision of land acquired by one owner (plus the spouse in the case of married co-owners), were consolidated. All transfers of 'farms' smaller than one hectare or with a per hectare market price greater than that commanded by the best quality farmland in KwaZulu-Natal were removed. The transfers remaining were then classified as either to 'white' or to 'previously disadvantaged entrants', with the latter being stratified into unique strata according to their method of financing/acquisition, namely government-assisted (grant financed), mortgage loan financed, cash purchase, or by private non-market transfers (mainly bequests and some donations and court orders) (see Graham & Lyne, 1999 for a full description of the stratification process).

Transfers to 'previously disadvantaged entrants' refer to transactions that transfer farmland from white owners to new owners who were previously excluded from land markets on the basis of racial segregation. However, this subset excludes a small number of transactions that shifted land from one disadvantaged owner to another – unless ownership transferred from males to females. The Deeds of Transfer do not explicitly record the race or gender of new landowners, and so the race and gender of individual entrants was established primarily on the basis of their names. While every effort was made to identify disadvantaged landowners, the authors accept that some of these new entrants may have been misclassified – the true rate of land redistribution in KwaZulu-Natal is, therefore, slightly understated. Note also that the Subdivision of Agricultural Land Act, 70 of 1970, prevents groups of individuals from owning undivided shares in farmland – except when property is jointly owned by a husband and wife married under national law. Where land had been acquired by corporate entities (for example, companies and trusts), information obtained directly from the entity or indirectly from the Registrar of Companies, Master of the Supreme Court, financiers, and the Provincial Department of Land Affairs was used to establish whether or not the land had transferred predominantly to disadvantaged people.

Table 4.1 reports the area of farmland acquired each year by disadvantaged owners in KwaZulu-Natal during the calendar years 1997–2001. This census estimate was used to compute the rate of land redistribution for each study year by expressing the area acquired by disadvantaged entrants as a percentage of the total area of farmland potentially available for redistribution in the province (5.31 million hectares = 4.13 million hectares owned by whites plus 1.18 million hectares of state farms, and game parks) at the time of political democratisation in 1994. The area of commercial farmland that transferred to white and disadvantaged owners rose markedly from 372 995 hectares in 1997 to 603 522 hectares in 1998, fell to about 300 000 hectares in both 1999 and 2000, and dropped further to 267 233 hectares in 2001. Annual farmland transfers to the disadvantaged group ranged from 17 345 to 36 109 ha/yr, representing between 5.8–11.8% of the farmland transferred annually, or between 0.33–0.68% *of the original area available for redistribution.*

Table 4.1: Estimated annual rates of farmland redistribution to disadvantaged owners in KwaZulu-Natal, 1997–2001.

Study year	*1997*	*1998*	*1999*	*2000*	*2001*
1. Area of farmland originally available for redistribution (ha)	5 308 559	5 308 559	5 308 559	5 308 559	5 308 559
2. Area of land transacted (ha)	372 995	603 522	306 433	300 799	267 233
3. Area of farmland acquired by, or for, disadvantaged people (ha)	22 934	17 772	36 109	17 345	27 324
4. Rate of land redistribution (%) ([3/1] x 100)	0.43	0.34	0.68	0.33	0.52

Although these rates are low, they are considerably higher than estimates made in previous years. In 1995, it was estimated that only 0.05% of the farmland available for redistribution transferred to previously disadvantaged people in the Northern Province (now Limpopo Province) of South Africa (Kirsten *et al.*, 1996: 218–223). In KwaZulu-Natal, the estimate for 1995 was 0.09% (Lyne & Darroch, 1997: 561–568.). Since 1997, about 121 484 hectares, or 2.3%, of the available farmland has been redistributed to disadvantaged owners in KwaZulu-Natal. Figure 4.1 graphs the annual rates of farmland redistribution reported in Table 4.1, and shows the cumulative rise in the total area of land that was redistributed. Of course, the area transferred says nothing about the *quality* of the redistributed land.

Table 4.2 presents the mean area of all farms acquired by white and disadvantaged entrants in KwaZulu-Natal, and – for those farms purchased – the mean

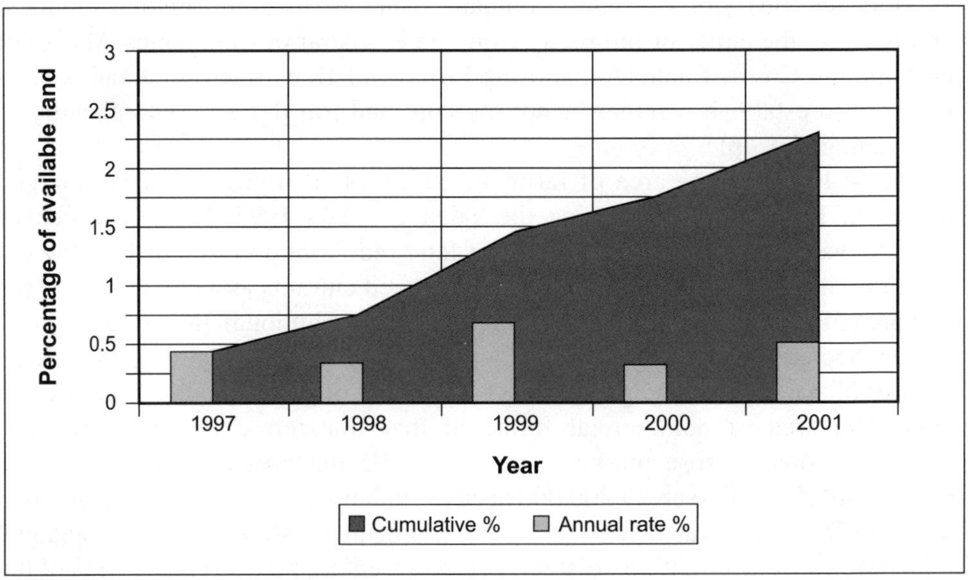

Figure 4.1: Estimated annual and cumulative rates of farmland redistribution to previously disadvantaged owners in KwaZulu-Natal, 1997–2001.

Table 4.2: Characteristics of farmland acquired by white and disadvantaged owners in KwaZulu-Natal, 1997–2001 (2000=100).

Farm characteristic	*Year*	*White group*	*Disadvantaged group*	*t-value*
Mean farm area (ha) – for all farms transacted (including non-market transfers such as bequests and donations)	1997	365	125	3.6***
	1998	1 007	100	2.4**
	1999	287	114	6.7***
	2000	268	109	5.7***
	2001	294	179	3.8***
Mean real farm price (R) – for all farms purchased	1997	1 193 882[a]	532 775	1.4
	1998	754 373	318 086	4.4**
	1999	879 400	312 339	3.4***
	2000	638 808	355 668	3.6***
	2001	652 318	382 006	3.3***
Weighted real land price (R/ha) – for all farms purchased	1997	2 554	2 796	
	1998	1 442	1 791	
	1999	2 761	1 678	
	2000	2 337	2 326	
	2001	1 993	1 660	

*** and ** denote statistical significance at the 1% and 5% level of probability, respectively
[a] Approximately R8=US$1 at the end of calendar year 2001

price of farms and the weighted price of land, during 1997–2001. All prices are expressed in real terms using the calendar year 2000 as the base year. The t-values reported in Table 4.2 test for differences in the mean characteristics of farms acquired by white and disadvantaged groups of entrants. The mean area (and hence price) of farms transacted is consistently smaller for the disadvantaged group than for the white group, but there is no consistent difference in the weighted average land price (total value of hectares traded in market transactions, divided by the total hectares of market transactions).

For farms of equal quality land, a lower per hectare price would be anticipated for larger farms as the fixed transaction costs and the value of fixed improvements are spread over a larger land area. This situation prevailed in 1997 and 1998 when private sugar millers working with Ithala Bank (as described in section 4.2.3.2) provided interest rate subsidies to 51 emerging farmers to buy medium-scale sugar cane farms on high quality land. The dip in 1999 values shows that farms purchased by disadvantaged buyers were not only much smaller, but also of poorer quality than those purchased by whites – partly because no farmers were settled under the graduated repayment scheme in that year. The quality gap, however, closed in 2000 and 2001, as the last group of 18 farmers under the graduated repayment scheme were settled, and because there were few new purchases of lower quality land by communities due to the moratorium on government (SLAG) grants.

The transactions involving disadvantaged entrants were further analysed according to the method of financing/acquisition – grant financed (government-assisted), private market transfers (mortgage loan financed or cash purchases), and private, non-market transfers. Table 4.3 disaggregates the land redistributed in KwaZulu-Natal each year during 1997–2001 into these four mutually exclusive strata. Figures 4.2 to 4.6 visually summarise this information, highlighting important differences in the quantity, market value and quality of land transferred by each mode of land redistribution. Figure 4.2 shows that there were far more private than government-assisted transactions redistributing land to disadvantaged people in KwaZulu-Natal during 1997–2001. In all, there were 905 private transactions, 471 of which were market transfers financed with mortgage loans (184) or cash (287). By contrast, only 89 transactions were financed with government (SLAG) grants. While the number of private purchases has been consistent over the study period (ranging from 79 to 120 annually), the number of government-assisted transactions declined from 32 in 1999 to nine in 2000 and rose to 22 in 2001, reflecting the imposition of the moratorium on new grants after July 1999, and its subsequent lifting in 2001.

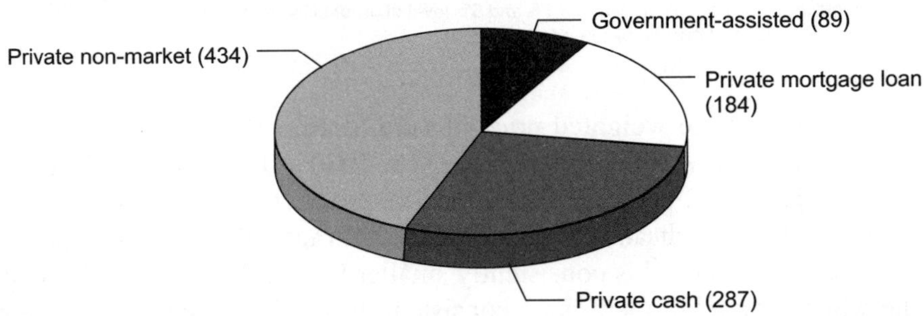

Figure 4.2: Total number of farmland transactions by mode of land redistribution to disadvantaged owners in KwaZulu-Natal, 1997–2001.

Table 4.3 and Figure 4.3 clearly indicate that private, non-market transactions (mainly bequests) transfer relatively small areas to new entrants (a mean of 37 hectares per transaction, and a total of 16 097 hectares). Farms purchased privately with cash averaged 85 hectares in area, and those financed with mortgage loans 196 hectares. Although groups of grant beneficiaries purchased the largest areas (mean of 513 hectares per transaction), the total area of farmland redistributed by private market purchases (60 266 hectares, being 36 148 hectares via private mortgage loans and 24 118 hectares via cash purchase) comfortably exceeded that redistributed by the government-assisted transactions (45 121 hectares), and consisted of higher quality land (greater weighted farmland price per hectare).

Figure 4.4 confirms that the area purchased with government grants peaked in 1999 (at 14 727 hectares) and then fell sharply (to 2 133 hectares) after the moratorium on new SLAG projects in 2000. Conversely, the area purchased

Table 4.3: Modes of land redistribution to, and characteristics of farmland acquired by, disadvantaged owners in KwaZulu-Natal, 1997–2001 (2000=100).

Farm characteristic	*Government-assisted*	*Private mortgage loan*	*Private cash*	*Private non-market*	*Total*
Number of transactions	89	184	287	434	994
Total area of land (ha)	45 121	36 148	24 118	16 097	121 484
Total market value of land (R million)	36.92	134.55	39.75		211.22
Mean area of farms (ha)	513	196	85	37	
Weighted farmland price (R/ha)	833	3 722	1 615		

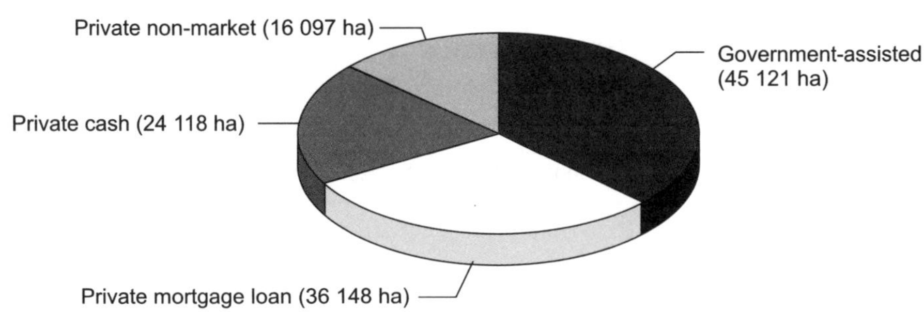

Figure 4.3: Total area of farmland by mode of land redistribution to disadvantaged owners in KwaZulu-Natal, 1997–2001.

privately (cash plus mortgage loans) increased steadily during 1997–99, stayed at the 1999 level in 2000, and then fell in 2001 when there were no sales of medium-scale sugar cane farms financed by private sector interest subsidies.

Figure 4.5 shows the impact of government-assisted and private land purchases on wealth redistribution in the commercial farming sector of KwaZulu-Natal during 1997–2001. Private cash and mortgage loan purchases redistributed nearly *five* times more land wealth (R174.3 million) than did government-assisted transactions (R36.9 million). This contrast in wealth is much greater than the difference in total area transferred by these modes of land redistribution. The implication is that agricultural land financed with government grants is of poor quality relative to that purchased privately (see Figure 4.6).

The census data presented in this section show that within the set of private purchases in KwaZulu-Natal during 1997–2001, the number of cash purchases (287) is high relative to the number financed with mortgage loans (184). This gap in numbers widened as the number of medium-scale sugar cane growers with access to interest rate subsidies under the graduated repayment scheme declined. These observations are consistent with Nieuwoudt & Vink's (1995: 509–517) argument

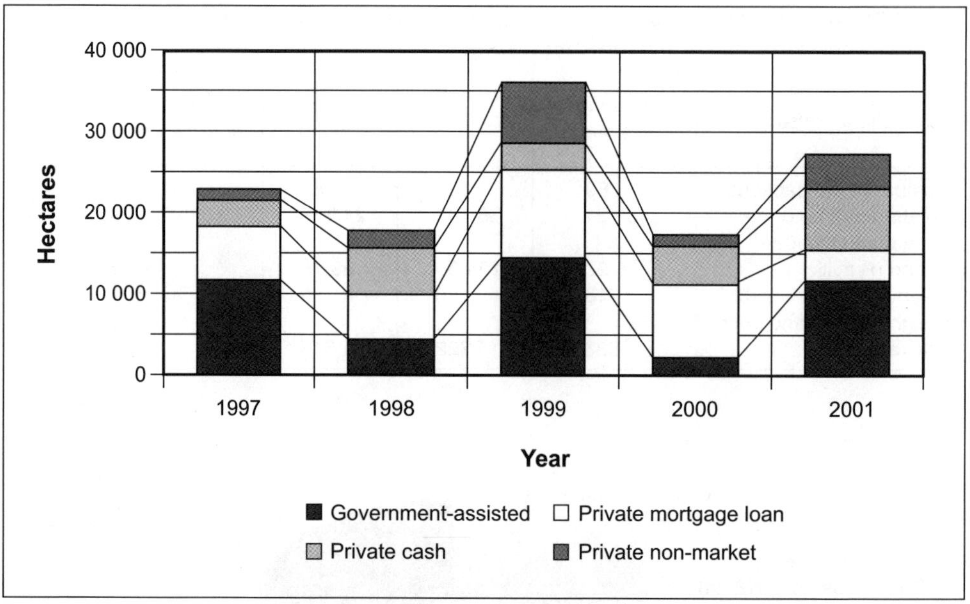

Figure 4.4: Annual area by mode of land redistribution to disadvantaged owners in KwaZulu-Natal, 1997–2001.

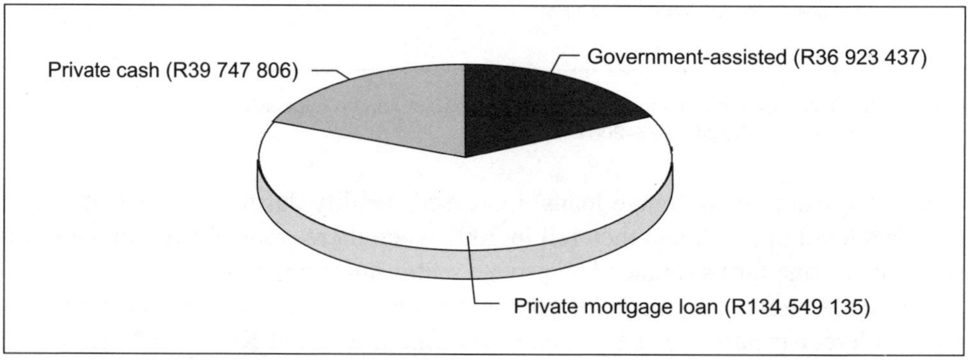

Figure 4.5: Total market value of farmland by mode of land redistribution to disadvantaged owners in KwaZulu-Natal, 1997–2001 (2000=100).

that buyers with limited equity cannot finance large, quality farms using a conventional mortgage loan owing to inflation-enhanced cash flow problems. Instead they pay cash for relatively cheaper farms. From Table 4.3 and Figure 4.6, it is evident that farms financed with cash are about 43% of both the size (ha) and quality (R/ha) of farms financed with mortgage loans. Moreover, it has recently been observed that within the group of medium-scale sugar cane growers, part-timers with off-farm income have outperformed their peers in terms of repaying their mortgage loans and reinvesting capital into their farms.

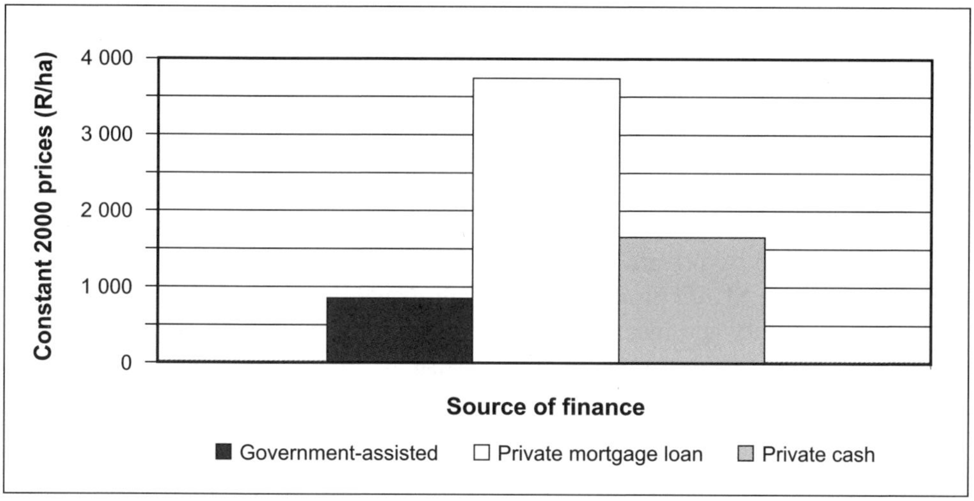

Figure 4.6: Weighted farmland price (R/ha) by mode of land redistribution to disadvantaged owners in KwaZulu-Natal, 1997–2001 (2000=100).

The relatively poorer performance of government programmes in redistributing farmland, especially high quality land, has been compounded by other problems. Under the SLAG programme, diverse groups of beneficiaries struggled to assign exclusive property rights to individual beneficiaries, or to design and enforce sound constitutions to manage communal resources. This institutional vacuum has already generated highly visible adverse outcomes on government land reform projects: commercial farmland has been needlessly lost to residential uses, private ranches have been converted into open access grazing resources (no restrictions enforced on livestock numbers), and insecure tenure has undermined investment in crop production (Lyne & Graham, 2001). This situation is expected to change now that the LRAD programme that aims to help settle farmers has replaced the SLAG programme.

4.4 CONCLUSION AND POLICY RECOMMENDATIONS

The annual census surveys of transfer deeds reported in this chapter show that a total of 121 484 hectares of farmland were acquired during 1997–2001 by historically disadvantaged people in KwaZulu-Natal. Private, non-market transfers (mainly bequests) accounted for 16 097 hectares of this land, government-assisted purchases for 45 121 hectares and private purchases (cash and mortgage loans) for 60 266 hectares. All told, this area of 121 484 hectares means that 2.3% of the farmland available for redistribution in KwaZulu-Natal transferred to disadvantaged owners during these five years. Clearly, the SLAG programme did not generate a rate of farmland transfer in KwaZulu-Natal consistent with the South African government's goal of redistributing 30% of white-owned farmland to disadvantaged people over 15 years.

The SLAG programme not only redistributed less land than did private purchases, but also transferred land of much lower quality, and about *five* times less total wealth, to beneficiaries whose tenure is relatively insecure. The weighted average farmland price for SLAG transfers during 1997–2001 was R833 per hectare, about *one-half* that for cash purchases (R1 615 per hectare) and *one-quarter* that for mortgage bond transactions (R3 722 per hectare). Over the same period, the SLAG programme redistributed farmland worth R36.9 million, compared to farmland worth R174.3 million transferred by private purchases. There is also further survey evidence that insecure tenure on SLAG projects in KwaZulu-Natal has impacted adversely on the beneficiaries' ability and incentive to finance seasonal inputs and improvements to cropland. Again, this outcome is not consistent with the expectation that land redistribution would result in a highly efficient small-scale farm sector.

The number of private transfers (905 – of which 184 were mortgage loans and 287 were cash purchases) also exceeded government-assisted transactions (89) redistributing land to disadvantaged people in KwaZulu-Natal during 1997–2001. A moratorium on new SLAG projects after July 1999 reduced annual transfers by this instrument in KwaZulu-Natal from a peak of 14 727 hectares in 1999 to just 2 133 hectares in 2000. Although the replacement LRAD programme is designed to extend larger grants to creditworthy farmers, there are some valid problems with the design and delivery of LRAD that must be addressed. For example, there is an implicit cap on LRAD grants imposed by commercial banks that usually require a debt/equity ratio of less than one when financing investments. Under optimal conditions when the lender is assured that the borrower will receive a grant, a prospective owner-operator would have to provide R100 000 in own equity in order to qualify for an LRAD grant of R90 000 and a commercial bank loan of R190 000. Under conditions that are less certain, and for previously disadvantaged farmers that lack savings, the implicit cap on LRAD grants would be much less generous.

Barriers to the subdivision of farmland imposed by the Subdivision of Agricultural Land Act, 70 of 1970, will prevent many emerging farmers from making private purchases. LRAD discounts the impact of Act 70, stating that 'permission to subdivide for sale under LRAD will be effective immediately upon the launch of LRAD' (Ministry of Agriculture and Land Affairs, 2001: 6). While this commitment is welcomed, it does not address other significant costs associated with the formal subdivision, registration and transfer of agricultural land. Moreover, it applies only to government-assisted transactions and, therefore, does nothing to improve market access for private buyers. LRAD relies heavily on private 'design agents' who will help prospective buyers to identify willing sellers, prepare their land-use plans and cash flow projections, negotiate sale agreements, arrange finance, and facilitate applications for LRAD. These agents are expected to recover most of their costs from a planning grant that is awarded once the project has been approved. However, this arrangement is fraught with uncertainty as there is no guarantee that government will approve a project. Consequently, existing landowners, property developers, estate agents and non-government organisations may be unwilling to act as design agents.

Although private purchases redistributed more land, and much more land wealth, to previously disadvantaged owners than did government-assisted purchases in KwaZulu-Natal during 1997–2001, these purchases together (105 387 hectares) accounted for less than 6% of the total area transacted (1 850 982 hectares). Clearly, the market has much larger potential to redistribute farmland than has been realised to date. The overwhelming response to innovative financial products administered by Ithala Bank (see section 4.2.3.2) and the LRCF (see section 4.2.3.5) to finance medium-scale farmers and equity shareholders suggests that access to the land market is constrained by inflation-induced cash flow problems associated with conventional mortgage loans, and by the costs, delays and uncertainty associated with the registration and transfer of affordable land subdivisions.

Obstacles preventing the repeal of the Subdivision of Agricultural Land Act, 70 of 1970, must be addressed without further delay. Scrapping this Act will make it easier for those poor and part-time farmers who are rationed out of the LRAD programme to obtain finance for smaller, more affordable farms. Likewise, the government should contain or reduce the inflation rate, and lower the statutory costs of subdividing and transacting farmland, to enable commercial banks to reach further down the income scale when financing prospective farmers and equity shareholders. Access to larger land reform grants for farm workers and aspiring farmers would also improve the outreach of private financiers because the poor cannot make significant contributions of their own to the purchase of equity and land. Ideally, the LRCF should be allocated a share of these grants and authorised to award them contingent upon the disbursement of a loan. Under these conditions, the LRCF will have to be capitalised at scale if it is to keep pace with growing demands for its loans from commercial banks.

Recent invasions of commercial farms in Zimbabwe highlight the urgent need for bold interventions to de-racialise the structure of commercial agriculture in South Africa. An effective alternative to the destructive 'fast track' policy adopted in Zimbabwe is to subsidise interest rates for a finite period on loans made to creditworthy land reform projects. The problems associated with cheap credit programmes are well documented and have encouraged the South African government to resist interest rate subsidies. Many of these problems, however, could be avoided by channelling *finite* interest rate subsidies *that decline over time* through commercial banks. This has the added advantage of drawing private sector finance and expertise into the land reform process. For example, the LRCF could either discount the wholesale interest rate that it charges commercial banks for its existing financial product, or it could allocate public and donor funds directly to commercial banks to support finite, diminishing interest rate subsidies on loans made to previously disadvantaged buyers of land and equity. These awards could be apportioned according to the value of transactions in land and equity that each lender expects to support during the forthcoming fiscal year. This process will give potential borrowers an opportunity to 'shop around' for favourable terms from a variety of decentralised and expert lending institutions.

A key area for future research would be to assess the extent to which land redistribution efforts countrywide may be constrained by the different approaches that the Provincial Departments of Land Affairs (PDLAs) use to apply the government's land reform strategies. For example, the KwaZulu-Natal PDLA has not supported any equity-sharing projects despite the relative success of these projects in other provinces. These discrepancies point to persistent differences in the willingness or capacity of staff in the PDLAs to implement relatively sophisticated projects. A national exchange programme involving mentors (drawn from government, non-governmental organisations and commercial banks) with experience in successful land reform projects is recommended to broaden views and transfer skills.

Acknowledgements

This chapter was made possible by support provided in part by the United States Agency for International Development (USAID) Agreement No. LAG-A-00-96-960016-00 through the Broadening Access and Strengthening Input Market Systems Collaborative Research Support Program (BASIS CRSP). All views, interpretations, recommendations and conclusions expressed in this chapter are those of the authors and not necessarily those of the supporting or co-operating organisations.

REFERENCES

BINSWANGER, H., DEININGER, K. & FEDER, G. (1993). Power, distortions and reform in agricultural land markets. In Berhman, J. & Srinivasan, T. (eds). *Handbook of Development Economics: Volume 3*. Amsterdam: Elsevier.

COUSINS, B. (2002). Controversy and consensus in South Africa's land reform programme. Invited paper presented at a World Bank regional workshop on land issues in Africa, 29 April – 2 May, Kampala.

DEININGER, K. & BINSWANGER, H. (1992). Are large farms more efficient than small ones? Government intervention, large-scale agriculture, and resettlement in Kenya, South Africa and Zimbabwe. Unpublished working paper, Department of Applied Economics, University of Minnesota.

DEPARTMENT OF AGRICULTURE. (2001). The strategic plan for South African agriculture. Department of Agriculture, Directorate of Agricultural Information Services, Pretoria.

DEPARTMENT OF LAND AFFAIRS. (1998). Feasibility study for the support of commercial land transfer projects. Department of Land Affairs, Pretoria.

ECKERT, J., HAMMAN, J. & LOMBARD, J. (1996). Perceiving the future: empowering farm workers through equity sharing. *Development Southern Africa* 13(5): 693–712.

GRAHAM, A. (2000). Land redistribution in KwaZulu-Natal: an analysis of farmland transactions recorded in 1997 and 1998. MSc thesis, Agricultural Economics, School of Agricultural Sciences and Agribusiness, University of Natal.

GRAHAM, A. & LYNE, M.C. (1999). Land redistribution in KwaZulu-Natal: an analysis of farmland transactions in 1997. *Development Southern Africa* 16(3): 435–445.

KIRSTEN, J., VAN ROOYEN, J. & NGQANGWENI, S. (1996). Progress with different land reform options in South Africa. *Agrekon* 35(4): 218–223.

KNIGHT, S.L. & LYNE, M.C. (2002). Perceptions of farmworker equity-share schemes in South Africa. *Agrekon* 41(4): 356–374.

KNIGHT, S.L., LYNE, M.C. & ROTH, M. (2002). Best institutional arrangements for farmworker equity-share schemes in South Africa. Unpublished working paper, Agricultural Economics, School of Agricultural Sciences and Agribusiness, University of Natal, Pietermaritzburg.

LRCF. (2000). The Land Reform Credit Facility: progress report for the period 4th November 1999 – 27th January 2000. Unpublished management report, Khula Enterprise Finance Ltd, Johannesburg.

LRCF. (2001a). The Land Reform Credit Facility: progress report for the period 1st October – 7th December 2001. Unpublished management report, Khula Enterprise Finance Ltd, Johannesburg.

LRCF. (2001b). Minutes of the sixth quarterly meeting of the LRCF Steering Committee, 22nd May. Khula Enterprise Finance Ltd, Johannesburg.

LRCF. (2001c). Implementation plan to establish the Land Reform Credit Facility (LRCF) on a permanent basis. Unpublished report, Khula Enterprise Finance Ltd, Johannesburg.

LYNE, M.C. & ORTMANN, G.F. (1996). Estimating the potential for creating additional livelihoods on commercial farmland in KwaZulu-Natal. In Lipton, M., Ellis, F. & Lipton, M. (eds). *Land, Labour and Livelihoods in Rural South Africa: Volume 2: KwaZulu-Natal and Northern Province*. Durban: Indicator Press.

LYNE, M.C. & DARROCH, M.A.G. (1997). Broadening access to land markets: financing emerging farmers in South Africa. *Development Southern Africa* 14(4): 561–568.

LYNE, M.C. & DARROCH, M.A.G. (2001). Land redistribution in KwaZulu-Natal, South Africa: four census surveys of farmland transactions, 1997–2000. Invited paper presented at a USAID BASIS CRSP synthesis workshop, 22–25 July, Magaliesburg.

LYNE, M.C. & GRAHAM, D. (2001). The impact of land redistribution on tenure security and agricultural performance in KwaZulu-Natal. *Agrekon* 40(4): 656–668.

MASHATOLA, M. (2001). Personal communication. Regional Economic Advisor: North Coast, South African Cane Growers' Association, Durban, South Africa.

MINISTRY OF AGRICULTURE AND LAND AFFAIRS. (2001). Land redistribution for agricultural development: a sub-programme of the land redistribution programme. National Department of Agriculture, Pretoria.

NATAL WITNESS. (2002). Land reform process has accelerated. 16 November.

NATIONAL DEPARTMENT OF AGRICULTURE (NDA). (2000). Abstract of Agricultural Statistics. The Directorate: Statistical Information, Pretoria.

NIEUWOUDT, W.L. & VINK, N. (1995). Financing of land purchase by small-scale farmers. *Development Southern Africa* 12(4): 509–517.

NGQANGWENI, S. & VAN ROOYEN, J. (1995). Equity sharing as a unique local agrarian reform experience: perceptions of farm workers. *Agrekon* 34(4): 211–214.

SIMMS, P. (1997). Land redistribution in South Africa: some practical lessons. In Bauer, L. (ed.). *Proceedings of the 11th International Farm Management Congress: Volume 1*. Alberta, Canada: International Farm Management Association and the Canadian Farm Business Management Council.

SHABANE, P.M. (2002). LRAD in KwaZulu-Natal. Unpublished paper presented at the Annual Symposium of the South African Society for Agricultural Extension: KwaZulu-Natal Branch, 5 December, Cedara.

SOUTH AFRICAN DATA PROFILE. (2002). World Development Indicators Database, World Bank, Washington DC.

SURPLUS PEOPLES PROJECT. (1999). An examination of the impact of share equity schemes on beneficiaries in the Western Cape and Mpumalanga. Unpublished report, Surplus Peoples Project, Cape Town.

SYKUTA, M. & COOK, M. (2001). A new institutional economics approach to contracts and cooperatives. *American Journal of Agricultural Economics* 83(5): 1271–1277.

TURNER, S. & IBSEN, H. (2000). Land and agrarian reform in South Africa: a status report. Unpublished report, Programme for Land and Agrarian Studies (PLAAS), University of the Western Cape, Cape Town.

WORLD BANK. (1993). Options for land reform and rural restructuring in South Africa. Unpublished World Bank report, World Bank, Southern Africa Department, Washington DC.

CHAPTER 5

Promoting Income and Employment Growth in Rural Areas through Smallholder Agriculture

Simphiwe Ngqangweni and Sheryl Hendriks

5.1 INTRODUCTION

Despite lack of convincing evidence that agriculture-led growth in South Africa's rural economies has the strength to drive economic reform in rural areas, the South African government has proclaimed rural agricultural development a priority (National Ministry of Provincial and Local Government, 2000; NDA, 2001; NDA, 1999; NDA, 1995; African National Congress, 1994). Numerous international studies (see Delgado *et al.*, 1998) have shown that agriculture-led growth has the potential to drive economic reform in developing countries through increased incomes (from both farm and non-farm activities) and improved employment rates. As most households in South Africa's communal areas have access to land and some knowledge of farming, it would seem reasonable to assume that encouraging agriculture-led growth could have widespread benefits for both farm and non-farm households. However, empirical estimation of *the strength of such growth* is crucial to test the potential for agriculture to drive economic reform in communal areas and to direct policy towards optimal supportive investment.

This chapter reviews the findings of a growth linkage study (Hendriks & Lyne, 2003a) conducted for two communal areas of KwaZulu-Natal – Swayimana and Umzumbe – and explores the potential benefits of exploiting the competitiveness of smallholder cattle and irrigated citrus in the Eastern Cape. The chapter begins with a brief overview of the benefits of growth linkages for African countries, while section 5.3 outlines the sampling and growth linkage methodology used in the chapter. Section 5.4 briefly describes the characteristics of the sample households in the two KwaZulu-Natal areas, and sections 5.5 and 5.6 present estimated expenditure elasticities and growth multipliers. The potential for optimal growth through conversion of non-tradables to tradables is then shown in section 5.7 using Ngqangweni's (2000) study of the competitiveness of currently non-tradable indigenous beef and irrigated citrus in the Eastern Cape. A concluding section discusses some policy implications of the chapter results, focusing on the conditions needed to optimise the potential for agriculture-led growth in communal areas of South Africa.

5.2 THE BENEFITS OF GROWTH LINKAGES FOR AFRICAN COUNTRIES

The potential for agriculture-led growth is estimated using growth multipliers from a regional model of typical household demands and intermediate demands between farm and non-farm tradable and non-tradable[1] sectors (Hazell & Röell, 1983). Growth multipliers measure the result of an exogenous income shock on extra income derived from stimulated regional demand, and hence production, in the local non-tradable sector (Delgado *et al.,* 1998). Growth results from a direct or indirect shift (to the right) of the regional supply curve. Direct growth is attained by stimulating the tradables sector through widespread adoption of new technologies that improve production of exportable goods, or by lowering the unit cost of marketing these goods. Indirectly, growth usually occurs when a non-tradable produced by many households becomes tradable. This may lead to increased opportunities for export production, while still meeting local demand.

Earlier African growth linkage studies suggested that growth linkages in Africa might be only half as powerful as those that occur in Asia (Hazell & Hojjati, 1995). However, Delgado *et al.* (1998) show that earlier estimates for African growth linkages may be underestated because of misclassification of commodities that are specifically non-tradable in the African context. The strength of growth linkages is dependent on the consumption patterns of those benefiting from an increase in agricultural income. If households spend most of the increased income on tradables, demand for locally produced non-tradables will not grow much. Delgado *et al.* (1998) review studies showing that non-agricultural employment opportunities in rural areas depend upon vibrant growth in local farm incomes. Without generating local purchasing power, employment in the non-tradable sectors will remain dependent on a steady flow of external remittances. If local production is responsive to this new local demand, the total amount of employment created indirectly through additional sales of non-agricultural goods and services can be twice the direct impact of the original influx of smallholder revenue.

Increased farm income has many spin-offs for the non-farm economy. Firstly, a growing agricultural sector raises agricultural wages, in turn raising the opportunity cost of labour in non-farm activities (Hazell & Haggblade, 1993). Secondly, an

[1] Non-tradables are goods that are neither exported outside the district of study, nor imported into this district, and for which there are no close substitutes. Services are by nature non-tradable. Non-tradable commodity prices are determined by local supply and demand, regardless of modest price movements outside the region of interest. Perishable and prepared foods are often non-tradables in rural areas. Tradability is a characteristic of the local market for a given item and not of the good itself. Non-tradable commodity examples are home and locally produced foods and gathered wild vegetables. Non-tradable non-farm goods and services include building materials, traditional medicines, child care, mini-bus transport, tribal dues and cattle held as investments.

increase in the opportunity cost of labour induces a shift in the composition of non-farm activity out of very labour intensive, low return activities to more skilled, higher investment, high return activities (Hazell & Haggblade, 1993). Thirdly, the income and employment generated by agricultural linkages are predominantly concentrated in rural areas and can partly stem urban migration (Delgado *et al.*, 1998). Fourthly, as most rural households are poor, they consume relatively more labour-intensive commodities like basic, starchy staples whereas higher income households consume relatively more capital intensive and imported goods (Machethe *et al.*, 1997). Fifthly, as incomes rise in the farm sector, farm households demand more goods from the non-farm sector (Islam, 1997). Finally, the kinds of non-farm goods and services demanded when rural incomes increase are typically products produced by small, labour-intensive enterprises focused on such sectors as transportation, hospitality and catering, entertainment, personal services, health, housing and residential construction, all of which lend additional opportunities for local employment (Hazell & Röell, 1983).

While increases in agricultural production are important for their own sake, non-farm activities are important for immediate and long-term household food security. As Hazell & Haddad (2001) explain, sustained increases in per capita incomes hinge on diversification into higher-value agricultural products and non-farm activities. Agro-industry and post-harvest systems can effectively raise rural incomes and absorb excess labour through local manufacturing, trade, construction, transportation, communications, value-adding, institutions, and services (Markets and Structural Studies Division, 2002; Islam, 1997). Strong household links to the rural non-farm economy help to reduce poverty and malnutrition as non-farm income provides liquidity to buy food, inputs and assets to increase production (Hazell & Röell, 1983). In turn, non-farm enterprise spurs farm profitability by providing markets for agricultural production and liquidity for purchasing farm inputs (Reardon *et al.*, 1994; Machethe *et al.*, 1997).

The benefits of growth linkages, namely raised rural incomes, increased non-farm activity and improved employment opportunities, are urgently needed to transform the South African rural economy and to provide livelihood opportunities for millions of poor households – most of which live in rural areas (May, 1998). Widespread increments to rural incomes have the potential to mobilise under-used resources by creating enterprise and employment opportunities in non-tradable farm and non-farm sectors. However, appropriate policy choices rest on knowing how much the elements that stimulate production of tradable commodities can increase the price elasticity of non-tradable output supply. It is, therefore, important to know if increased employment can be supported by food supplies, the types of employment possible from alternative production technologies, and the efficiency (ability to address unemployment) of various policies, strategies, investments and other support measures (De Janvry, 1994; Mellor, 1976: 171). Consequently, policies that effectively promote agricultural exports from the local area also promote increased employment not only directly but possibly also indirectly by creating local purchasing power for demand-constrained items (Delgado, 1997).

5.3 SAMPLING AND GROWTH LINKAGE METHODOLOGY

A stratified, multi-stage sampling procedure was used to draw a representative sample of 99 KwaZulu-Natal households (see Hendriks 2002 for a detailed description of the sampling and estimation methods) – 49 in Swayimana and 50 in Umzumbe – for consumption surveys in the latter half of 1997. Due to the unavailability of suitable respondents during successive rounds of data collection, and missing data relating to expenditure, the effective sample size was reduced to 93 households. Structured interview schedules with different recall periods were used to collect household expenditure data on foods, non-food items and services through each of three rounds of data collection. Information on household composition, income sources, savings, assets, agricultural enterprises, inputs, harvests, and sales was also collected. Farm inputs were excluded from total household expenditure in order to avoid double counting. The opportunity costs of subsistence products were estimated using farm-gate selling prices.

Since savings rates were negligible, household expenditure was almost equivalent to income. Household expenditure was, therefore, used as a proxy for income, as suggested by Alderman (1993), Devereux (1993), Puetz (1993), and Hazell & Röell (1983). Per capita household expenditures were used to control variations in household sizes, as household size directly influences total household expenditure. The expenditure data were annualised to account for the different recall periods used in the study. Monthly hire purchase installments were multiplied by 12 months and the cash prices of durables were converted into annual flows using the capital recovery method (Monke & Pearson, 1989: 139). Infrequent expenditures on occasions such as feasts for weddings, funerals and births were also converted into annual flows, on the assumption that each such event occurred every 10 years.

A variant of the Working-Leser model (Delgado *et al.*, 1998; Hazell & Röell, 1983) was estimated in share form by ordinary least squares as:

$$S_i = b_i + a_i / E + c_i \log E + \Sigma_{\mathrm{j}}) (\mu_{\mathrm{ij}} Z_{\mathrm{j}} / E + \lambda_{\mathrm{ij}} Z_{\mathrm{j}} \quad (1)$$

where $S_i = E_i / E$ is the share of commodity *i* in total per capita expenditure, and Z is a vector of *j* household characteristics and district dummy variables to control differences between households and survey units. The marginal budget shares (MBS_i), average budget shares (ABS_i), and expenditure elasticities (ξ_i) for each category of expenditure were calculated using mean values. Following Delgado *et al.* (1998), the equations used for these calculations were:

$$MBS_{i,} = \delta E_i / \delta E = b_i + c_i (1 + \log E) + \Sigma_j \lambda_{ij} Z_j \quad (2)$$

$$ABS_i = S_i, \text{ and} \quad (3)$$

$$\xi_i = MBS_i / ABS_i. \quad (4)$$

The t-statistics computed for ABS_i, MBS_i and ξ_i are presented in section 5.5 to test the null hypothesis that the MBS and ABS estimates do not differ from zero, and that the estimated ξ_i do not differ from unity.

Following analyses to compare district budget shares and consumption elasticities, sample households were divided into two 'wealth' groups to compare the effect of increased income on consumption between these groups. For both studies, a comparison of relative wealth groups showed no significant differences in expenditure behaviour, hence the results are not presented here. Finally, annual consumption was aggregated across commodities to compare the effect of income changes on predicted consumption of farm and non-farm tradables and non-tradables.

Growth multipliers were computed from the MBS_i estimated for district and wealth group *non-tradables* to determine if raised household incomes would generate additional income in the local economies. The simplified semi-input-output model (Hazell, 1984, cited by Haggblade *et al.*, 1991) was used to compute the multipliers. The multiplier (M) measures the effect of an initial one South African Rand increase in household incomes on local incomes through increased consumption and production of non-tradables (Delgado *et al.*, 1998). The simplified multiplier is represented as:

$$M = \frac{1}{1 - \beta_n(1 - s)} \tag{5}$$

where $\beta_n = MBS_i$ and s = average annual savings rate (1.59% of mean household expenditure).

5.4 CHARACTERISTICS OF THE KWAZULU-NATAL STUDY AREAS AND SAMPLE HOUSEHOLDS

Sample households were drawn from the adjacent tribal wards of Gcabane and KwaHlongwa near Umzumbe, and the inland ward of Swayimana. Gcabane and KwaHlongwa were treated as one 'district' due to comparable political, environmental and sociological characteristics. Swayimana households were on average[2] slightly larger and poorer than Umzumbe households. In Swayimana, 44% of households fell below the national poverty line (May, 1998), compared to 20% of Umzumbe households. Both districts lacked infrastructure, although 12% of Swayimana households and 8% of households at Umzumbe were supplied with electricity. Fifteen of the surveyed households, all in Swayimana, had on-site piped water, while only two households (both in the Umzumbe sample) had a telephone in their homes. The local cellular phone network could not be accessed from either district at the time the survey was conducted. Swayimana households were on average further from a tarred road (35.35 km) than were Umzumbe households (16.86 km). Only three households, all in the Umzumbe district, owned a vehicle.

[2] Although these sample means – and those that follow – were computed as if all sample households had equal probability of selection, they are not necessarily biased estimates of the district means as the primary sampling units (PSUs) were of roughly equal size within each stratum.

The two districts are located in prime agricultural areas, with Swayimana households having larger crop areas. A total of 77% of Swayimana households and 66% of Umzumbe households farmed crops and/or livestock. Maize, amadumbes (*taro*) and beans were the major crops grown in both districts. Swayimana households produced more maize than Umzumbe households, while Umzumbe households produced relatively more beans. The types of vegetables and fruit grown by households in the two districts were similar. Vegetable crops included cabbage, swiss chard, tomatoes and pumpkins. Avocados were grown in both districts, and bananas at Umzumbe. Three sample households in each district farmed sugar cane. Cattle were owned by 63% of Swayimana households, compared to 43% of Umzumbe households, while most households (96% and 80% for Swayimana and Umzumbe, respectively) kept poultry. Agriculture's share of household income (including cash sales and the opportunity cost of home production) was only 16.3% and 6.5%, respectively for Swayimana and Umzumbe, despite the good agricultural potential of these districts.

5.5 EXPENDITURE ELASTICITIES FOR DISTRICTS, WEALTH GROUPS AND FARM AND NON-FARM TRADABLES AND NON-TRADABLES

Estimated expenditure elasticities for households in Swayimana and Umzumbe are presented in Tables 5.1 and 5.2. These estimates show little possibility of income-induced growth in the overall demand for food (expenditure elasticities of 1.09 and 0.98 for Swayimana and Umzumbe, respectively), except for horticultural products (7.91 and 4.56), and to a lesser extent, for meat and poultry products (0.97 and 1.04). For these two categories, the expenditure elasticities indicate likely proportional demand growth for meat and poultry products, and more than proportional demand growth for horticultural products as income increases.

Households in both districts would substantially increase their expenditure on durable items (furniture and appliances), with rising incomes (expenditure elasticities of 2.17 and 2.20 for Swayimana and Umzumbe, respectively). This suggests that should income rise, the demand for imported consumer durables may grow faster than the demand for locally produced commodities, directing part of any additional income out of the local economy. This leakage will tend to lower the strength of rural growth linkages.

Large positive elasticity estimates (2.46 and 2.72, for Swayimana and Umzumbe respectively) for housing expenditure have positive implications for the local economy, as building materials (for example, concrete blocks), construction and repair services tend to use local resources, especially labour. Estimated expenditure elasticities for transport were high (8.30 and 2.52, respectively for Swayimana and Umzumbe). Wealthier and poorer households display similar expenditure behaviour for most products, but the demand growth generated by poorer households for durables, housing and horticultural products is greater than for wealthier households.

Table 5.1: Expenditure elasticities for significant aggregate commodity groups in Swayimana and Umzumbe, 1997 (n=93), (Hendriks & Lyne, 2003b).

Expenditure group	*F-statistic*	*Swayimana (n=46)*			*Umzumbe (n=47)*		
		ABS	*MBS*	*Elasticity*	*ABS*	*MBS*	*Elasticity*
Consumer expendables	6.74	0.02	0.02	0.76	0.03	0.02	0.71
		(4.23)	(13.41)	(9.52)	(6.68)	(19.65)	(18.50)
Durables	1.91	0.02	0.05	2.17	0.06	0.12	2.20
		(0.76)	(6.84)	(8.29)	(1.99)	(18.23)	(22.32)
Food	1.83	0.70	0.76	1.09	0.59	0.58	0.98
		(9.84)	(44.81)	(8.48)	(8.36)	(34.28)	(1.31)
Housing	2.86	0.02	0.04	2.46	0.06	0.17	2.72
		(0.50)	(5.09)	(6.79)	(1.69)	(19.12)	(27.16)
Social obligations	2.95	0.06	0.02	0.31	0.06	–0.02	–0.29
		(2.11)	(2.73)	(13.61)	(2.32)	(–2.84)	(–28.14)
Transport	1.97	0.01	0.05	8.30	0.03	0.08	2.52
		0.44	*15.37*	*30.40*	*2.25*	*23.65*	*32.06*

Note: Figures in parentheses are t-statistics for the null hypothesis that the predicted ABS and MBS are zero, and that the expenditure elasticity is unity

The results of the expenditure analysis for farm and non-farm tradables and non-tradables are presented in Table 5.3. Estimated expenditure elasticities for farm tradables are positive and less than unity (0.53) for the relatively poorer and more remote Swayimana households, and 0.41 for Umzumbe households. Expenditure on farm tradables is the largest portion of household expenditure (estimated ABS of 0.65 and 0.53, for Swayimana and Umzumbe respectively). The expenditure elasticities for non-tradable farm goods of 1.23 and 1.22 for Swayimana and Umzumbe respectively, indicate likely demand growth for meat and poultry products, and considerable demand

Table 5.2: Expenditure elasticities for significant food commodity groups in Swayimana and Umzumbe, 1997 (n=93), (Hendriks & Lyne, 2003b).

Expenditure group	*F-statistic*	*Swayimana (n=46)*			*Umzumbe (n=47)*		
		ABS	*MBS*	*Elasticity*	*ABS*	*MBS*	*Elasticity*
Staples	5.33	0.34	0.06	0.17	0.25	–0.03	–0.11
		(8.30)	(5.73)	(64.95)	(6.04)	(–2.77)	(–62.84)
Maize	5.32	0.13	0.01	0.10	0.07	0.00	0.04
		(5.30)	(2.22)	(44.70)	(2.69)	(0.42)	(24.29)
Rice	1.55	0.07	–0.01	–0.12	0.09	–0.01	–0.07
		(3.24)	(–1.65)	(–34.10)	(4.28)	(–1.23)	(–42.94)
Roots	6.89	0.05	0.01	0.19	0.03	–0.01	–0.24
		(5.97)	(4.62)	(45.64)	(3.46)	(–3.45)	(–40.18)
Wheat	2.97	0.09	0.04	0.44	0.06	–0.02	–0.32
		(5.57)	(10.57)	(29.43)	(3.75)	(–4.92)	(–46.19)
Horticultural products	2.65	0.05	0.42	7.91	0.09	0.42	4.56
		(1.18)	(38.87)	(76.37)	(2.06)	(39.10)	(68.62)
Meat and poultry products	2.11	0.08	0.08	0.97	0.05	0.06	1.04
		(4.49)	(18.07)	(1.44)	(2.84)	(12.36)	(1.18)

Note: Figures in parentheses are t-statistics for the null hypothesis that the estimated ABS and MBS are zero, and that the expenditure elasticity is unity

Table 5.3: District expenditure analysis for tradable/non-tradable commodity groups, Swayimana and Umzumbe, 1997 (n=93), (Hendriks & Lyne, 2003a).

Expenditure category	*F-statistic*	*Swayimana (n=46)*				*Umzumbe (n=47)*			
		Observed	*Estimated*			*Observed*	*Estimated*		
		ABS	*ABS*	*MBS*	*Elasticity*	*ABS*	*ABS*	*MBS*	*Elasticity*
Farm tradables	2.72	0.64	0.65 (15.93)	0.34 (27.08)	0.53 (88.49)	0.53	0.53 (19.24)	0.22 (42.18)	0.41 (85.60)
Farm non-tradables	1.83	0.09	0.10 (1.48)	0.12 (7.55)	1.23 (3.08)	0.05	0.05 (2.83)	0.06 (14.47)	1.22 (6.00)
Non-farm tradables	3.55	0.15	0.15 (8.70)	0.37 (70.52)	2.58 (76.98)	0.30	0.29 (4.32)	0.57 (46.53)	1.94 (64.15)
Non-farm non-tradables	2.27	0.12	0.12 (3.71)	0.17 (19.16)	1.47 (8.27)	0.13	0.12 (3.44)	0.15 (21.15)	1.24 (15.29)

Note: Figures in parentheses are t-statistics for the null hypothesis that the predicted ABS and MBS are zero, and that the expenditure elasticity is unity

growth for horticultural products following an increase in household incomes. Increased demand for these currently demand-constrained, non-tradable, farm goods could, therefore, generate higher turnover and employment on local farms.

Consumption income elasticities were also greater than one, which is consistent with other South African studies (Van Zyl *et al.*, 1991; Nieuwoudt & Vink, 1989). Rural households are thus twice as likely to spend additions to income on commodities other than food. Expenditure elasticities estimated for non-tradable non-farm commodities are lower than for tradable non-farm commodities in KwaZulu-Natal. These results suggest some potential for income-induced growth in the local economy, mainly through greater demand for local housing materials, building and repair services, transport, and catering and hospitality services.

The estimated expenditure elasticities for each category of commodities did not differ much between districts or wealth groups. Since the estimated MBS_i did not differ much between districts or wealth groups, it appears that growth linkages in communal areas are less likely to depend on location or on which rural households receive increased incomes. Reasonably strong growth linkages seem possible for KwaZulu-Natal, driven mainly by more than proportional increased local demand for farm non-tradables and non-farm non-tradables. However, these linkages are likely to be weakened by increased expenditure on non-farm tradables, a leakage from the local economy.

5.6 GROWTH MULTIPLIERS FOR KWAZULU-NATAL STUDY AREAS

Estimated growth multipliers presented in Table 5.4 show the resultant additions (from consumption and production of farm and non-farm non-tradables) to gross income of local households following an initial income shock of one (added) South

Table 5.4: Growth multipliers for Swayimana and Umzumbe, 1997 (n=93), (Hendriks & Lyne, 2003a).

Commodity	*Whole sample*	*Swayimana*	*Umzumbe*	*Wealthier decile*	*Poorer households*
Tradables	1.00	1.00	1.00	1.00	1.00
Value added from farm non-tradables	0.09	0.12	0.06	0.10	0.09
Value added from non-farm non-tradables	0.19	0.20	0.18	0.21	0.18
Total value added	0.28	0.32	0.24	0.31	0.27
Multiplier	1.28	1.32	1.24	1.31	1.27

African Rand. This shock could come from any positive outside effect such as improved technology, better infrastructure or institutional support that increases profit in the production and marketing of rural tradables. The results show that over the whole sample, a one Rand increase in income could lead to total value added from spending on these demand-constrained items of 28 cents.

The total value added by a one Rand income shock in Swayimana households is 32 cents for farm and non-farm tradables, slightly higher than the 24 cents estimated for the less remote Umzumbe households. The total value added for the wealthier income group is only four cents higher than the multiplier for the poorer households, implying negligible differences between income groups in the way in which they would spend additional income.

The results suggest that agriculture has limited potential to drive growth in South Africa's communal areas, and that growth linkages could be weaker than for other African economies. The estimated whole sample multiplier of 1.28 is far lower than the local multipliers for African countries reported by Delgado *et al.* (1998) that range from 1.71 to 2.67. South African rural households are generally less remote and less isolated from urban areas than are rural households in other parts of Africa. The expected value added from farm non-tradables is roughly half that from non-farm non-tradables. The estimates are lower than those from other sub-Saharan African countries (Delgado *et al.*, 1998), where the farm contribution is usually larger than the non-farm contribution.

Although the findings do not represent particularly strong growth linkages, it is clear that increased incomes could generate additional growth through consumption linkages. The results raise two important policy issues. Firstly, the multiplier indicates *potential* growth linkages because the supply of non-tradables may not be perfectly elastic. An implicit assumption underlying these results is that increased demand for non-tradable goods and services will be met by new production of these items. In other words, the supply response of non-tradables is assumed to be elastic, although this is not necessarily the case in communal areas where supply elasticity is often low. This is because, by definition, new demand for these items cannot be met from imports. Secondly, while potential growth linkages are driven by non-tradables, these non-tradables are unlikely to generate the initial income shock, as they are demand-constrained.

5.7 COMPETITIVENESS OF NON-TRADABLE COMMODITIES IN THE EASTERN CAPE

Tradable commodities that can compete with rival products could provide the necessary income to create effective demand for non-tradables that, via the multiplier effect, could bring underused resources into production. In rural households in the Eastern Cape, citrus and beef are largely non-tradable commodities simply because smallholder farmers currently do not extensively farm these commodities for export out of local communities. However, intensified farming systems and technology advances (such as the introduction of irrigated citrus) could lead to commercialisation. Ngqangweni (2000) constructed Policy Analysis Matrices (PAMs) (Monke & Pearson, 1989) for selected citrus and beef enterprise case studies to estimate the efficiency of smallholders' resource use and, hence, their competitiveness (or comparative advantage) in these enterprises. Production and marketing data were collected through focus group interviews, farmer recall, and interviews with organisations involved in smallholder extension and marketing in the Mpofu district in 1997.

Due to a lack of reliable secondary data, no attempt was made to describe *representative* smallholder farms and activities. Rather, the objective was to give insights into what was *possible*, given the observed activities of samples of relatively successful smallholders currently involved in farming and selling significant portions of their output in the market. Smallholder farm budgets (using prevailing technologies) were developed on a per unit basis, and compared with secondary data on transportation costs, prices, and shadow prices, to assess partial equilibrium indicators of competitiveness. Social cost values for land, labour and capital, were estimated using the closest proxy of their opportunity costs. These opportunity costs determine whether an activity makes efficient use of resources.

Six indicator ratios derived from the PAMs measure the competitiveness of different agricultural activities given current technology and government policies within and between regions. They rank the comparative advantage of various smallholder enterprises and identify possible areas of investment to increase the growth of national income. The first indicator, the domestic resource cost (DRC), is a measurement specific to a given technology, a given end market, and a given location of production. It condenses into a single ratio the relationship between the true cost of producing one unit of the item in question and the return in selling it. A DRC of less than one is usually thought to indicate efficient production based on the existence of comparative advantage. The DRC for a particular smallholder activity is equal to the value of domestic inputs used, priced using social prices, divided by product revenues at social prices less the cost of tradable inputs priced at social prices. A DRC less than unity, therefore, shows that the opportunity cost (the cost of production as valued by the foregone most profitable alternative uses of the inputs) of the domestic resources used is less than the value-added earned from the sale of those resources. Put simply, it indicates how well the activity uses resources to earn value. Comparing across activities, the activity with the lowest DRC is the one that earns the most value with the least value of inputs. Although DRCs are usually employed in the context of international trade, and refer to the saving or earning of foreign exchange, the methodology also applies to regions within a country, provided the commodity in question is tradable.

Non-tradables are commodities (or resources) whose equilibrium local market price is too high to permit profitable export to 'outside' (non-local) markets, but too low to justify transporting the goods into the local area, given prices 'outside'. Furthermore, true non-tradables are not good economic substitutes for other tradables (their prices are not correlated with tradables). DRC measurements can only be interpreted as showing comparative advantage in the cases of tradables, although they are computed for all goods. A DRC of less than one for a tradable suggests that it would be profitable to concentrate more resources in that activity, but the same cannot be said of the DRCs for non-tradables. For them, increased local production will lead to falling local prices as the local market is saturated, and the computed DRC will rise quickly.

The valuation of domestic factors of production for black smallholder farmers in the Eastern Cape assumed that opportunity costs were comparable to those of their white commercial farm counterparts. Given the history of smallholder repression, it would probably be incorrect to compare the two groups of farmers in this way, but the analysis does show how efficient smallholder production would be if subjected to the same cost assumptions applying to commercial farmers. Table 5.5 contrasts outcomes from two scenarios: smallholder private costs versus white commercial farm costs (Lyne, 2000; see also Ohene-Anyang, 1997). The 'private-level' columns show data on land and labour costs, which approximate those reported or observed from the smallholders' point of view (private costs), and a resultant DRC. Under the 'commercial level' scenario, the opportunity costs of land and labour that approximate rates for commercial farmers, and a resultant DRC, are shown.

Under the given set of assumptions relating to opportunity costs, smallholders in the Eastern Cape seem to produce the chosen commodities profitably and efficiently at commercial cost rates. With a DRC ratio of 0.62, the indigenous beef activity has the better indicator of efficiency, followed by citrus at 0.65. These assertions could not be considered conclusive without conducting a sensitivity analysis of the assumptions regarding opportunity costs.

Table 5.5: Contrasting outcomes from smallholder 'private level' versus 'commercial level' assumptions, Mpofu district, Eastern Cape, 1997.

Activity	*'Private level' opportunity cost estimates*			*'Commercial level' opportunity cost estimates*		
	Land (R/ha)	*Labour (R/day)*	*DRC*	*Land (R/ha)*	*Labour (R/day)*	*DRC*
Indigenous cattle	12	10	0.37	30	20	0.62
Irrigated citrus	2 900	30	0.79	2 000	20	0.65

Smallholder cattle farmers in Mpofu district leased land from the state at a nominal rate of R12 per hectare that was lower than the rate on commercial farms of R30 per hectare. This is land transferred from the then Republic of South Africa to the former homelands, which subsequently became property of the South African Native Trust as stipulated by the Natives Trust and Land Act of 1936. The hired labour employed was paid a wage of R10 a day, which was considered as the market-related rate in the study area.

In the sensitivity analysis reported in Table 5.6, the private level resource costs borne by the indigenous beef smallholder were taken as the first possible opportunity cost scenario. An additional factor was recorded as part of the first scenario, i.e. an off-take rate of 16% was assumed. Under this scenario, smallholder farmers would enjoy a comparative advantage as shown by a DRC of 0.37. If higher opportunity costs at the commercial rates are applied, efficiency falls but the DRC is still below one at 0.62 at a higher and more likely off-take rate of 16% (a higher off-take rate would make more sense for indigenous cattle considering its physical advantages). Smallholders will enjoy comparative advantage in indigenous beef production only up to a certain level of opportunity costs of land and labour. No threshold level was measured, but it seems that if land and labour opportunity costs were to increase to R60 a hectare and R20 per day, respectively, rural communities would no longer gain from smallholder use of resources for raising indigenous beef.

Citrus farmers in the study area directly incur a cost of R2 900 (rental of state land) per hectare for their land. Labour costs were double those of other activities chosen for analysis, at R30 per day. Taking these observed conditions as the first (lower limit) scenario in the sensitivity analysis, smallholder citrus would still enjoy a degree of comparative advantage, with a DRC of 0.79. Further investigation

Table 5.6: Sensitivity analyses of DRC estimates for indigenous cattle and citrus, Mpofu district, Eastern Cape, 1997.

Scenario			*Opportunity cost beef*		*Cattle off-take rate (%)*	*DRC beef*	*Opportunity cost citrus*		*Citrus crop export (%)*	*DRC citrus*
	Land price	*Labour price*	*Land (R/ha)*	*Labour (R/day)*			*Land (R/ha)*	*Labour (R/day)*		
1	Low	Low	12	10	16	0.37	2 900	30	60	0.79
2	Low	Low	12	10	14	0.44	2 900	30	65	0.74
3	Low	High	12	20	16	0.46	2 900	20	60	0.74
4	Low	High	12	20	14	0.46	2 900	20	65	0.69
5	Medium	Low	30	10	16	0.59	2 000	30	60	0.75
6	Medium	Low	30	10	14	0.59	2 000	30	65	0.69
7	Medium	High	30	20	16	0.62	2 000	20	60	0.69
8	Medium	High	30	20	14	0.61	2 000	20	65	0.65
9	High	Low	60	10	16	0.84	1 000	30	60	0.70
10	High	Low	60	10	14	0.84	1 000	30	65	0.65
11	High	High	60	20	16	0.86	1 000	20	60	0.64
12	High	High	60	20	14	1.90	1 000	20	65	0.60

revealed that opportunity costs of land and labour for smallholder citrus in the study area were R2 000 per hectare and R20 per day. This scenario significantly improves the activity's efficiency, and the DRC of between 0.65 and 0.69 for smallholder citrus at these rates and for citrus crop export rates between 60% and 65% makes it less advantageous than indigenous beef.

All 12 scenarios reported in Table 5.6 for citrus showed potential for comparative advantage for smallholder citrus, regardless of the dominance of land and labour opportunity costs, and export quantity. However, the impact of a good citrus harvest and a high export percentage cannot be ignored. Keeping the opportunity costs of land and labour constant, and adjusting export percentage up or down by 5%, had at least an equal effect on the DRC. The only major threat to the output side is the incidence of hailstorms that negatively affect the export percentage. Other factors such as smallholders' long-term cash flow and debt status should be factored in when interpreting these results.

The results show that smallholders are at least competitive in the chosen commodities, suggesting some potential for indigenous beef and citrus in this sector to act as engines of growth in the rural areas. In the context of broader rural restructuring, it is important to know whether investment in this sector has any potential for broader income and employment growth, and could hence be justified. Such investment seems justified to the extent that the sale of these competitive tradable commodities would, first, create the effective demand necessary to purchase non-tradable rural goods, and second, bring underutilised rural resources into use.

5.8 DISCUSSION

Growth linkages in the selected areas of KwaZulu-Natal studied in this chapter appear to be weaker than in most Asian and other African countries. Nevertheless, increased farm incomes could generate additional growth through consumption linkages, but may be weaker than those presumed by government. For local economies to benefit from growth linkages, it is important (a) that the local supply of non-tradables is price elastic, and (b) that the initial income shock is *widespread*. Economic growth is generated through an initial exogenous income shock (such as technological change or improved infrastructure), resulting in extra income derived from stimulated regional demand and production in the local non-tradable sector. In under-productive, but high potential farming areas where most households have knowledge of farming and access to land, this initial income shock would best come from increased agricultural production but this is unlikely without substantial investment in public goods. As growth linkages are relatively low even in areas of good agricultural potential, public investment would be best targeted at areas of high agricultural potential.

For agricultural growth to reduce poverty in South Africa, a number of conditions are required in designing and promoting appropriate agricultural development policies in the country's communal areas. Some conditions necessary to increase income from farm tradables, improve the price elasticity of supply for non-tradables, and stimulate enterprise and employment opportunities in the non-tradable farm and non-farm sectors are now discussed. Firstly, increased agricultural production can increase income from farm tradables, but would require adoption of new technology, including irrigation, new and diverse crop varieties, new production methods, and possibly also more land- and labour-intense production systems. Much smallholder farm produce is currently non-tradable due to the bulkiness and perishability of produce.

Tradable products with comparative advantage for local producers are the mostly likely engines of sustained rural growth. Growing and exporting a tradable agricultural good to outside the region brings in more income without depressing local prices. Although these possibilities have not been sufficiently explored, Ngqangweni's (2000) study suggests that certain smallholder activities may potentially be privately and socially profitable. However, improved production incentives are needed. More research by smallholder farmer collectives and/or government is needed to identify commodities for which smallholders have a comparative advantage. This will help to better target policy to strengthen smallholder production and market access.

Secondly, it is possible that the price elasticity of supply of non-tradables would increase as farm incomes rise and non-farm enterprises emerge and grow, accelerating demand growth for non-tradable commodities. Increased local demand for non-tradables should be translated to increased production of these items. Moreover, the demand for non-farm products is dependent on the income from farmers and farm workers, and initial impetus for growth is not likely to come from the non-farm sector. However, the Hendriks & Lyne (2003a) study results show that the non-farm sector is likely to contribute more to growth multipliers for KwaZulu-Natal communal areas than the farm sector itself. This concurs with Ravallion & Datt's (2002) finding that non-farm output changes have a more powerful, longer-term impact on the prospects of escaping absolute poverty in India.

Most of the non-farm sector consists of small industries that exploit rural market niches, particularly for bulky products such as wood for fuel, building material and many farm and non-farm services. Should non-farm incomes rise, there may be many opportunities for non-farm enterprise growth in rural areas due to the high costs of transporting similar products and the inaccessibility of urban-based services. Again the comparative advantage of non-farm enterprises should also be analysed, as agricultural development depends on provision of services from the non-farm sector, while the non-farm sector depends to some extent on the supply of raw materials from agriculture and demand for inputs and services from the farm sector.

Thirdly, to transform non-tradables into tradable commodities, transaction costs need to be reduced, leading to better prices for producers, increased access to technology and increased market opportunities. High transaction costs result from weak physical and institutional infrastructure (Matungul *et al.*, 2001). While public policy should, therefore, focus on alleviating these constraints in communal areas where most of South Africa's rural poor live, smallholders and the private sector also have roles to play in reducing transaction costs, by for example, setting up marketing agreements between commercial and communal farmers. Government should also perhaps play a greater role in reducing and/or bearing some of the costs of collective action, and ensure that contracts are enforced in an unbiased and predictable way by local regulatory institutions such as tribal courts. Public extension agents could, for example, facilitate land rental transactions by identifying and maintaining lists of willing lessors and lessees; supplying copies of *proforma* lease agreements, and facilitating and witnessing rental contracts.

Fourthly, business opportunities in the non-tradable farm and non-farm sectors would lead to increased employment for rural populations. Poor households tend to consume relatively labour-intensive commodities, while small rural enterprises tend to be relatively labour intensive, providing employment opportunities within rural areas. In addition, multiple-cropping, agricultural diversification and intensification will most likely lead to increased aggregate demand for agricultural labour, bidding up wages for new entrants into an expanding non-farm sector (Ravallion & Datt, 2002). Although agricultural labour is an easy entry point for unskilled workers, López (2002) points out that higher incomes for unskilled workers is a short-term approach to poverty alleviation as skill development does not necessarily occur, limiting earning capacity of unskilled workers. An increase in the opportunity cost of labour induces a shift in the composition of non-farm activity out of the very labour-intensive, low return activities, to more skilled, higher investment, high returns activities (Hazell & Haggblade, 1993) which provide long-term solutions to abject poverty. However, unskilled workers need exposure to education and training opportunities. This would require extensive investment in enterprise development and training provision and support for farm and non-farm enterprise, technology, production and marketing skills.

Public, non-governmental and community-based extension services play a critical role in mobilising, empowering and supporting emerging rural enterprises (Aliber, 2001; Kepe *et al.*, 2001). Public extension needs to adopt a client-orientated, problem solving approach, which in turn, will require changes in incentive structures and new partnerships with non-governmental organisations, private sector firms and farmers (Hazell & Haddad, 2001). In addition, public extension services should address issues beyond agricultural production (Ndlela, 2002: 86), and together with non-governmental and community-based organisations, offer training and both technical and business support for a range of farm and non-farm enterprise opportunities.

Finally, increased employment opportunities and wages for unskilled workers are only possible if policy and market situations are conducive to agricultural growth. For example, if credit market failures curtail farmers' access to liquidity, the supply of agricultural products will remain inelastic and rural populations will not reap the full benefits of public investment in new technology, infrastructure and training. The same is true of the land market. Without an efficient rental market for farmland, expansion of farming enterprises is not possible. Agricultural-led growth requires new technology and improved infrastructure (both physical and institutional) to reduce unit production and transaction costs, improve product prices, to broaden access to land and credit markets, and to convert non-tradables into tradables.

5.9 CONCLUSION

Although the KwaZulu-Natal case study presented here shows that agriculture-led growth has potential to drive economic reform in the two communal areas studied, the strength of the multipliers is lower than government expects. In under-productive but high potential areas such as these, the income shock is most likely to come from increased agricultural production, but it not likely without substantial investment in public goods.

REFERENCES

AFRICAN NATIONAL CONGRESS. (1994). *The Reconstruction and Development Programme.* Braamfontein: Umanyano.

ALDERMAN, H. (1993). Obtaining useful data on household incomes from surveys. In Von Braun, J. & Puetz, D (eds). *Data Needs for Developing Countries: New Directions for Household Surveys.* Washington DC: International Food and Policy Research Institute: 193–201.

ALIBER, M. (2001). Study of the incidence and nature of chronic poverty and development policy in South Africa: an overview. Unpublished paper, Programme for Land and Agrarian Studies, School of Government, University of the Western Cape.

DE JANVRY, A. (1994). Farm-non-farm synergies in Africa: discussion. *American Journal of Agricultural Economics* 76: 1183–1185.

DELGADO, C.L. (1997). The role of smallholder income generation from agriculture in sub-Saharan Africa. In Haddad, L. (ed.). *Achieving Food Security in Southern Africa.* Washington DC: International Food Policy Research Institute.

DELGADO, C.L., HOPKINS, J. & KELLY, V.A. (1998). *Agricultural Growth Linkages in Sub-Saharan Africa.* Washington DC: International Food Policy Research Institute.

DEVEREUX, S. (1993). Collection of production and income data: a commentary. In Von Braun, J. & Puetz, D. (eds). *Data Needs for Developing Countries: New Directions for Household Surveys.* Washington DC: International Food and Policy Research Institute: 202–206.

FOOD AND AGRICULTURE ORGANISATION. (2001). Financing term investments in agriculture. Draft report prepared for the World Bank, Food and Agriculture Organisation, Rome.

HAGGBLADE, S., HAMMER, J. & HAZELL P.B.R. (1991). Modelling agricultural growth multipliers. *American Journal of Agricultural Economics* 73: 361–374.

HAZELL, P. & HADDAD, L. (2001). *Agricultural Research and Poverty Reduction.* 2020 Brief 70. Washington DC: International Food Policy Research Institute.

HAZELL, P.B.R & HAGGBLADE, S. (1993). Farm-non-farm growth linkages and the welfare of the poor. In Lipton, M. & Van der Gaag, J. (eds). *Including the Poor: Proceedings of a Symposium*

Organized by the World Bank and the International Food Policy Research Institute. Washington DC: World Bank: 190–204.

HAZELL, P.B.R. & HOJJATI, B, 1995. Farm non-farm growth linkages in Zambia. *Journal of African Economies* 4(3): 406–435.

HAZELL, P.B.R. & RÖELL, A. (1983). *Rural Growth Linkages: Household Expenditure Patterns in Malaysia and Nigeria.* Washington DC: International Food Policy Research Institute.

HENDRIKS, S.L. (2002). Expenditure elasticities and growth linkages for rural households in two study areas of KwaZulu-Natal. PhD thesis, University of Natal, Pietermaritzburg.

HENDRIKS, S.L. & LYNE, M.C. (2003a). Growth linkages for two communal areas of KwaZulu-Natal. *Development Southern Africa* (Forthcoming).

HENDRIKS, S.L. & LYNE, M.C. (2003b). Expenditure elasticities for two communal areas of KwaZulu-Natal. *Development Southern Africa* 20(1): 105–127.

ISLAM, N. (1997). *The Non-farm Sector and Rural Development: Review of Issues and Evidence.* 2020 Vision, Brief 47. Washington DC: International Food Policy Research Institute.

KEPE, T., NTSEBEZA, L. & PITHERS, L. (2001). Agri-tourism spatial development initiatives in South Africa: are they enhancing rural livelihoods? Paper number 65, Natural Resource Perspectives, Overseas Development Institute, London.

LÓPEZ, R. (2002). Agricultural growth and poverty reduction. Theoretical Note Module 3 - Poverty reduction. Food and Agriculture Organization Roles of Agriculture Project. http://www.fao.org/es/esa/roa/forum (Accessed 16/01/03).

LYNE, M.C. (2000). Personal Communication. University of Natal, Pietermaritzburg.

MACHETHE, C.L., REARDON, T. & MEAD, D.C. (1997). Promoting farm/non-farm linkages for employment of the poor in South Africa: a research agenda focused on small-scale farms and agro-industry. *Development Southern Africa* 14(3): 377–394.

MARKETS AND STRUCTURAL STUDIES DIVISION. (2002). Promoting growth and diversification through markets for high-value crop and animal products. http://www.ifpri.org/themes/mp07.htm (Accessed 16/03/01).

MATUNGUL, M.P., LYNE M.C. & ORTMANN, G.F. (2001). Transaction costs and crop marketing in the communal areas of Impendle and Swayimana, KwaZulu-Natal. *Development Southern Africa* 18(3): 347–364.

MAY, J. (1998). Poverty and inequality in South Africa. Report prepared for the office of the executive Deputy President and the Inter-Ministerial Committee for Poverty and Inequality. Summary Report. http://www.gov.za/reports/1998/pirsum.htm (Accessed 01/10/99).

MELLOR, J. (1976). *The New Economics of Growth: A Strategy for India and the Developing World.* Ithaca, NY: Cornell University Press.

MONKE, E.A. & PEARSON, S.R. (1989). *The Policy Analysis Matrix for Agricultural Development.* Ithaca, NY: Cornell University Press.

NATIONAL DEPARTMENT OF AGRICULTURE (NDA). (1995). *White Paper on Agriculture.* Pretoria: Department of Agriculture Directorate Agricultural Information Services.

NDA. (1999). Agricultural policy in South Africa: a discussion document. http://www.nda.agric.za (Accessed 07/10/99).

NDA. 2001. *The Strategic Plan for South African Agriculture.* Pretoria: Department of Agriculture Directorate Agricultural Information Services.

NATIONAL MINISTRY OF PROVINCIAL AND LOCAL GOVERNMENT. (2000). *The Integrated Sustainable Rural Development Strategy.* Pretoria: Ministry of Provincial and Local Government.

NDLELA, N.R. (2002). Provincial Department of Agriculture Extension Service and the needs of women's clubs in Hlanganani, KwaZulu-Natal. MSocSc thesis, University of Natal, Pietermaritzburg.

NGQANGWENI, S.S. (2000). Promoting income and employment growth in the rural economy of the Eastern Cape through smallholder agriculture. PhD thesis, University of Pretoria.

NIEUWOUDT, W.L. & VINK, N. (1989). The effects of increased earnings from traditional agriculture in southern Africa. *South African Journal of Economics* 57(3): 257–269.

OHENE-ANYANG, E. (1997). Comparative advantage and policy incentives for commercial maize and wheat production in South Africa. MSc thesis, University of Pretoria.

PUETZ, D. (1993). Improving data quality in household surveys. In Von Braun, J. & Puetz, D. (eds). *Data Needs for Developing Countries: New Directions for Household Surveys*. Washington DC: International Food and Policy Research Institute: 173–185.

RAVALLION, M. & DATT, G. (2002). Why has economic growth been more pro-poor in some states of India than others? *Journal of Development Economics* 68: 381–400.

REARDON, T., CRAWFORD, E. & KELLY, V. (1994). Links between non-farm and farm investment in African households: adding the capital market perspective. *American Journal of Agricultural Economics* 76: 1172–1176.

STATISTICS SOUTH AFRICA. (1998). Unemployment and employment in South Africa. http://www.statssa.gov.za/u&e/Prelim_p.htm (Accessed 01/10/99).

VAN ZYL, J. & KIRSTEN, J. (1998). Defining small-scale farmers in the South African context. *Agrekon* 37(4): 560–571.

VAN ZYL, J., MACHETHE, C., SARTORIUS VON BACH, H. & SINGINI, R. (1991). The effects of increased earnings from traditional agriculture in Lebowa. *Agrekon* 30(4): 276–278.

VAN ZYL, J. & VINK, N. (1992). Effects of farmer support programmes on consumption and investment. *Development Southern Africa* 5(2): 196–207.

CHAPTER 6

The Development of Small-Scale Sugar Cane Growers

A Success Story?[1]

Richard Bates and Patrick Sokhela

6.1 INTRODUCTION

The South African sugar industry is located in the KwaZulu-Natal, Mpumalanga and Eastern Cape provinces of South Africa. Approximately 50 000 growers produce sugar cane on 434 507 hectares (SASA, 2002b). In certain areas of South Africa the growing cycle of sugar cane is longer than 12 months and therefore not all land is harvested in a season. In the 2001–02 production season 329 591 hectares were harvested to produce 21.2 million tons of sugar cane crushed in 15 sugar mills. The average yield was 64 tons of sugar cane per hectare and a total of 2.4 million tons of sugar were produced, of which about one-half was exported (SASA, 2002a: 3). The gross income from the sale of sugar and by-products was R5.7 billion.

Small-scale growers (SSGs)[2] comprise 96% of growers in the industry and they produced 14.4% of the sugar cane crop on 19.7% (85 418 hectares) of registered sugar cane land during the 2001–02 season (SASA 2002b). These small-scale growers are mainly black[3] growers who produce sugar cane in communal (traditional tenure) areas. Fifty-six per cent of the growers are located in the Zululand area of KwaZulu-Natal, and deliver sugar cane to the Amatikulu, Felixton, Entumeni and Umfolozi mills. Significant production of sugar cane by SSGs in Mpumalanga and the Eastern Cape provinces only commenced in the 1980s.

[1] Appreciation is expressed to the South African Sugar Association and the South African Cane Growers' Association for assistance provided. Opinions expressed and conclusions arrived at are those of the authors and are not necessarily those of the individual Associations.

[2] SSGs are sugar cane growers who produce less than 2 100 tons of sugar cane per season. This tonnage would require a maximum of ±40 hectares (see note 3).

[3] This chapter is primarily concerned with black SSGs as they have been the most marginalised in terms of their access to resources compared to other racial groups. Bates (1996) found that 90% of small-scale black growers had less than 4 hectares of sugar cane land. The average area was 2.5 hectares.

For much of its history, the sugar industry has continuously exercised production controls taking into account local demand and that of an export market prone to price volatility. This has meant that expansion of production has had to be undertaken in a controlled manner. A driving force in the South African sugar industry is the maintenance of production and, when the opportunity exists, its expansion. Furthermore, the maximisation of throughput of sugar mills is critical in reducing the cost of sugar production.

About 14% of the mills' sugar cane throughput is now sourced from SSGs. Sugar cane production by SSGs increased over the period 1946–93, and in particular during the years 1973–93. The increase in production appears to be correlated to an increase in the total area planted and not necessarily to an increase in productivity per unit area (Bates, 1996: 76).

Promoting the development of SSGs by the sugar industry is unique in South African agriculture. SSGs were identified as a 'sector' at an early stage and their special needs were recognised in the development of the industry.

SSGs derived a gross income of approximately R487 million from sugar cane production in 2001–02. The impact on livelihoods in rural areas in which sugar cane is produced has been significant, not only in terms of the impact of monetary income, but also in terms of the development of infrastructure (roads, etc.) and of economic multiplier effects.

The development of SSGs has not been without its problems and a number of issues will need to be addressed in the future. These, however, should be seen in the light of changing economic, social and political conditions. Interventions that achieved identified objectives some years ago may be inappropriate today and will require reviewing.

This chapter traces the development of SSGs and, where appropriate, identifies problems that have constrained development. This approach is not to detract from an historically significant development intervention by a large agro-industry, but is adopted to provide a model that could be applied elsewhere.

Rural areas in South Africa require the opening up of development opportunities as there are high levels of unemployment in the economy and the option of migration to urban areas currently does not offer the solution that it did previously to rural residents. Agriculture and its associated businesses and industries have a role to play in this regard and the sugar industry's example should provide some guidelines as to appropriate interventions.

6.2 SMALL-SCALE SUGAR CANE PRODUCTION

Commercial sugar cane production by black (Zulu) growers was first recorded in the 1860s. In 1934, the South African Board of Trade and Industries directed that the sugar industry should give attention to 'the position in the industry of small cane growers and the need of according them special treatment in the future organisation of the industry' (Huntley, 1966: 6).

Following a further inquiry in 1946, the Board noted that small-scale farmers were not producing sugar cane according to the yield potential in areas they occupied. It noted that attempts had been made to encourage small-scale farmers to improve production, but these had not been successful; and that a serious handicap that small-scale farmers experienced was a 'lack of knowledge of cane growing' and a 'lack of transport and draft animals' (Board of Trade, 1947: 60). The Board urged the government to provide extension services and assistance with mechanisation.

In the early 1970s, the sugar industry embarked on a major expansion programme, part of which involved the development of the small-scale grower sector. The total number of SSGs increased from approximately 4 000 in 1972–73 to 50 000 in 1994–95 (Bates, 1996: 54), with the number remaining more or less constant since the mid-1990s.

In 1973, the Small Grower Financial Aid Fund (FAF), renamed Umthombo Agricultural Finance (UAF) in 2001, was established and provided the stimulus for an increase in the number of and production by small-scale sugar cane growers. The number of SSGs, primarily with financial assistance from FAF, increased to 50 561 in 2001. Sugar cane production by SSGs increased from approximately 350 000 tons in 1972–73 to more than 4 million tons in 1997–98, before declining to approximately 3 million tons in 2001–02 (SASA, 2002b). In 1994–95 the sugar industry discontinued recording deliveries by racial groupings so it is not possible to determine the status of deliveries of black SSGs since then. The decline in SSG production has been a cause for concern and some of the reasons for this will be discussed further in this chapter.

6.3 PROFILE OF SMALL-SCALE GROWERS

The area of black SSG land increased from 1.3% of the sugar industry's total area in 1952 to 20% in 1992–93, while the share of production rose from 1.3–7%. Small-scale grower production in 2001–02 accounted for 14% of total production.

The South African government established an assistance programme for SSGs in 1956 providing finance for ploughing and the purchase of fertilizer and seed cane. A total of 1 060 new SSGs on 4 409 hectares were assisted (SASJ, 1968: 623), and the area under SSG registered land tripled between 1952–53 and 1954–55 from 2 388 to 7 616 hectares.

From 1963–67 small-scale growers were issued sugar cane quotas to plant 6 840 hectares. The total expansion approved by the industry amounted to 37 727 hectares, and the SSG portion of the expansion amounted to 22%. The period 1973–93 witnessed the most concerted effort in the expansion of small grower production with the establishment of FAF to provide credit to SSGs. At the same time, sugar-milling companies increased their inputs into the small-scale sector. The registered area of SSG production increased from 13 408 hectares in 1972–73 to 96 792 hectares in 1992–93, a 7.2 fold increase. Unfortunately, the progress of this group has become difficult to track due to a lack of separate racial records. UAF and mill records, however, indicate decreasing deliveries by black SSGs.

Although the SSG registered area increased significantly during the period 1989–90 to 1992–93, their total deliveries declined from the 1984–85 season at an average annual rate of 5%. Notwithstanding the severe drought which affected the 1992–93 season, the general decline in production raised serious questions. It was observed that the average small-scale grower's sugar cane yield of between 22–35 tons/ha/yr was well below that of the overall sugar industry average of 50–55 tons/ha/yr. In addition, the total production per SSG also appeared to be declining. This was seen as being linked to an apparent decline in the average area planted by individual SSGs (Bates, 1996: 80).

It was found in an analysis of a sample of SSGs that their production was skewed: approximately 60% of the growers accounted for 35–40% of SSG production; and the top 20% of SSGs accounted for 45–50% of production (Bates, 1996: 86).

The sugar cane price increased in nominal terms in the period 1960–2002. However, in real terms it decreased between 1960–72, increased from 1973–83 and then continued to decline for the most part to the present. Bates (1996) suggested that the real movement in the sugar cane price played an important role in the motivation of SSGs to enter into or withdraw from sugar cane production. In addition to the declining real price of sugar cane, SSGs are faced with increasing costs of production, the long-term nature of the crop, and productivity issues that include poor management and low efficiency of contractual services.

Small-scale sugar cane growers identified a number of sugar cane production problems in a survey undertaken in the Umbumbulu area of KwaZulu-Natal (Bates, 2002). They considered, amongst others, the following as important problems:

(a) high loan interest rates which led to a request for grant funding;
(b) sugar cane land area too small;
(c) lack of and/or poor condition of roads to their fields;
(d) lack of firebreaks, which gave rise to runaway fires in young sugar cane;
(e) lack of management of the sugar cane harvesting programmes;
(f) inadequate extension services; and
(g) insufficient contractual services for ploughing and harvesting.

The above problems require greater analysis. For example, the problem with high interest rates probably arises from the reduced returns to sugar cane production. The interest rates charged by UAF are the lowest that SSGs could obtain.

In addition to the above, the SSG contractors are facing difficulties in servicing the sector. Contractors maintain that the contracting rates, which are established by grower's committees, are insufficient to meet their requirements. They also maintain that they do not have access to credit to assist them in their businesses. Other problems they face revolve around the acquisition of inputs and the maintenance of machinery.

As already mentioned, the average area of sugar cane land cultivated by SSGs is regarded as too small and access to additional land is a desire expressed by many growers. The current traditional communal land tenure system constrains efforts to address this problem. Per unit area productivity also needs attention, and SSG average production per unit area, as already recorded, is low. The small unit size and the prevailing form of traditional land tenure are considered to be major problems impeding improved productivity of the SSG sector.

With respect to road infrastructure, considerable development has occurred in sugar cane producing areas. Maintenance of roads, however, was noted as a problem. The question of who should be responsible for linking SSG sugar cane fields to the road infrastructure has always been a vexed one and growers feel this should not be their responsibility.

The sugar industry has provided extensive input into sugar cane extension services and has developed close linkages with the respective provincial Departments of Agriculture. The apparent problem of inadequate extension services probably relates to the number of growers requiring servicing and the costs thereof. It has also been noted that the majority of SSGs have a proportionally less than equal share of SSG production: 20% of SSGs produce 45–50% of the SSG sugar cane. If analysed further, a greater number of problems would probably be found to lie with 60% of growers with small production as opposed to the more productive growers. This was apparent in the Umbumbulu survey referred to earlier.

The withdrawal of sugar mill contracting services has probably exacerbated problems in a number of SSG producing areas. Up to 2002, some sugar mills provided contractual ploughing and ratoon management services, which led to rapid development in the SSG producing areas. Their withdrawal, however, has led to a vacuum in the provision of support services and is probably a factor contributing to the decline in SSG production. The issues surrounding the provision of large-scale contractual services are controversial and need further research. Bates (2002) reports that in SSG areas where small-scale contractual services were promoted, total production has remained more or less constant, whereas it has declined in areas serviced by large-scale contractors. Use of small-scale contractual services, however, led to a further array of problems (challenges) being identified by SSGs.

This section has indicated that significant development of SSGs has taken place since 1973 and that a number of structural problems have arisen since the early 1990s. A major contributing factor to these problems has been the change in the sucrose pricing structure with a move from a two-tier system to a single pricing system in the late 1990s and, in 2000–01, a move from a system based on the sucrose percentage content of sugar cane to one based on the recoverable value (RV) payment system which takes account of the sucrose content less elements of sucrose which cannot be or are not converted into sugar during the milling process. Under unaltered SSG farming methods, the impact of these changes has meant lower returns for SSGs. It should be noted that the sugar industry provided compensatory payments to enable growers to make adjustments to the new systems. Up to now, however, SSGs would appear not to have adapted successfully to the changes.

A major impact of the change from the two-tiered payment system was a change in how sugar milling companies viewed SSG production. The two-tiered system gave a preferential price to 'A' pool production for both growers and millers. All SSG production was paid according to 'A' pool prices and hence there was an incentive for sugar millers to promote this production. Once the incentive was removed, a contraction of sugar miller inputs and services occurred.

The foregoing changes would appear to have had a negative impact on UAF loan redemptions, with higher levels of bad debts being experienced by UAF in recent years. In 2002, this led to UAF having to review its operations. UAF has not completed this process and is currently researching how it can operate under vastly changed circumstances – massively reduced sugar mill involvement, squeezed sugar cane margins and declining SSG production.

SSGs receive a 'flat rate' refund of their VAT (value added tax) payments incurred on the purchase of production inputs and a rebate on the cost of diesel used in their production based on the number of tons of sugar cane they produce. The justification for this is that, due to their scale, they are unable to take advantage of these benefits as they do not retain the required records and their revenue is below the threshold for registration with the tax authorities. The VAT and diesel rebates were negotiated by the South African Cane Growers' Association (Canegrowers).

Canegrowers is carrying out a project, the Sugar Logistics Improvement Project (SLIP), to address sugar production efficiencies. SSG problems will be addressed in this project.

SSG sugar production has increased during the previous 30 years. The sector has increased in significance in terms of total production, but this production remains significantly less than large-scale production. Despite this increase, the sector has laboured under a number of problems, not least of which have been those associated with structural changes in the pricing of sugar cane and the squeeze on margins arising from increasing input costs and, in many instances, poor management. Section 6.4 shows that the sugar industry has directed extensive resources to promoting the development of the sector.

6.4 SUGAR INDUSTRY INTERVENTIONS IN SMALL-SCALE GROWER DEVELOPMENT

The South African sugar industry continues to support the development of rural people in areas associated with the industry. Amongst a number of interventions, the following are noteworthy:

(a) Umthombo Agricultural Finance;
(b) SSG extension and training;
(c) the Small Grower Development Trust; and

(d) the New Freehold Farmer programme (although farmers in this programme are not classified as SSGs, the programme offers an opportunity for SSGs to progress into larger-scale production).

The sugar industry's interventions have contributed to:

(a) the creation of wealth in rural areas;
(b) increased sustainable employment;
(c) improved farming efficiency; and
(d) promoting Black Economic Empowerment.

Umthombo Agricultural Finance will be dealt with in the following section, and brief details of other important interventions are discussed below.

The South African Sugar Association's Experiment Station (SASEX) aims, through its research and information transfer activities, to promote the attainment of maximum economic returns for both commercial and small-scale sugar cane growers. This is achieved through the development of improved sugar cane varieties, and providing expertise and advice that optimises farm productivity while encouraging development of environmentally sound agricultural practices to promote and sustain the sugar industry. Outcomes from all routine research and development activities have relevance to both SSGs and commercial growers (SASA, 2002b).

Small-scale grower needs are catered for by joint extension projects entered into between SASEX and the provincial Departments of Agriculture in KwaZulu-Natal, Mpumalanga and the Eastern Cape. SASEX's resources are combined in partnership with provincial governments and other service providers. SASEX provides the technical expertise and extension experience and the partners provide the bulk of the funding and the required field staff.

The Small Grower Development Trust (SGDT) was founded in 1992 by SSGs themselves. Its objective is to promote economic empowerment of SSGs and also to develop viable and independent cane growing communities and businesses (SACGA, 2002). The sugar industry has contributed R21.6 million to the Trust to date. This funding has been supported by appropriating the premium on USA sugar quota proceeds for the purpose of supporting SSG development activities.

Since its formation, the SGDT has focused on the training of elected representatives of SSG organisations (institutional development) and the funding of administration and operational activities of Mill Cane Committees (these are SSG representative structures established in each mill area and will be discussed in section 6.6).

The SGDT recently supported the Cane Contractor Support Programme (CCSP), a project to improve small-scale contracting services. The CCSP identified problems in the SSG sector, which confirmed those already identified above. The project also highlighted a major problem with poor SSG organisational structures.

In respect of Black Economic Empowerment (BEE), the sugar industry's interventions are contributing to South Africa's national transformation programme, the Reconstruction and Development Programme. BEE is expected to contribute to 'job creation, rural development, urban renewal, poverty alleviation, land ownership, specific measures to empower black woman, skills and management development, education, meaningful ownership, and access to finance for households and for the purpose of conducting business' (BEE.Com, 2001: 2).

In terms of land redistribution, a number of sugar millers introduced a scheme in 1996 to promote the redistribution of commercial sugar cane land. The scheme involved the sale of miller-cum-planter (mill estate) land using an innovative financial package.

To make the purchase of land affordable for new freehold farmers, sugar millers invested 18% of the capital raised from the sale of the land with Ithala Bank to subsidise the interest rate on the loans on a declining basis for seven years. The interest rate subsidy is intended to improve the viability of the new freehold farmers in the early years following the purchase of the land. There are currently 120 new freehold farmers on 12 000 hectares.

6.5 UMTHOMBO AGRICULTURAL FINANCE

Umthombo Agricultural Finance (UAF), previously known as the Financial Aid Fund (FAF), was established in 1973 after the sugar industry had emerged from a long cycle of depressed export prices. An amount of R5 million was appropriated from funds in excess of the requirements of the sugar industry's Price Stabilisation Fund. A further R1 million was granted to FAF from the sugar industry's Development Fund in 1978. Additional funding, as and when required, was raised by means of loans from financial institutions. FAF was established to provide assistance to small-scale sugar cane growers who did not have access to normal credit facilities and who lacked the capital, equipment, experience and basic essentials to become viable and efficient farmers.

UAF continues to provide loans and financial services to economically viable SSGs involved in sugar cane production with the purpose of enhancing capacity for development and establishing self-sustaining entrepreneurs. The objectives of UAF can be summarised as follows:

(a) to develop financial instruments (lending and savings) in order to improve accessibility and outreach which will lead to growers becoming independent of subsidies, while at the same time promoting strong incentives for repayment;
(b) to develop policies, procedures and lending methodologies, which will meet the specific requirements of the target market, while ensuring minimal risk exposure to Umthombo;
(c) to interact with other role players in the development and rural finance sectors; and
(d) to ensure that financial facilities are provided at the lowest possible cost to clients (UAF, 2001: 2).

The operational and executive control of UAF vests in the Council of the South African Sugar Association (SASA). A UAF Committee, comprising representatives from SASA, the South African Cane Growers' Association and the South African Sugar Millers' Association Limited, is responsible for UAF policy and procedures and the guidance of its management.

A Consultative Forum, comprising representatives from Mill Group Local Committees, was established to enable liaison with stakeholders on matters affecting them. This Forum makes recommendations to the UAF Committee. Mill Group Local Committees, comprising SSG, large-scale farmer and mill representatives, are responsible for the administration of UAF matters at mill area level.

Prior to 2001, UAF (FAF) relied on mill field staff to carry out a major portion of its loan administration. Mill staff have since been withdrawn and UAF has introduced its own mill area offices and loan staff. UAF currently (2002–03) has a budgeted staff complement of 35, 18 of whom are stationed in mill areas, and eight are Loans Officers.

UAF provides the following loan products:

(a) Sugar cane establishment loans, to pay for ploughing, planting and weed control contractual services, as well as the purchase of fertilizer, seed cane and, in a limited number of cases, equipment. These loans have a maximum term of eight years.
(b) Ratoon management loans, for paying the costs of fertilizer and weed control for ratoon crops. The maximum term is two years.

Applicants for loans are required to obtain assistance with the completion of their applications from UAF Loans Officers. Detailed procedures are laid out in the UAF Policy and Procedure Manual. Applicants are required to pay an application fee and to take out a credit life insurance policy covering the repayment of the loan should they die during the term of the loan.

A Mill Area Loans Committee, made up of a UAF Loans Officer, a mill representative and a grower facilitator, screens loan applications and makes recommendations regarding the approval or not thereof. The Umthombo Management Loans Committee, comprising the UAF Manager, the UAF Loans Administration Manager, the UAF Operations Manager and the UAF Finance and Administration Manager, approve or reject loan applications.

Applicants are checked for creditworthiness. The Mill Area Loans Committees are required to make appropriate recommendations and the UAF Loans Officer is required to check the following with regard to an applicant:

(a) his/her character;
(b) his/her capacity to repay the loan;
(c) collateral; and
(d) his/her capability for producing sugar cane.

The majority of SSGs farm on traditionally occupied land. They are unable to alienate the land in any way which implies that it cannot be used as collateral for loans. UAF requires a cession over a borrower's crop proceeds to be registered at the receiving sugar mill as security for the repayment of a loan. This provides a degree of security, but is dependent on the borrower farming effectively; there being no adverse climatic conditions; and that the borrower does not undertake repayment avoidance activities such as delivering on another grower's delivery number or him/herself having any other undeclared delivery numbers.

Loan monitoring is carried out by a Loans Grading Committee, comprising the UAF Operations Manager, UAF Finance and Administration Manager, UAF Regional Managers, UAF Loans Administration Manager and the responsible mill area Loans Officer. Results from the Loans Grading Committee are used to manage loan recovery and provide information for loan provisioning and write-offs.

Up to 2000, the loan interest rate was calculated on a compound basis on a declining loan balance. In establishing loan interest rates, an attempt was made to take the opportunity cost of capital, the cost of lending, risk and a margin, including an allowance for inflation, into account. In reality, the interest rate did not cover all these requirements as the UAF Committee based final decisions on representations made by SSGs. The interest rates were 'soft' and regularly below both the Land Bank's prevailing rate and commercial banks' prime lending rates. The interest rate is currently 16% per year.

UAF has to date advanced a total of R344.4 million in loans; of this amount R83.2 million is outstanding in current loans. A total of R18.7 million has been written-off in bad debts.

The average bad debt write-off to date is 5% of total loans advanced. UAF, however, is currently adjusting to the withdrawal of sugar mill contracting services and is anticipating a negative impact on the recoverability of loans from SSGs previously reliant on these services.

The total number of loans advanced has declined over the period 1998–2002. The number of loans approved for this period, excluding 2000–01, amounts to approximately 1 400 per season and the average value of individual loans falls in a range of between R8 000–R10 000 per loan.

UAF provides a savings scheme (the retention savings scheme), which was established to assist SSGs in saving a portion of their sugar cane income to purchase fertilizer and to finance weed control operations for their ratoon crops. Small-scale grower savings have shown substantial growth since 1985.

The saving scheme was seen as a way of improving UAF's security and reducing risk by ensuring that SSGs would have sufficient funds available to maintain a high level of productivity. Funds are retained by UAF in trust for SSGs and these funds, in terms of deposit-taking regulations, cannot be used to fund lending as they would have been in a normal banking institution. Savings have not been mobilised in the true sense of a savings scheme.

SSGs with outstanding loans are required to participate in the savings scheme. Once a loan has been repaid a grower can opt to withdraw from the scheme. Growers without loans can participate voluntarily in the savings scheme.

Small-scale grower savings are based on their productivity. The amount saved per ton of sugar cane delivered is determined by Mill Group Local Committees (MGLC) in each area according to the agronomic needs of the area.

During the season 2001–02 a total of R49.7 million was deposited and R53.9 million was withdrawn for ratoon management purposes. The current activity in the savings scheme is approximately four times that in the loan scheme, which is significant in that SSGs are increasingly relying on own resources and avoiding debt. The sugar industry is the only agricultural industry in South Africa that provides a savings mechanism for small-scale farmers.

6.6 SMALL-SCALE GROWER ORGANISATIONAL STRUCTURE

All sugar cane growers are members of a representative organisation, the South African Cane Growers' Association (Canegrowers). The Association has a pyramidal structure with growers being members of grower organisations at mill area level. These grower organisations, 'member organisations', come together in Local Grower Councils at mill area level. The Local Grower Councils have equal representation of large- and small-scale growers. Large- and small-scale growers are also equally represented up to the level of the Board of Directors of the South African Cane Growers' Association.

In respect of 'member organisations', SSGs are represented by 16 Mill Cane Committees (MCCs) at sugar mill area level. MCCs draw their membership from Local Farmers' Associations and Sub-Committees, of which there are approximately 270 representing SSGs in the South African sugar industry.

The current structure of grower representation at the lower levels, i.e. mill group associations, planters' associations, farmers' associations, local associations and sub-committees, are split according to historical racial divisions. From the Local Grower Council level upwards there is a 50:50 representation of small- and large-scale growers. Canegrowers is addressing this with emphasis on dismantling the numerous lower structures and integrating them into fewer, more efficient representative organisations. The process has not been without tensions, with a SSG structure attempting a breakaway. This latter movement would appear to have involved a minority of SSGs, but has had a negative impact on the process.

The above process should lead to a strengthened voice for Canegrowers. The issue, however, of balancing the varying requirements of small- and large-scale growers, as discussed below, will necessitate sensitive stewardship.

According to Sokhela (1999: 2), MCCs 'are the most influential representative (bodies) of the SSGs in that they are directly accountable to the Local Farmers' Associations (LFAs) and they elect the individuals who serve as directors in the South African Cane Growers' Association'.

Currently, the SGDT finances the major part of the MCC administration costs. Since its establishment, the SGDT has advanced R27.2 million to these costs. SSGs do pay a levy from their sugar cane proceeds, but these are insufficient to meet the total costs. The SGDT anticipated that the subsidisation of MCC administration and operational costs would be phased out over a period of 10 years. This has not been achieved, with SSG contributions only amounting to R2 million, which is approximately 5% of the total expenditure. Given the continuation of the current level of services and the current level of SSG production, the phasing out of this high subsidy is probably not possible.

Although there are structures in place, Sokhela (1999: 19) reported that 'small-scale growers have needs which are not adequately met through the structures designed to service large-scale growers. Their particular limiting circumstances have made it necessary for Canegrowers and the sugar industry in general to be proactive in modifying organisational structures and functioning'.

The lack of education of a majority of SSGs is a major problem with respect to SSG representation in a sophisticated agricultural industry. In a survey, Sokhela (1999: 64) reported that 43% of MCC representatives sampled (n=136) were 'basically illiterate' (having an educational level of less than Grade 5). His sample indicated that 79% of MCC representatives had less than a Grade 7 level of education and that none of the sampled respondents had achieved a Grade 12 or higher level of education. A minimum of Grade 6 is considered necessary in terms of being functionally literate. 'Because of low educational levels of the MCC members, the sugar industry ends up negotiating with a weak "partner". The industry as a whole is weakened' (Sokhela, 1999: 65).

Sokhela (1999: 42) also found that SSGs are concerned more with broader development issues than those being addressed by the sugar industry. The sugar industry's interventions have contributed a great deal to the economic upliftment of SSGs. However, their community development needs and educational requirements now appear to become more important than the mere production of sugar cane.

It is a moot point whether the sugar industry should be addressing the broader development issues. Sokhela (1999: xv) stated that 'small-scale growers must realise that the sugar industry is not responsible for providing improved education and training and support services either for sugar production or for activities outside of the sugar industry'. 'Small-scale growers should control their destiny and be committed to self-management of SSG development in the long-term. This may be achieved by developing effective organisational structures, proper interaction with support service providers and less dependency on the sugar industry.'

Canegrowers and the SGDT introduced membership and leadership development programmes in 2002 to address the problems identified by Sokhela.

6.7 CONCLUSION

The South African sugar industry has undertaken development of small-scale farmers on a scale not attempted by any other agricultural sector in South Africa. Small-scale growers have produced sugar cane since the establishment of the industry in the nineteenth century. Small-scale farmers, by definition, are marginalised in terms of their access to resources (Shaner *et al.*, 1982: 16) and the South African sugar industry has attempted to address difficulties faced by SSGs by providing financial and other services to assist with their development. Amongst other interventions, Umthombo Agricultural Finance has advanced a total of R344 million to approximately 48 000 SSGs who earned a gross income totalling R2 238 million between 1997–2002.

This chapter has highlighted development initiatives undertaken by the sugar industry and indicated some challenges in the development of the SSG sector. With recent changes in the industry, SSG development faces some major problems, not the least being a review of how credit and savings services can be provided in a viable and sustainable manner. UAF was established on the basis of a large input from sugar mills, and now that this has been substantially reduced or discontinued a new approach is required.

In addition, the viability and sustainability of the SSG sector, as a whole, needs attention in terms of a changed and changing national and globalised agricultural economy. The sector cannot be expected to expand while faced with low productivity, declining returns and increasing costs. Further to this, the unit size and land tenure issues need to be addressed.

The sugar industry has a structured and representative sugar cane grower organisational system, which enables the SSG voice to be heard and taken into account in industrial decision making. Historic divisions in the structure are being addressed, with movement towards consolidation of the many foundational member organisations into fewer representative organisations being promoted. The lack of adequate education of a majority of SSGs is recognised as problematic and will require attention to avoid unnecessary misunderstanding clouding the advancement of SSGs.

The reasons for the sugar industry investments in SSG development can be questioned in a number of respects, but the impact which the development has had on rural livelihoods in areas associated with the sugar industry is important. Significant development of infrastructure has taken place, and this has stimulated other economic activities. It was estimated at one stage that the government's investment in infrastructure equated rand-for-rand with the level of loans advanced by the industry, i.e. R344 million.

The production from the SSG sector forms an important component of the economy in a number of rural areas in South Africa and reached a peak in the 1990s. Since then there has been a decline in production as a result of substantive changes to the pricing structure, to restructured interventions by the sugar milling sector, and the declining profitability of sugar production.

Despite the above, a significant development impact has occurred. Is it a success story? Are lessons learnt and is there a future? The answer to the first question is probably a qualified yes. If problems currently faced are addressed satisfactorily the qualification could be dropped. If not, a balance sheet will have to be drawn up to provide an unqualified answer.

With regard to the second question, there are many lessons, positive and negative, and as indicated there is a future if the necessary solutions to current problems are addressed timeously. The SSG sector faces adjustment to changed circumstances but, given past experience, the sugar industry should be in a position to adapt and continue developing the sector so that sugar cane production continues contributing to sustaining livelihoods in rural areas.

REFERENCES

BATES, R.F. (1996). An evaluation of the South African sugar industry's Small Cane Growers' Financial Aid Fund. PhD thesis, University of Natal, Pietermaritzburg.

BATES, R.F. (2002). Survey of small-scale growers in the Umbumbulu area. Unpublished. Durban.

BEE.COM. (2001). *The Black Economic Empowerment Commission Report.* Johannesburg: Skotaville Press.

BOARD OF TRADE. (1947). *The South African Sugar Industry: Review.* Report No. 298. Pretoria: Government Printer.

HUNTLEY, J.K. (1966). The Sugar Industry Central Board: its origin and functions. Unpublished. South African Sugar Association, Durban.

SHANER, W.W., PHILIP, P.F. & SCHMIHL, W.R. (1982). *Farming Systems Research and Development: Guidelines for Developing Countries.* Boulder, Colorado: Westview Press.

SOKHELA, M.P. (1999). Enhancing the contribution of small-scale growers in the sugar industry. PhD thesis, University of Natal, Pietermaritzburg.

SOUTH AFRICAN CANE GROWERS' ASSOCIATION (SACGA). (2002). Personal communication. Durban

SOUTH AFRICAN SUGAR ASSOCIATION (SASA). (1993). Financial Aid Fund: mission, objectives and procedures. Unpublished. South African Sugar Association, Durban.

SASA. (2002a). *Industry Directory 2002/2003.* Durban: South African Sugar Association.

SASA. (2002b). Personal communication. Durban.

SOUTH AFRICAN SUGAR JOURNAL (SASJ). (1968). Article on grower development. South African Sugar Association, Durban.

UMTHOMBO AGRICULTURAL FINANCE (UAF). (2001). Policies & procedures manual. Unpublished. South African Sugar Association, Durban.

CHAPTER 7

Agricultural Marketing in South Africa[1]

Herman van Schalkwyk, Jan Groenewald and André Jooste

7.1 INTRODUCTION

Agricultural marketing in South Africa virtually went through a full circle in the twentieth century – from a *laissez-faire* system to one of very rigid market controls and back to a system with fewer controls and less government intervention than in most countries of the world.

The movements toward control in the earlier parts of the twentieth century were in step with prevailing thought and action of the time – a movement away from *laissez-faire,* toward intervention in economic life.

The First World War and its economic aftermath added fuel to the fire. Two post-war depressions, in 1922 and the early 1930s, caused severe hardship for agriculture; and agricultural product prices generally declined more rapidly than other prices, reducing farmers' purchasing power. Many countries did not regard the maintenance of agriculture as a purely economic consideration, but also as a necessity for social, historical and political reasons.

In the South African interior, commercial farming had only emerged from subsistence agriculture after the discovery and eventual development of mining and cities of diamonds and gold in the 1870s and 1880s. Market-directed production was still in its infancy when it was severely interrupted by the Anglo-Boer War (1899–1902), with recovery occurring only later during the first decade of the twentieth century. The expansion of railways enabled farmers in the interior to produce for export markets. However, many farmers, still in only the second generation of commercial production, were for various reasons unable to cope with the harsh economic situation: poor training, a lack of managerial expertise, resistance to change, an inheritance system leading to non-viable farming units, biologically maladjusted farming systems, shortage of credit, and injudicious use of credit obtained (Grosskopf, 1932: Chapters 2, 5 and 8; Leppan, 1931: 36–39). Farmers were thus ill-equipped to deal with the

[1] A large section of the first part of this chapter has been taken from Groenewald (2000)

so-called World Agricultural Crisis (Marais, 1936; De Swardt, 1983). 'Overproduction' became an often-used term in South Africa – until the 1980s – although its economic meaninglessness was exposed at an early stage (Davis, 1931).

The poor state of agriculture made it necessary for the state to act and the question revolved around the type of measures to use. Two opinions became prominent: free competition produces adverse effects (De Swardt, 1983), and powerful monopolistic tendencies prevail in the trade of agricultural products (McLoughlin, 1938). The latter view corresponded with a standard version of a large number of small farmers, unorganised, having to cope with a limited number of powerful, monopolised or cartelised traders.

The first step to ameliorate this state of affairs was the promulgation of the Co-operative Act of 1939. One important stipulation was extension of limited liability, similar to that existing in respect of companies, to co-operatives. Expectations were high that improved facilities and a militant co-operative approach would solve many, if not most problems (De Swardt, 1983), but this was not to be. Efforts to control marketing and raise (and maintain) prices at levels above equilibrium are doomed to fail (Kohls & Downey, 1968). The next step was compulsory co-operative marketing of wine and tobacco, and maize exports were subsidised after 1931 (De Swardt, 1983).

A Commission of Inquiry into Co-operation and Agricultural Credit strongly advised against compulsary co-operatives, price control and one-channel marketing by means of statutory boards. It advised the government to strengthen available effective institutions and stick to the competitive system (Frankel, 1934; De Swardt, 1983). Its report (U.G. 16 of 1934) was accepted by the government, but only two years later political pressure caused the same government to accept the principle of one-channel marketing through statutory control boards.

The issue became the subject of much debate with the majority of the critics professional economists, while the protagonists were typically involved with existing control systems or Organised Agriculture. A noteworthy contribution was an article by Richards (1936), entitled 'The "New Despotism" in agriculture' – a title which much later was proven rather accurate. According to Richards, the proposed Marketing Act disregarded basic dynamics of economic life in that it ignored economic evolution (now largely called innovation), the nature of productive (value added) services in marketing, complexity and the necessity of risk-bearing in marketing. These may all be summarised in terms of ignoring entrepreneurship and adjustment to a changing social, economic and technological environment. Richards made the following prediction (Richards, 1936: 417):

'The Marketing Bill, if passed, will not solve the agricultural problem, which is as bad as it is because of the very methods it is proposed to extend; it will make present maladjustments infinitely worse. It is cumbersome and unworkable, and will most probably lead to

(a) the growth of an agricultural monopoly;
(b) increased production of unwanted products;
(c) increased prices to consumers;
(d) a rising cost of living and costs of production all round;
(e) heavy losses to the State, and ultimate failure'.

However, this was an age when the public at large, and politicians in particular, appear to have had an almost unlimited confidence in the ability of bureaucrats to regulate affairs satisfactorily, provided they had legal sanction, and the Marketing Act was passed by parliament in 1937.

7.2 THE NATURE OF THE ACT AND CONTROL SCHEMES

The Marketing Act enabled the Minister of Agriculture to promulgate marketing schemes, administered by control boards. Each scheme would be responsible for an agricultural product or group of products, and the control board could exercise certain defined rights. The majority of members on a control board had to be producers' representatives appointed by the Minister and the other parties represented were to be traders, processors, consumers and the Department of Agriculture. A National Marketing Council was to advise the Minister on matters concerning the actions of control boards. The board's powers were wide-ranging and included:

(a) one-channel marketing, in which only the board or its agents were legally entitled to buy, sell or store a product;
(b) the fixing or regulation of prices;
(c) conduction of pools, and also the transfer of proceeds among pools;
(d) registration of traders and processors (including the right to refuse and to withdraw registration);
(e) registration of producers;
(f) prohibition of the erection and/or use of facilities for mass handling or storage of products;
(g) fixing of transport tariffs;
(h) enforcement of marketing quotas; and
(i) imposition of levies on the products.

The main aims of the boards were stability of prices and reductions in price spreads between producers and consumers.

Five types of control schemes developed under the Marketing Act:

(a) *One-channel fixed price schemes*: The total production was bought, marketed and sold by the control board (or its agents) at prices as determined by the Minister or the board.

(b) *One-channel pool schemes*: The control board was the only buyer and seller and proceeds of sales were pooled with the shares of individual producers equivalent to their contributions.
(c) *Surplus removal schemes*: The control board used various methods to remove products from the market if, in its opinion, there was a surplus. In some years, bananas were destroyed, along with prohibitions on the marketing of lower grades of potatoes. Meat surpluses were stored at a high cost which later led to the use of quota or permit restrictions on the marketing of livestock.
(d) *Supervisory schemes*: The board attempted to install order in marketing, for example, through negotiations between buyers and sellers under the auspices of the board.
(e) *Publicity schemes*: Only one product (karakul pelts) was marketed in this way and publicity was the board's only function.

By 1959–60, approximately 64% of the value of agricultural production consisted of products controlled under the Marketing Act (Van Rensburg, 1962). The Act was amended a few times, for example, in 1968, 1974 and 1975. The amendments were largely concerned with the closing of loopholes and with extending control. The Marketing Act, together with other specific legislation, put in place a range of schemes affecting a major portion of the agricultural industry. These are set out in Table 7.1.

Table 7.1: Schemes under the Marketing Act of 1968 and other interventionist legislation, 1997.

Product	*Nature of legislation*	*Percentage of gross value of production*
A Marketing Act:		
(i) Maize; winter grain	Single-channel, fixed price	21.0
(ii) Fresh deciduous fruit; citrus; dried fruit; bananas; wool; mohair; tobacco; oilseeds; chicory; rooibos tea; lucerne seed; buckwheat	Single-channel pools	15.0
(iii) Red meat; eggs; dairy; potatoes; grain sorghum; dry beans	Price support/surplus removal	28.0
(iv) Canned fruit; cotton; karakul pelts	Supervisory/price regulatory/ promotion schemes	0.6
Subtotal		64.6
B Other legislation:		
(i) Sugar cane	Surplus removal, fixed price + quotas	5.0
(ii) Wine grapes	Single-channel, fixed price + quotas	4.0
(iii) Ostriches	Single-channel, prohibition of export of breeding material	0.8
(iv) Tea	Arrangement re local tea uptake	0.3
Subtotal		10.1
Total		74.7

Source: Kassier (1999)

7.3 PROBLEMS WITH THE MARKETING ACT

Government interference in the South African economy was not restricted to agriculture and many industries, for example, steel, synthetic fuel, railways, etc. were nationalised. In addition, the government supported the development of secondary industry, notably in the manufacture of machinery, machinery components, chemicals and metal products. The once-popular infant industries argument undoubtedly provided a rationale for some of these developments, while ideological and, consequently, defence considerations also became important motivations. These industries were protected by various devices, such as import controls, import duties and preferential credit arrangements. The products of most such industries were produced at higher costs, and hence, higher selling prices than those at which these could be imported. Some other sectors and industries used these manufactured products as inputs, and costs of production in these sectors were therefore also increased. Agriculture was one of these sectors. According to Le Clus (1982), prices of a wide spectrum of agricultural inputs rose by between 35–74% more rapidly than would be the case had these been imported duty-free. The result was that even the price fixing, quota and export promotion efforts of the agricultural marketing control institutions were inadequate to keep farm product prices at a par with costs. Although producers' prices of farm products rose faster in South Africa than in the great majority of developed market economies (the main markets for agricultural exports), costs of farming inputs in South Africa rose so fast, that the parity of product prices *vis-à-vis* inputs in South Africa declined considerably faster than those in by far the majority of countries in respect of which the Food and Agricultural Organisation (FAO) had data of this nature (Groenewald, 1982).

Warnings that South African agriculture was heading towards a severe financial crisis (Louw, 1979; Groenewald, 1979, 1980) went unheeded and eventually, the mid-1980s became a period of financial setbacks, including many insolvencies.

7.3.1 Restrictive registration and economic concentration

Restrictive registration as practiced by control boards inevitably brought economic concentration in its wake. Examples of such concentration were numerous: two major companies and one central co-operative had a joint market share of 84% in the beef and 94% in the pork markets (Lubbe, 1992a). Fifteen out of a total of 328 registered traders handled 60% of dry bean sales (Van Zyl & Kirsten, 1993). The Commission of Inquiry into the Regulation of Monopolistic Conditions Act (Republic of South Africa, 1977) also found intense concentration in some agricultural processing and marketing activities, as shown in Table 7.2.

In addition, all the boards operating one-channel schemes were statutory monopolies. Under such conditions, no or little competitive pressure existed to enforce improved performance and efficiency. The Board on Tariffs and Trade (1992) concluded that concentration engendered by control under the Act made an important contribution to high marketing margins of food in South Africa.

Table 7.2: Concentration in agro-industries, 1976.

Industry	*Number of firms in industry*	*Percentage of total turnover in hands of three or fewer largest firms*	*Number of firms controlling at least 70% of total turnover*
Abattoirs	7	89.1	3
Condensed milk, milk powder and other milk products, except ice cream	14	81.8	2
Dried fruit packaging	2	100.0	1
Vegetable raw oil, oilcake and oilcake meal	5	98.5	2
Cigarettes, cigars, tobacco, snuff, etc.	15	98.4	2
Cotton gins	6	84.4	2

Source: Republic of South Africa (1977)

7.3.2 Surplus removal and quotas

Surplus removal can occur in three ways: buying up for storage; destruction of products; or exports. A surplus could hardly develop with profitable export opportunities and the third alternative could only occur at a loss, involving wealth transfer from local consumers and taxpayers (should the state subsidise such exports) to local producers and foreign consumers.

The destruction of products was rarely used, probably because of the public outcry when the Banana Board tried this remedy. The press pointed out that there could not be moral justification for such a step as long as there were people who lived under conditions of severe poverty. The Potato Board used another *modus operandi,* namely, prohibition of the marketing of lower grades of products. It was shown that, with different grades having different demand elasticities, such a policy caused grade combinations offered on the market to deviate from the optimum both in terms of producers' incomes and consumers' satisfaction (Myburgh *et al.,* 1986).

Buying up products for storage involves costs which may be high if refrigeration and freezing are involved and there is also a risk of product deterioration. Consumers are, for example, not willing to buy frozen meat at the same price as fresh meat. The Meat Board operated a floor price scheme by which carcasses were auctioned, and those which did not reach the floor price were purchased by the Board, put into cold storage and sold later, mostly as frozen meat. The costs were borne by levies. In order to operate the scheme, a prohibition was placed on the introduction of carcasses in proclaimed areas (covering all the metropolitan areas), thus forcing urban meat to come from urban abattoirs to which live animals were transported, often over long distances. This deviated substantially from what would be optimal abattoir locations and added substantially to costs (Eales, 1979) and, therefore, was also to the detriment of consumer and producer alike. Over time, the authorities found it impossible to police the illegal movement of live animals to informal slaughter facilities in townships (Myburgh, 1992; Karaan & Myburgh, 1992).

Floor prices can, if set at a realistic level, reduce price fluctuations over the short term. If floor prices are set too low, they become useless and hardly have any effect. If, however, they are set at levels which would exceed equilibrium prices over sustained periods, continuous 'surpluses' result with storage cost implications.

In practice, floor prices for some products, particularly red meats, were set at levels high enough to cause perennial surpluses and mounting storage costs. In order to reduce such damages, the Meat Board embarked on two types of quota programmes: quotas to livestock agents who decided by themselves how much they should buy and from whom; and marketing permits to producers (a permit being only semantically different from a quota). A permit or quota specified how many animals a producer or trader was allowed to deliver to a specific abattoir within a specified period of time. In the issue of permits and quotas, preference was given to producers or traders who could supply a steady stream of animals, which obviously discriminated against producers who were not able to do so – mainly extensive graziers and smaller producers. An opinion survey indicated that the great majority of producers would be in favour of more, smaller abattoirs in production areas rather that the few, large abattoirs in metropolitan areas. Farmers complained that the quota/permit system prevented them from marketing stock in the seasons associated with the best prices. They were of the opinion that the system favoured the feedlots, big farmers and those close to the market and discriminated against graziers, small farmers and those distant from the market (Elliott *et al.*, 1987). The system certainly discriminated against small-scale black cattle farmers (Nkosi & Kirsten, 1994).

Another study also found that feedlots and larger producers received preferential treatment (Lubbe, 1992a). The feedlots became concentrated, with the five largest feedlots having a market share of 64% and the two largest feedlots belonging to two of the dominant firms in the meat trade (Lubbe, 1992a). The quota/permit system certainly did not benefit those farmers who were intended to be the main beneficiaries. Various authors (Paarlberg, 1962, 1964; Groenewald & Nieuwoudt, 1979; Nieuwoudt, 1976) have mentioned some intrinsic weaknesses of quotas:

(a) they are often not effective in controlling output, since producers change factor proportions;
(b) they become capitalised in fixed asset values, bestowing a windfall gain on present producers and a cost of doing business to new and future producers;
(c) they lend themselves to nepotism; they can cause inefficiency and cause an industry to become uncompetitive; and
(d) they lead to rigidities.

When the quota system is disbanded, producers can also suffer capital losses (Nieuwoudt, 1978).

Lubbe (1992b) found that the Meat Board's scheme aggravated, rather than reduced cyclical price fluctuations in red meat prices. Thus, the price stabilisation motive was not achieved, and instead producers suffered from inadequate market access during over-supply periods, deteriorating natural resources because the system forced them to overstock at times, and increased income variability. Consumer interests were harmed by artificially high prices during over-supply periods (Lubbe, 1992b). The system also did not succeed in increasing net prices to producers: the price-decreasing effects of Meat Board frozen meat sales exceeded the price-rising effects of floor price purchases (Venter & Van Zyl, 1996). The only people to really benefit from the meat scheme – apart from the bureaucrats involved – were those who gained in terms of quasi-monopoly power.

7.3.3 One-channel fixed price schemes

One-channel fixed price schemes were in vogue for maize and winter grains. In the case of an export product, such as South African maize, a domestic price higher than export parity can be maintained only if local consumers and/or the government (and thus tax payers) make up for the loss, which is what happened. In 1954 it became clear to the Maize Board that future exports would be done at a loss and consequently, the Board established a stabilisation fund. The unused balance of previous export profits was deposited into this, and from then on, contributions were made to the fund from levies paid by producers and local buyers, and the government also contributed from time to time (Scholtz & Tomlinson, 1979). De Swardt (1962), who was the Secretary of Agricultural Economics and Marketing, stated that in cases where products were exported at a loss, expansion of production and exports should be discouraged and prices should not be raised to either producers and/or consumers. This is, however, exactly what was done. Net producers' prices and the Maize Board's domestic selling prices rose steadily from 1970–71 to 1986–87; production expanded more rapidly than domestic consumption (Abstract of Agricultural Statistics, 1996); and exports rose as shown in Table 7.3.

During the 1980s the government terminated subsidisation of the Maize Board's stabilisation fund. In contrast to the prognostication of De Swardt (1962), the Maize Board financed its export losses by raising domestic prices in those years in which large exportable surpluses were produced. The Maize Board's local selling prices were higher in times of surplus and lower in the two years in which droughts caused production to be smaller than local use (Groenewald, 1989). In particular, lower-income black consumers, for whom maize is a staple food, were adversely affected. Given local demand conditions, this policy was certainly shortsighted: animal consumption had been growing considerably more rapidly than human consumption, with the share of the latter declining from 63% in 1961–62 to 41% in 1989–90. The demand for maize as animal feed is relatively elastic, with an elasticity of –2.2, while the human demand is relatively inelastic (elasticity = –0.149) (Van Zyl, 1990). It was shown that in the long term, lower maize prices would, in the form of an expanded local market, have benefited maize farmers more than the policy of the Maize Board (Van Zyl, 1990).

Table 7.3: Mean annual maize exports in four consecutive five-year periods.

*Period**	*Mean annual maize exports (million tons)*
1965/66–69/70	1.60
1970/71–74/75	2.50
1975/76–79/80	3.20
1980/81–84/85	3.40

Source: Abstract of Agricultural Statistics (1996)
* Export data between 1986–87 and 1990–91 were suppressed

The Maize Board policies transferred welfare, with maize producers gaining, and local consumers as well as stock farmers who were buying maize as animal feed, losing. The welfare transfers were measured by Wright (1992). In Table 7.4 significant net welfare losses are shown.

Table 7.4: Welfare effects of maize price policies (R million).

Year	*Welfare gain in maize production*		*Welfare gain in maize consumption*	
	White	*Yellow*	*White*	*Yellow*
1986	255	263	–358	–310
1987	564	507	–573	–522
1988	634	479	–613	–597
1989	564	239	–517	–549
1999	206	178	–554	–595

Source: Wright (1992)

The higher-than-import-parity prices of wheat were accompanied by subsidies on bread. In this way, the burden on poorer consumers in particular was made more bearable, although the bread subsidy was discontinued in 1990.

One-channel fixed-price marketing became unsustainable.

7.4 THE MARKETING OF AGRICULTURAL PRODUCTS ACT OF 1996

Various other factors, such as globalisation, the establishment of a democratic dispensation and a shift of political power from commercial farmers to consumers also contributed towards a relaxation of stringent interventionist measures. For instance, fixed prices were made more 'market related' and quotas in the wine and sugar cane industries were suspended.

By the beginning of the 1990s, there were clamours from consumer and business groups to terminate the Marketing Act. They argued that the Marketing Act furthered the interests of white commercial farmers to the detriment of black farmers, and of consumers. The degree of monopolisation engendered was a source of concern, as were arguments concerning nepotism. There were calls by some farmers to be allowed to market freely, for example, in fruit exports, which in the case of deciduous and citrus fruits was subject to compulsory one-channel exports. Some farmers' organisations, businessmen and officials who had obtained vested interests in the system understandably campaigned for the retention of the Marketing Act.

The escalating public debate caused the Minister of Agriculture to appoint a Committee of Inquiry into the Marketing Act (Kassier, 1993). This Committee recommended that all the statutory schemes established under the Marketing Act should be abolished with suitable sunset arrangements negotiated with each of the industries; a statutorily guaranteed monopoly in the private sector would not be acceptable. Furthermore, a more representative Marketing Council was advocated to advise the Minister on matters relating to agricultural marketing and trade.

The report of this Committee, as well as that of the Agricultural Marketing Policy Evaluation Committee, were instrumental in the repeal of the Marketing Act and the formulation of new legislation, the Marketing of Agricultural Products Act of 1996.

The new Act provides for the establishment of the National Agricultural Marketing Council (NAMC) to be appointed by the Minister of Agriculture from nominations by any 'directly affected group' and recommendations by the agriculture portfolio committees of parliament. The NAMC has its own staff, is financially autonomous, and the chairperson reports directly to the Minister.

The objectives of the new legislation are:

(a) increasing market access for all market participants;
(b) promoting efficiency in the marketing of agricultural products;
(c) optimising export earnings; and
(d) enhancing the viability of agriculture.

A policy of no intervention could have negative effects on food security, job opportunities or fair labour practices. To balance this, the Act contains strict prescriptions and procedures to be followed for any proposed intervention and all interventions have to be reviewed every two years. All new interventionist measures have to promote the objectives of the Act, and all 'directly affected groups' (a register of which is kept by the NAMC) are given an opportunity to comment on the proposed intervention.

The main differences between the previous Marketing Act of 1968 and the Marketing of Agricultural Products Act of 1996 are summarised in Table 7.5.

The last control boards ceased to exist on 5 January 1998. Some of the boards had considerable assets at the time of closure which were transferred to industry trusts or companies not for gain.

Table 7.5: Differences between the Marketing Act of 1968, and the Marketing of Agricultural Products Act of 1996.

1968 *Inherently interventionist*	*1996* *Minimum intervention*
Increased productivity	Increased marketing efficiency
Reduction of marketing margins	Optimum export earnings
Increased consumption and food self-sufficiency	Food security at household level
Maximum commercial producers on land	More emphasis on small-scale farmers
Economic farming units; minimum farm size	Increased sustainability of agriculture
Non-participative and bureaucratic introduction of intervention	Participative, transparent and all-inclusive
Stabilising product prices	Producers must themselves stabilise income
Intervention inclusive of single channel; pools, surplus removal, fixed prices, quotas; price support; promotion; general and special levies, registration, records and returns	Limited to levies; export control; pools; registration; records and returns
Requested by producers or introduced by Minister	Requested by any directly affected group of persons or firms
Consultation not always necessary although certain quantified producer support required	Consultation process prescribed by Act inclusive of all directly affected groups
No political process to approve levies apart from Minister	Levies need to be approved by both parliamentary portfolio committees and the Minister
No maximum period and no interim testing of intervention	All statutory measures to be introduced for fixed period and tested at least every two years

The wine industry and ostrich industry have also been deregulated while the sugar industry has abolished cane quotas. However, the sugar agreement, with the division of proceeds dispensation between growers and millers, is still in place as growers feel very strongly that they need some form of protection against the millers, who are effectively monopoly buyers.

7.5 IMPACT OF MARKETING REFORM

The overall impact of the marketing reform has resulted in significant private sector response, with the major reactions including:

(a) The appearance of a large number of new fruit exporters to compete with the previous single-channel exporters, Outspan and Unifruco (now Capespan).
(b) The establishment of an Agricultural Markets Division (AMD) of the South African Futures Exchange (SAFEX).

(c) Farmers are becoming more involved in managing their risks by means of storage (especially in the case of maize, forcing the co-operatives which own the vast majority of grain silos to become more commercially orientated), forward contracts and diversification.
(d) An acceleration has occurred in the establishment of new enterprises in the food and agricultural sectors.
(e) Agricultural land prices have fallen overall, in real terms, by at least one-third since 1990 (Bayley, 2000).

Agricultural support, expressed as the Producer Subsidy Equivalent (PSE), has dropped dramatically from 12.14% in 1995 to 2.7% in 1998. The 1998 PSEs were around 7, 16 and 22% for Australia, Canada and the USA respectively with the only lower figure being quoted for New Zealand at 0.8% (Gouse *et al.*, 2000).

According to Vink & Kirsten (2000) deregulation has benefited agriculture by, *inter alia*:

(a) reducing food price inflation;
(b) increasing the general level of investment in agriculture; and
(c) increasing foreign investment in input supply, agro-processing and marketing related to agriculture

7.6 CURRENT MARKETING PROBLEMS

In a world characterised by static overall demand, productivity improvements and increasing competitiveness are not by themselves a panacea for solving South Africa's problems of unemployment, poverty and economic progress. What is needed is an expansion of total demand and markets, as only when the economy is growing rapidly can it be certain that the success achieved by one enterprise is not at the expense and detriment of another.

Figure 7.1 shows the per capita GDP from 1980 until 2001. Economic growth continues to elude South Africa and, of greater concern, South Africa's position is slipping relative to other countries.

7.6.1 Agricultural information

The marketing boards played an important role in the collection and dissemination of agricultural data. Their abolition caused the supply of data to decrease and in some cases to be discontinued, despite a substantial increase in the need for data by decision-makers. Before the agricultural control boards were abolished, agricultural outlook conferences (AGROCON) were held annually since 1977, but with the disappearance of control boards these conferences also ceased.

For most products, Section 21 companies and producer organisations replaced the data collection and dissemination functions of the former marketing boards.

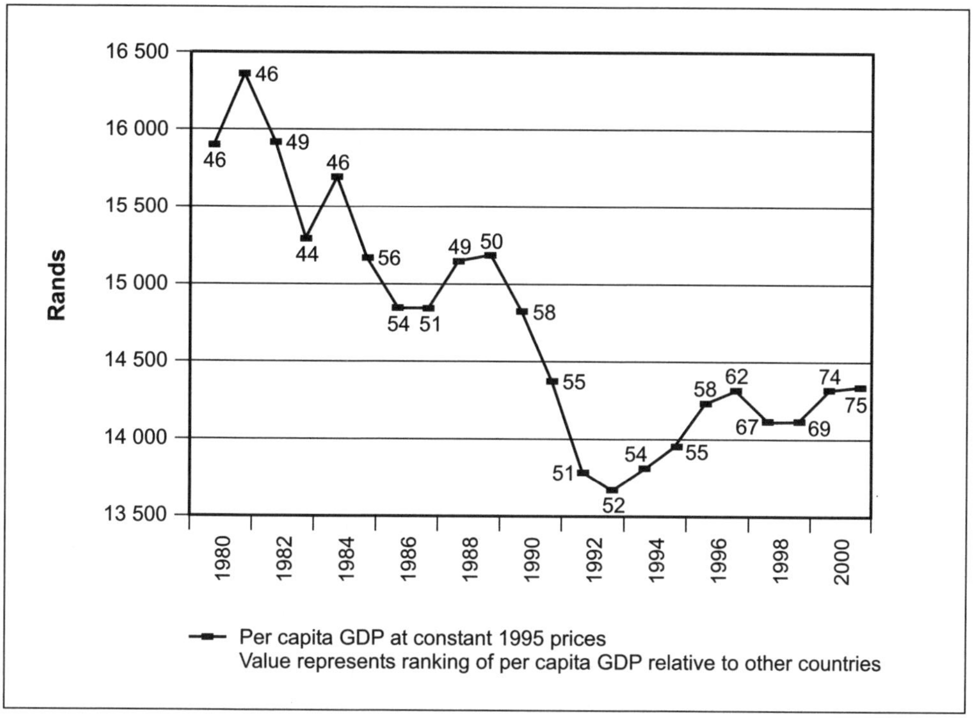

Figure 7.1: Per capita GDP in South Africa (1980–2001).

However, according to Willemse (1996) and Van Schalkwyk & Swanepoel (1997), these organisations serve a particular constituency with a specific interest.

Research on data needs was funded by the National Department of Agriculture, and the results were published (Frick & Groenewald, 1999a, 1999b, 2001; Groenewald & Frick, 2001), but the Department has not yet perceptibly acted in this regard. Agricultural data play a very important part in the decision-making process by both private and public decision-makers and the current situation is disturbing when one considers that good economic information improves the competitiveness and the efficiency of markets.

7.6.2 The small farming sector

The dualism that characterises South African agriculture and economic life is very evident in agricultural marketing. The established commercial sector and the areas in which commercial agriculture preponderates, are served by a sophisticated agricultural marketing system and by good infrastructure supporting agricultural production and marketing. This infrastructure includes roads, railways, telecommunications, postal services, marketing and financial services. The picture is very different with the small farming sector and in the areas where this sector is most common, including the former 'homelands'. Marketing services to these small farmers are often poor (sometimes non-existent), as are roads, telecommunications, physical marketing infrastructure and financial services.

No development can be expected in large areas of South Africa, and no upliftment of the rural poor can occur in the absence of significant improvements in the marketing set-up serving these areas and people. In the absence of an efficient marketing system for the farmers' products and inputs, they do not have either the opportunity or the incentive to become productive. This can probably be best summarised by Schultz (1968: 99): 'In my judgement, the real culprit causing the poor performance of agriculture in the less developed countries is the lack of economic opportunities in agriculture, opportunities that are rewarding to farmers. It is the lack of viable opportunities that is the crux of the matter.'

7.7 THE WAY FORWARD

The South African agricultural sector has now experienced a full circle to and from heavy state intervention to market liberalisation, free markets and the opening up for international trade. The question is often raised as to which of the two approaches provides the solution. The answer is that neither of the approaches proved to be successful in the creation of economic growth. A new paradigm concentrating on making South Africa's agricultural sector more competitive is needed. Currently, the main focus is on price. A new approach should, amongst others, involve that supporting industries be developed and research and training be focused and co-ordinated. It is also necessary to increase the competitiveness of the agricultural sector by expanding agriculture's capability to pursue process innovations and technology upgrading, increasing product and service quality, and expanding market opportunities.

Compared to countries such as Australia and New Zealand, the support to the agricultural sector in South Africa is inadequate and more should be done with regard to:

(a) training and education of farmers and prospective marketers in marketing;
(b) appropriate extension services;
(d) the development of export markets and the identification of niche markets;
(d) research with regard to the possible effects of international trends on the South African agricultural sector;
(e) the availability of adequate and reliable technical information;
(f) forecast of agricultural commodity prices;
(g) improvement of services to small-scale farmers; and
(h) training of young people (at school level) in business, technical and accountancy skills.

The *critical underlying factor* in order to implement any successful strategy as discussed is the *availability of timely and accurate market and economic information.*

Deficiencies in the marketing of agricultural produce and producers' procurement of inputs and services, particularly in the former 'homelands', have certainly slowed the achievement of growth objectives and have been major disincentives for improved productivity.

Some of the efforts to develop agriculture, including marketing development, in these areas appear to have been western-oriented and to have ignored the socio-cultural-economic conditions of the previously disadvantaged rural producers. This has to be rectified. The traditional tendency to co-operate should, in particular, be accommodated and used.

More attention and recognition should go into rural markets. Market information must be improved, and extension personnel should be retrained for this purpose. The promotion of high-value crops and of value adding should be considered (Van Renen, 1997). A new approach is needed to encourage vendors and hawkers to distribute small farmer produce, both locally and in urban centres. A fresh look at co-operatives is need.

The authorities also have the task of developing and maintaining the infrastructure needed to stimulate marketing development and the other activities that depend on marketing, particularly production. Decisions on this infrastructure must be made in consultation with the communities involved.

REFERENCES

BAYLEY, B. (2000). *A Revolution in the Market: The Deregulation of South African Agriculture.* Oxford: Oxford Policy Management.

DAVIS, E. (1931). Some aspects of the marketing of farm products in South Africa. *South African Journal of Economics* 1: 167–187.

DE SWARDT, S.J.J. (1962). Problems of adaptation in agriculture. *Agrekon* 1(1): 13–23.

DE SWARDT, S.J.J. (1983). Agricultural marketing problems in the nineteen thirties. *South African Journal of Economics* 51: 1–28.

EALES, R. (1979). A long-term physical abattoir development plan for South Africa. PhD thesis, University of the Witwatersrand.

ELLIOTT, M.B., NIEUWOUDT, W.L. & LYNE, M.C. (1987). An opinion survey on aspects of beef marketing in South Africa. *Agrekon* 26(3): 62–65.

FRANKEL, S.H. (1934). Some comments on price and marketing control in South African agriculture. *South African Journal of Economics* 2: 324–331.

FRICK, A. & GROENEWALD, J.A. (1999a). Agricultural statistics needs of agribusiness in South Africa. *Agrekon* 38: 526–534.

FRICK, A. & GROENEWALD, J.A. (1999b). The need for agricultural information in the new South Africa. *Agrekon* 38: 241–254.

FRICK, A. & GROENEWALD, J.A. (2001). The need for agricultural information and data: researchers and policy makers. *Agrekon* 40: 104–117.

GOUSE, M., KIRSTEN, J.F., TREGURTHA, N. & TSWAI, J. (2000). The producer support estimate (PSE) for South African agriculture for 1996, 1997 and 1998. Paper presented at the 38th Conference of the Agricultural Economics Association of South Africa, 27–29 October, Sun City.

GROENEWALD, J.A. (1979). Inleiding tot algemene bespreking. *Tegnikon* 27(3): 29–32.

GROENEWALD, J.A. (1980). Economic considerations in the determination of agricultural development priorities by the extension worker. *South African Journal of Agricultural Extension* 9: 15–20.

GROENEWALD, J.A. (1982). Changes in the parity position of South African agriculture. *Agrekon* 21(2): 8–14.

GROENEWALD, J.A. (1989). Review of the report of the Committee of Enquiry into Alternative Marketing Arrangements for Maize. *Development Southern Africa* 6: 467–488.

GROENEWALD, J.A. (2000). The Agricultural Marketing Act: a post-mortem. *South African Journal of Economics* 68(3): 364–402.

GROENEWALD, J.A. & FRICK, A. (2001). South African farmers' needs of agricultural data. *South African Journal of Economic and Management Sciences* NS 4: 316–331.

GROENEWALD, J.A. & NIEUWOUDT, W.L. (1979). Welfare aspects of state intervention in agriculture. *Finance & Trade Review* 13: 59–81.

GROSSKOPF, J.F.W. (1932). *Plattelandsverarming en Plaasverlating*. Stellenbosch: Carnegie Kommissie en Pro Ecclesia – Drukkery.

KARAAN, A.S.M. & MYBURGH, A.S. (1992). Food distribution system in the urban informal markets: the case of red meat marketing in the Western Cape townships. *Agrekon* 31: 289–293.

KASSIER, W.E. (1993). News Release by the Chairman. Committee of Inquiry into the Marketing Act. Pretoria.

KASSIER, W.E. (1999). The process of agricultural market deregulation in South Africa with special reference to fresh fruit. Paper delivered at the Fruit Producer's Federation, Auckland, New Zealand.

KOHLS, R.L. & DOWNEY, D.W. (1968). *Marketing of Agricultural Products*. New York: MacMillan.

LE CLUS, C.F. (1982). Die ekonomie van mielieverbouing. *Mielies/Maize* 19: 7A–11A; 16A.

LEPPAN, H.D. (1931). *Agricultural Policy in South Africa*. Johannesburg: Central News Agency.

LOUW, A. (1979). Finansiële bestuur in die boerdery: riglyne en strategie. *Tegnikon* 27(3): 14–23.

LUBBE, W.F. (1992a). The Red Meat Marketing Scheme: an evaluation in a dynamic environment. PhD thesis, University of Pretoria.

LUBBE, W.F. (1992b). Price stabilisation policies: has the Meat Scheme benefited beef producers in South Africa? *Agrekon* 31: 74–80.

MARAIS, J.S. (1936). 'n Landboubeleid vir die toekoms. *South African Journal of Economics* 4: 207–220.

MCLOUGHLIN, J.R. (1938). A defence of control in the marketing of agricultural products. *South African Journal of Economics* 6: 291–302.

MYBURGH, A.S. (1992). Die bepaling van die realiteite van voedselbemarking in die Kaapse plakkersgemeenskappe. *Agrekon* 31: 285–288.

MYBURGH, A.S., LOMBARD, J.P. & SINCLAIR, M. (1986). Supply limitations according to product quality. *Agrekon* 25(1): 35–38.

NKOSI, S.A. & KIRSTEN, J.F. (1994). The marketing of livestock in South Africa's developing areas: a case study of the role of speculates, auctioneers, butchers and private buyers in Lebowa. *Agrekon* 32: 230–237.

NIEUWOUDT, W.L. (1976). Rents of land and production quotas in agriculture. *South African Journal of Economics* 44: 194–196.

NIEUWOUDT, W.L. (1978). Production quotas with special reference to meat: reply. *South African Journal of Economics* 46: 175–178.

PAARLBERG, D. (1962). Discussion: contributions of the New Frontier to agricultural reform in the United States. *Journal of Farm Economics* 44: 1179–1183.

PAARLBERG, D. (1964). *American Farm Policy*. New York: John Wiley & Sons.

RICHARDS, C.S. (1935). Subsidies, quotas, tariffs and the excess cost of agriculture in South Africa. *South African Journal of Economics* 3.

RICHARDS, C.S. (1936). The 'New Despotism' in agriculture: some reflections on the Marketing Bill. *South African Journal of Economics* 4: 469–504.

SCHOLTZ, A.P. & TOMLINSON, F.R. (1979). Comments on the problem of creating stability in the maize industry. *Agrekon* 18(3): 1–5.

SCHULTZ, T.W. (1968). US malinvestments in food for the world. In Heady, E.O. (ed.). *Alternatives for Balancing World Food, Productivity and Needs*. Ames: Iowa State University Press.

VAN RENEN, E. (1997). The BATAT marketing drive: improving market access for small-scale farmers. *Agrekon* 36: 648–655.

VAN RENSBURG, H.J. (1962). Control under the Marketing Act. *Agrekon* 1(1): 35–42.

VAN SCHALKWYK, H.D. & SWANEPOEL, L. (1997). How free is information in the new South African market? 9 May.

VAN ZYL, J. (1990). Prysbeleid in die mieliebedryf: effekte van pryssensitiwiteit, interafhanklikheid en relatiewe inkomsteverskuiwings. *Agrekon* 29: 108–112.

VAN ZYL, J. & KIRSTEN, J.F. (1993). Bemarkingsmoontlikhede van droëbone in Suid-Afrika: meer of minder beheer? *Agrekon* 32: 110–117.

VENTER, W.J. & VAN ZYL, J. (1996). Die effektiwiteit van die vloerprysskema as stabilisasiemeganisme in die primêre beesbleisbedryf. *Agrekon* 35: 141–148.

VINK, N. & KIRSTEN, J. (2000). Deregulation of agricultural marketing in South Africa: lessons learned. Monograph, Free Market Foundation, Sandton.

WILLEMSE, J. (1996). Information on South African agriculture: frustrations and opportunities. Paper presented at the AGROCON '96 Conference, January, Pretoria.

WRIGHT, P.D. (1992). A study in political economy: the South African maize and dairy industry. PhD thesis, University of Natal.

CHAPTER 8

Market Access for Small-Scale Farmers in South Africa

Moraka Makhura and Madime Mokoena

8.1 INTRODUCTION

The democratisation of South Africa in 1994 brought hope to African farmers who expected that the political transformation would be accompanied by positive changes in the agricultural sectors (Nyamande-Pitso, 2001). The improvement of small-scale agricultural production and the emerging farmers' increased participation in the mainstream of the economy were the pillars of the Reconstruction and Development Programme (RDP) proposed by the African National Congress (ANC). This programme aimed to increase household food security and alleviate poverty, especially in the rural and communal areas where most of the small-scale farmers are situated (ANC, 1994).

The ANC government has introduced several policies to try to improve the status of small-scale farmers in South Africa. The land reform programme was intended to make land accessible and to ensure security of tenure for the rural poor, thereby improving the capacity of their small-scale agricultural production in a sustainable manner (Department of Land Affairs, 1997; ANC, 1994). Market deregulation and trade liberalisation policies aimed to ensure market access for agricultural products, presumably from both emerging and commercial entrepreneurs, in domestic as well as international markets. The Broadening Access to Agriculture Thrust (BATAT) was another key attempt to enable small-scale farmers to participate in commodity markets.

Despite these policies, small-scale farmers in South Africa still experience problems in trying to access resource and commodity markets. This chapter discusses the extent to which these farmers typically market their produce, the aim being to identify common constraints that they face in gaining market access, and to suggest how to alleviate these constraints. Market access is the ability of farmers to use agricultural markets for their produce, and it assumes market availability. Market access and market utilisation are both viewed in the chapter as reflecting market participation. The basic issue is how small-scale farmers can access and participate in available commodity markets in South Africa, in order to promote their viability (Bayley, 2000).

Section 8.2 addresses policy issues that create an appropriate environment for market availability. Section 8.3 discusses the extent to which emerging farmers in South Africa participate in markets, while section 8.4 identifies constraints on commodity market access. Section 8.5 makes suggestions for addressing these constraints, and a concluding section discusses some of the policy implications of the material in this chapter.

8.2 POLICY PROVISIONS FOR MARKET AVAILABILITY

This section outlines the major steps that the South African government has followed to try to ensure the availability and accessibility of markets for established commercial and emerging farmers. Trade liberalisation and market deregulation processes are outlined and their implications for small-scale farmers are discussed.

8.2.1 Trade liberalisation measures

Since the advent of democracy in 1994, South Africa has become more actively involved in international trade issues and bodies. Currently, the country is a member of international organisations at regional as well as multilateral levels. Most of these organisations have trade packages as part of their agreements, while others influence trade issues. At the regional level, South Africa is a member of the Southern African Customs Union (SACU) and the Southern African Development Community (SADC). The principle objective of the SACU Agreement of 1969 is to maintain free interchange of goods between member countries and to apply the same tariff and trade regulations to imports from outside the common customs area on a basis that sustains the economic development of all the member countries.

At the multilateral level, South Africa, as one of the founder members of the General Agreement on Tariffs and Trade (GATT), became a full member of the World Trade Organisation (WTO) in 1995. On a bilateral basis, South Africa has signed a Trade, Development and Co-operation Agreement with the European Union (EU-SA TDCA). In addition, as a beneficiary of the United States government's Africa Growth and Opportunity Act (AGOA), the country benefits from the US's Generalised System of Preferences (GSP).

South Africa's membership of the above organisations has brought changes to its domestic agricultural sector that led to shifts from protection and regulation policies to market liberalisation and deregulation polices. This situation has improved the access of the country's agricultural products into international markets, while at the same time allowing agricultural products from the foreign countries to penetrate its commodity markets (for example, reduced tariffs on meat imports).

8.2.2 Marketing deregulation measures

The South African agricultural sector has emerged from a history of protection and support at various levels of production. Administered prices, import controls, compulsory single-channel local and export marketing and export subsidies, amongst other instruments, were used by previous governments to intervene extensively in agriculture. This intervention was aimed at supporting commercial farm incomes, promoting food self-sufficiency, and stabilising prices (Van Schalkwyk, 1997).

Until recently, the marketing of some 80% of South Africa's agricultural commodities was controlled by commodity-specific, producer-dominated, agricultural marketing boards, which operated under the Marketing Act of 1968. The boards also collected levies for various purposes (such as research), and some were responsible for issuing permits in terms of which agricultural products could be imported and exported (NAMC, 1999). Some of the former homeland areas also had their own agricultural marketing legislation. Initiatives in terms of this legislation generally took place in close co-operation with the various marketing boards, but implementation of this legislation was patchy, and only a relatively small number of homeland farmers were directly affected. This was particularly so where farming households produced predominantly for their own consumption and were net consumers of staple foods (NAMC, 1999).

The marketing arrangements affected all aspects of agriculture, including prices of, access to, and use of natural resources, finance, capital, labour, local markets, and trade (Jooste & Van Zyl, 1999). Groenewald (2001) states that the one-channel schemes proclaimed under the Marketing Act of 1968 became close to being unmanageable in the face of local and international pressure. This resulted in a gradual process of deregulation and the introduction of the new Marketing of Agricultural Products Act of 1996 (Van Dijk, 2000; Bayley, 2000) which, it was hoped, would improve access to markets for agricultural products.

8.2.3 Implications for small-scale farmers of changes in marketing policy

South Africa's trade regime changed considerably after the country became a signatory of the Marrakech Agreement in 1994 (Vink *et al.*, 2002). Agriculture was brought into the multilateral trade rules at the conclusion of the Uruguay Round and the establishment of the WTO in 1995. Although hailed as one of the achievements of the Round, it does not focus on addressing the development needs and concerns of the small-scale subsistence farmers in developing countries, particularly in Africa. The WTO Agreement on Agriculture tends to emphasise commercial agriculture as opposed to the small-scale farming practised by many farmers in South Africa and the rest of Africa (Ndirangu, 2002).

Increasing globalisation of the South African economy has, therefore, raised some challenges for the small-scale farmers, as they try to access finance to produce quality products that meet the required standards, and also try to find markets to sell these products at a competitive price, with no government subsidies in place (Nyamande-Pitso, 2001). The WTO Agreement on Agriculture tends to favour farmers in developed countries. It does not allow farmers in Africa to receive any form of support other than that listed under the Green Box and the *de minimis* percentage of 10% of agricultural production. Developed countries, which provided considerable help to their farmers before the WTO was established, have continued to assist them. These nations are permitted to invoke the special safeguard clause, which allows imposition of additional duties in the case of increased imports without proof of injury in the domestic market. In contrast, African countries do not have recourse to this clause as they agreed to ceiling binding during the Uruguay Round.

After the conclusion of the Uruguay Round negotiations, the farm sector in Africa became more exposed to external forces and subsidised exports from developed nations which has had adverse effects on production and food security. South Africa's classification as a developed country prevents small-scale farmers from benefiting from other preferential treatment offered to some African countries by the international community. For example, through the Cotonou Agreement (previously called the Lomé convention), the EU has a policy of non-reciprocal, duty-free market access of agricultural products from the African, Caribbean and Pacific (ACP) countries. Commercial farmers in the ACP countries also benefit from this arrangement, but small-scale farmers in South Africa are excluded due to the country's developed status. As a result, our small-scale farmers must compete with the commercial farmers in terms of producing high-quality and value-added products required by international markets (Mokoena, 2002).

It seems that the South African government's efforts to improve access to domestic and international markets through trade liberalisation and market deregulation have not markedly affected small-scale farmers. The benefits (and costs) of these measures still impact mainly on commercial farmers, and there is still a need to improve market access for small-scale farmers. According to Kirsten & Vink (2002), this is probably the main reason why questions are being raised about the current system of marketing agricultural products. One potential solution may be for South Africa to start negotiating trade agreements that will accommodate participation by small-scale farmers who produce the desired quality and type of commodities.

8.3 CURRENT EXTENT OF COMMODITY MARKET ACCESS OF SMALL-SCALE FARMERS IN SOUTH AFRICA

The access of small-scale farmers to fresh produce markets, livestock markets such as auctions, and grain markets is still limited. This section discusses a sample survey of 158 farming households conducted in 1997 among emerging farmers in Limpopo (Makhura 2001), and compares the results, where possible, with those of a study of constraints on crop marketing in the Impendle and Swayimana districts of KwaZulu-Natal (Matungul *et al.*, 2001). The aim is to show the typical extent of market access by these farmers, and the typical constraints to market access which they face.

8.3.1 Market sales

Market participation is based on the ability of households to sell all or some of their produce for cash. In the Limpopo household sample, agricultural incomes are generated when farmers sell high-value commodities and food crops. The high value crops include horticulture (fruit and vegetables) and livestock (large stock, small stock and poultry). The food crops include maize and other field crops. Table 8.1 shows that 19% of the households sell horticultural crops, 17% sell livestock, and about 20% sell maize. Furthermore, 22% of the households sell other field crops (wheat, groundnuts, beans, melons, and sunflowers). It follows that almost 50% of the households sell agricultural produce on the cash market.

Table 8.1: Market participation by smallholder farmers in Limpopo Province, South Africa, 1997.

Source of sales	*Annual mean (R)*	*% of households that sell products*
Horticulture	1 663	19
Livestock	492	17
Maize	293	21
Other field crops	459	22
Total agricultural income	2 907	50

Source: Makhura (2001)

Matungul *et al.* (2001) report that all Impendle sample households, and 83% of sample households in Swayimana, market some agricultural products. Yearly crop income ranged from R80 to R11 510 per household. Some Limpopo farmers sell horticultural commodities and livestock, while others sell horticultural and field crops (including maize). Some farmers use both dryland and irrigation practices for commercial production, while others market livestock, field crops and maize. In Impendle and Swayimana, multi-cropping (especially maize, madumbis, potatoes and dry beans) is the norm.

8.3.2 Value of exchange and subsistence production

Households often exchange agricultural products for processed products as an alternative form of market access, where value is added to the smallholder's product without change in ownership. For example, farmers agree with a co-operative, milling company or trading store to deliver their maize in exchange for maize meal. The costs incurred per 80 kg bag in the Limpopo Province involve transport (about R10), milling costs (R22), and storage costs (about R18) that are paid when farmers collect the maize meal. Almost 50% of the Limpopo households exchanged maize for maize meal. Similar arrangements, particularly with local traders and consumers, apply to other products. Livestock farmers in almost 25% of the Limpopo households barter one type of livestock for another to be used for different purposes. Table 8.2 shows that about 49% of households make exchange arrangements.

Table 8.2: Participation by households in other markets (%), Limpopo Province, South Africa, 1997.

Value categories	*% of households that participate*
Sell for cash	50.3
Market exchange	48.7
Annual value of own produce consumed	
Not consumed	30.6
R1 – R1 000	42.0
R1 001 and greater	27.4

Source: Makhura (2001)

Another alternative to cash sales is home consumption, estimated by the value of own produce consumed at the purchase price. According to Table 8.2, the annual value of own production that is consumed by households exceeded R1 000 for 27.4% of households. The yearly mean value of maize consumed in the Limpopo sample households is about R384, that of other crops R142, and of livestock R177. In total, the mean value of household consumption is R705. The total yearly value of consumption, agricultural sales, and exchanged goods is about R4 547 on average. This is equivalent to 21% of total household income of R21 365. Household consumption is 15.5% of agricultural income and 3% of total household income (R705 out of R21 365).

8.3.3 Participation in different commodity markets

Small-scale farmers in the Limpopo Province participate mainly in four markets:

(a) Horticultural market

Households participating in this market are considered to be more commercially inclined due to the nature of their product. Horticultural crops are generally perishable and require immediate disposal, implying that the farmers who produce horticulture crops do so with the intent to sell them. Some 19% of the sample households sell all, or a portion of, their fruits and vegetables to market outlets ranging from informal markets to large urban-based, fresh produce markets. They also use other channels such as direct-to-consumers, and selling to factories (Mathye *et al.*, 2000; Mathye, 2001). Typically, many of the households producing fruits and vegetables also have access to a dryland plot where they produce maize and/or other field crops.

(b) Livestock market

Households selling to livestock markets are considered commercially-orientated since livestock production is a high-value enterprise. About 17% of the Limpopo sample households sell all or some of their cattle, sheep, goats and poultry at auctions, to the local community, and directly to other markets. Most livestock producers also have access to arable lands for crop production, particularly dryland crops.

(c) Maize market
Households selling maize are not usually seen as commercially-orientated, as maize is mainly regarded as a subsistence crop. As in the KwaZulu-Natal districts, only when consumption needs are met, will farmers consider selling some maize. Another reason for this classification is that maize is a low-value commodity. As such, maize selling may be a spill-over of access to markets for other commodities. This is illustrated by about 41% of the Limpopo sample households selling maize, as well as horticultural and other field crops, while only about 25% of the maize sellers also sell livestock.

(d) Other field crop markets
The indicator of market participation of other field crops used in the Limpopo study aggregates several crops such as wheat, beans, coriander, and grain sorghum. Some of these crops have a higher commercial orientation than others. Most of the farmers selling these commodities also sell other commodities – about 39% and 36% of the households also sell maize and horticultural crops, respectively. Relatively fewer farmers (about 18%) also sell livestock.

8.4 DETERMINANTS OF, AND CONSTRAINTS ON, MARKET ACCESS

Access by small-scale farmers to commodity markets in South Africa is determined and limited by a number of factors. Studies conducted by the NAMC, Makhura (1994, 2001) and Matungul *et al.* (2001) focus on the marketing constraints for these farmers. Some of these issues are related to policy, but most are location-, farmer-, and/or commodity-specific, and some are more pertinent to certain provinces than others:

(a) Transport: Most small-scale farmers do not have their own means of transport and are dependent on contractors, taxis or neighbours. Hired transport is in some cases relatively expensive because of the small quantities involved. In other cases, transport contractors are not prepared to service the rural areas because of the very bad condition of the roads. Furthermore, these farmers are often poorly serviced by public transport.

(b) Assembly and storage of products: This involves both the transport of crops to storage facilities and the provision of suitable storage facilities within reasonable distance from producers. In many rural areas, storage facilities are non-existent, unsuitable, or centralised to the extent that crops need to be transported over long distances. A problem is that low volumes may not justify storage facilities.

(c) Market infrastructure: A system of centralised National Fresh Produce Markets makes it necessary to transport products over long distances, at considerable cost and sometimes with loss of quality.
(d) Road infrastructure: In some provinces, particularly Limpopo, KwaZulu-Natal and the Eastern Cape, roads are in bad condition.
(e) Discrimination: Some farmers experience discrimination at livestock auctions and fresh produce markets. Their products are usually traded last and often fetch lower prices due to their deteriorating quality caused by delays.
(f) Market information: There is a lack of market information in the rural areas and also a lack of means to effectively disseminate information. With the advent of liberalised marketing, small-scale farmers will increasingly need market information to survive in a very competitive marketing environment.
(g) Bargaining power: Small-scale farmers lack organisation, bargaining power and knowledge to make effective use of their membership in the marketing trusts that were formed after the marketing boards in South Africa were abolished (Kirsten & Vink, 2002). A stronger public representation is, therefore, required in such trusts to enhance synergy, co-ordination and accountability. The lack of bargaining power also applies to the purchase of inputs.
(h) Institution responsibility: There is a lack of institutional responsibility focused on ensuring market access for emerging farmers. Loose responsibilities are shared amongst the NAMC, National Department of Agriculture (NDA), and the Provincial Departments of Agriculture (PDAs) (Kirsten & Vink, 2002). While dedicated institutions may be required, farmers stand to gain if these shared responsibilities are better co-ordinated.

Both the Limpopo and KwaZulu-Natal studies identify transaction costs as adversely affecting market participation by small-scale farmers. The first of these transaction cost factors is access to information. Farmers who presently do not participate in the markets might respond positively if they could have reasonable access to information about markets, via close proximity to markets, appropriate contacts with extension officers, or radio and television. These conditions are particularly relevant for high-value commodities such as horticultural crops. In this regard, the government could appoint extension officers who specialise in marketing – this would require that these officers receive formal college education and in-service training.

Farmers are also likely to participate in product markets if they can access productive assets such as land and livestock, and precautionary assets such as non-farm income. The land reform programme in South Africa is currently not redistributing land at the envisaged rate (see Chapter 4 of this book). Potential ways to improve this situation would be to provide incentives to motivate high-value commodity production, and to create access to grazing rights for small-scale livestock farmers. It is hoped that the new Land Redistribution and Agricultural

Development (LRAD) programme will lead to greater access to productive resources. Non-farm income helps to provide liquidity to purchase inputs and contractor services needed to produce and market agricultural outputs (Matungul *et al.*, 2001: 360). The provision of credit (avoiding the pitfalls of past programmes) as an alternative source of liquidity may also be considered.

Furthermore, households participate in commodity markets when they have most of their members involved in production activities rather than being dependants. To facilitate the commercialisation of small-scale farmers, it thus seems pertinent to determine how different household members influence market participation. For example, since women are more involved in livestock markets and less in horticulture markets in the Limpopo sample, how can opportunities be enhanced for them to participate in livestock enterprises, and what factors limiting their participation in horticultural enterprises could be removed? This may require changes in the legal status of women as contractual partners in business transactions. Finally, farmers are more willing to sell their crops and livestock when markets are more accessible in terms of good roads or farmers have access to their own transport.

8.5 TOWARDS IMPROVING MARKET ACCESS FOR SMALL-SCALE FARMERS IN SOUTH AFRICA

Kirsten & Vink (2002) suggest an audit of all institutions to find the appropriate system within government and its agencies for promoting market access by small-scale farmers. In addition, improved co-operation with provincial departments and the implementation agents is needed. The NAMC could establish a market-access desk, as a contact point for farmers, and promote the dissemination of relevant marketing information. There is also a need to consider appropriate measures for small-scale farmers regarding the allocation of funds, exemptions from obligations, and benefit from promotional activities. A 'developing' agriculture brand could be developed and backed up with appropriate marketing locations, technologies, products and systems. Co-operation in marketing could be considered, for example, via contract production whereby processors provide financial, technical and marketing services to smallholders. Lessons from the South African sugar industry will be useful in this regard.

According to the NAMC, constraints to market access could also be alleviated by:

(a) designing an improved market information system provided by the PDAs that is adapted to the communication facilities available to small-scale farmers;
(b) promoting small-scale processing and value-adding (to generate higher profit margins and reduce transport costs);
(c) promoting high-value crops such as certain herbs and spices, cashmere and certain kinds of fruit, vegetables and flowers;

(d) providing infrastructure such as market places, improved transport, road and telecommunications, storage facilities (sources of raw material for rural-based processing facilities); and
(e) training farmers and extension officers in the interpretation and use of market information, technical aspects of the production of non-traditional crops, storage techniques, grading and packaging of products, and small-scale processing/value-adding techniques.

Makhura (2000, 2001) suggests the following *additional* components of a marketing strategy to improve market access for smallholders:

(a) The information system should link Agricultural Marketing (extension) Officers with market information centres at district or service centre level, and enhance the link with farmers by improving their access to and the use of cellular phones.
(b) The appropriate transport system should promote emergent transport contractors, the opening of road networks, and the development of collection points. The role of both the public and private sector is important here. Members of local communities should be encouraged to provide transport services to ferry products to market centres or collection points.
(c) Asset accumulation should be promoted by, for example, addressing both the acquisition of, and ownership rights to, land by small-scale farmers. Matungul *et al.* (2001: 361) recommend institutional innovations such as encouraging a rental market for land to enable households wanting to farm to increase their land area and thus their potential income from crop sales.
(d) Diversity in market access should be promoted in order to give farmers alternative marketing channels and thereby lower transaction costs (Matungul *et al.*, 2001).
(e) The potential role of price guarantees and price insurance for new entrants into agricultural product markets should be considered (Makhura *et al.*, 1996).

8.6 CONCLUSION

Despite the introduction of policies to enable agricultural market development, most small-scale farmers in South Africa still do not participate in product markets. These farmers have limited technical expertise and only a partial understanding of market requirements. They have also not benefited to the extent that their commercial counterparts have from past agricultural support measures in this country, or in other developed countries, before the conclusion of the Uruguay Round of world trade discussions. As a result, small-scale farmers in South Africa still have to manage basic issues like pre- and post-harvest marketing, and how to enter local and global markets, while their commercial counterparts focus on issues relating to how to improve competitiveness.

Present trade, marketing, and institutional policies in this country make limited provision for the unique interests of emerging small-scale farmers. Furthermore, the current institutions involved in promoting market access are not co-ordinated. The various constraints to market access can be addressed through a combination of public intervention and private sector involvement. Potential solutions include overcoming infrastructure backlogs, improving market information, institutional reform (such as promoting a land rental market), supporting small-scale food processing and value-adding, and encouraging closer links between small-scale and established commercial farmers. Farmer-Public-Private-Sector Partnership is likely to be crucial in addressing the problem of market access of small-scale farmers in South Africa.

REFERENCES

AFRICAN NATIONAL CONGRESS. (1994). *The Reconstruction and Development Programme: A Policy Framework*. Johannesburg.

BAYLEY, B. (2000). *A Revolution in the Market: The Deregulation of South African Agriculture*. Oxford: Oxford Policy Management.

DEPARTMENT OF LAND AFFAIRS. (1997). *White Paper on South African Land Policy*. Pretoria: Government Printer.

GROENEWALD, J.A. (2001). Impediments to delivery of policy for growth with equity in South African agriculture. Unpublished report, Africa Institute for Policy Analysis and Economic Integration (AIPA).

JOOSTE, A. & VAN ZYL, J. (1999). Regional agricultural trade and changing comparative advantage in South Africa. Technical Paper No. 94, SD Publication Series, USAID.

KIRSTEN, J.F. & VINK, N. (2002). A note on a future agricultural marketing dispensation in South Africa. Unpublished submission to the NAMC, Pretoria.

MAKHURA, M.T. (1994). Theoretical and econometric models of commercialization behaviour of households in the developing areas of South Africa. MSc thesis, Pennsylvania State University.

MAKHURA, M.T. (2000). Promoting market participation in small-scale agriculture in the Northern Province. An outline paper presented at a seminar organised by NIEP, LAPC and University of Pretoria on market access. Agrivillage, Northern Province Department of Agriculture, Pietersburg.

MAKHURA, M.T. (2001). Overcoming transaction costs barriers to market participation of smallholder farmers in the Northern Province of South Africa. PhD thesis, University of Pretoria.

MAKHURA, M.T., COETZEE, G. & GOODE, F.M. (1996). Commercialisation as a strategy for reconstruction in agriculture. *Agrekon* 35(1): 35–40.

MATHYE, M.M. (2001). An analysis of transaction costs in marketing of mangoes and bananas in Mhinga-Shikundu and Phaswana areas of the Northern Province. Unpublished report, University of Pretoria.

MATHYE, M.M., MAKHURA, M.T. & KIRSTEN J.F. (2000). Transaction costs in the market for bananas. Paper presented at the 38th Annual Conference of the Agricultural Economics Association of South Africa, 27–29 September, Sun City.

MATUNGUL, P.M., LYNE, M.C. & ORTMANN, G.F. (2001). Transaction costs and crop marketing in the communal areas of Impendle and Swayimana, KwaZulu-Natal. *Development Southern Africa* 18(3): 348–363.

MOKOENA, M.R. (2002). An overview of South Africa's international and regional trade relations: implications on the agrarian reform. Working document, Human Sciences Research Council.

NATIONAL AGRICULTURAL MARKETING COUNCIL (NAMC). (1999). Report on the investigation into market access. Report on Phase 1, February, Pretoria.

NDIRANGU, N. (2002). Africa worse off after Agreement on Agriculture. Paper presented at a Roundtable on Food and Trade: The WTO Development Change, 4–5 November, Ottawa, Canada.

NYAMANDE-PITSO, A. (2001). The experience of black business in the import/export market. Paper presented at the 8th Annual Agriculture Management Conference, 30 October, Midrand, South Africa.

VAN DIJK, G. (2000). GATT/WTO: changes, functions and structure. Working Document, National Department of Agriculture.

VAN SCHALKWYK, H.D. (1997). Agricultural trade policy. Working document, University of the Free State.

VINK, N., KIRSTEN, J.F. & TREGURTHA, N. (2002). South Africa's changing agricultural trade regime. Unpublished article, University of Stellenbosch.

CHAPTER 9

Water Usage and Irrigation Policy

Gerhard Backeberg

9.1 INTRODUCTION

The natural resource base, and in particular the water resource base, forms the ultimate support of economic activity in agriculture. Together with entrepreneurial capacity, it comprises the most important competitive strength of the South African political economy (Spies, 1996; Liebenberg, 1996). Although natural resources of high quality are limited, they are not a constraint to economic growth if they are used as productively as possible. This applies particularly to water use for irrigated and dryland cropping, and to water use by natural grazing. The challenge is to formulate policy, design institutions and implement policy instruments that will unlock the productive potential of people who are using water resources (Backeberg, 1996).

The purpose of this chapter is to give a brief overview of water use, to review the unfolding policy process, and to highlight some key issues of people-centred development which require attention for future action.

9.2 WATER USE

Water use for agricultural production can best be described as rainfall dependent, surface-water and groundwater dependent. The average annual rainfall in South Africa is nearly 500 mm on a land area of 102.8 million hectares used for crop and livestock farming as well as forestry. Approximately 62% of rainfall is used for maintenance and production by natural vegetation, woodlands and forests, while 12% is used as evapotranspiration for crop production on drylands. It is estimated that 8.5% and 6% of rainfall respectively contributes to run-off and deep percolation of water, which becomes available as surface-water and groundwater (Bennie *et al.*, 1998: 1–3). The use of surface- and groundwater for irrigation will be analysed in more detail (Backeberg & Odendaal, 1998: 49–53).

According to the Overview of Water Resource Availability and Utilisation in South Africa (DWAF, 1997a), the maximum annual surface-water yield is 33 290 million m^3 while the annual water requirement in 1996 was 20 045 million m^3 (see Table 9.1). The total requirement for irrigation is estimated at 10 740 million m^3

per year, or on average 53.6% of total water requirements. However, this varies from below 40% to above 80% of total water requirements for the Eastern coastal and inland Karoo regions respectively (see Table 9.1). In addition, the maximum annual yield of groundwater is 5 400 million m^3 of which 2 000 million m^3 per annum is exploited and 80% is used for irrigation (Water Research Commission 1996: 15–16). The water used annually for irrigation is equivalent to 2.4% of rainfall.

Table 9.1: Water resource potential and requirements for regions with catchments of similar hydro-meteorological characteristics, 1996.

Region	*Maximum yield*	*Water requirements*	*Irrigation and afforestation requirements***	
	$10^6 m^3/yr$	*$10^6 m^3/yr$*	*$10^6 m^3/yr$*	*%*
Northern[1]	2 566	3 373*	1 861	55.2
Eastern inland[2]	4 834	2 320	1 826	78.7
Eastern coastal[3]	13 199	5 604	2 217	39.6
Southern coastal[4]	1 793	1 768	1 350	76.4
South western[5]	3 095	2 396	1 570	65.5
Karoo[6]	6 014	2 555	2 173	85.0
Central[7]	1 789	2 029*	1 347	66.4
Total/Average	33 290	20 045	12 344	66.7

Source: DWAF (1997a)

* Shortfalls are supplied via different transfer schemes

** Water requirements for afforestation are estimated at 1 604 million m^3 per year or 8% of total requirements, which gives irrigation requirements of 10 740 million m^3 or 53.6% of total requirements

1 Crocodile/Limpopo and Olifants basin

2 Nkomati and Maputo basin

3 Mfolozi, Thukela, Mgeni/Mzimkulu, Mzimvubu and Mbashe/Kei basin

4 Sundays, Gamtoos and Gouritz basin

5 Bree/Berg, Olifants/Doring and Buffels basin

6 Lesotho, Orange to Vaal and Orange basin below Vaal confluence

7 Vaal basin

The above-mentioned water requirements have been revised for the purpose of the National Water Resources Strategy (NWRS) with 2000 as the reference year. These revisions are based on improved assessments of streamflow and they also account for environmental water requirements as a reduction in the yield of water resources, rather than as an abstractable requirement as applies to other water-use sectors. Water requirements have been standardised to a 98% assurance of supply and are now sub-divided according to the 19 newly defined water management areas. The total water requirement is thus calculated as 13 280 million m^3/yr, of which 7 836 million m^3/yr, or 59% of the total, is required for irrigation (DWAF, 2002: 22–24).

The total area under irrigation in South Africa is 1 290 132 hectares and the potential for expansion is estimated at 283 350 hectares. Water is relatively more limiting than land since the projected available water according to official estimates only allows for expansion of 178 000 hectares (DWAF, 1995). Of the total current area under irrigation, 86.5% is classified as very suitable or suitable (also known as class 1 and 2 soils), and 82.5% of the potential area falls within the same classes (see Table 9.2). The total potential irrigable land of 1 573 482 hectares is approximately 10% of the arable land of 15.8 million hectares (NDA, 1996: 8). However, it is noteworthy that satellite imagery captured in 1994–95, which has an overall mapping accuracy of approximately 80%, indicates that the total area under irrigation in South Africa is 1 589 884 hectares (Nell & Van den Berg, 2001). This shows that the projected potential for expansion has already been exceeded.

Included in this total irrigated area is an area of 47 486 hectares (or 3.7% of the total) which is used for foodplot and smallholder farming on a large number of irrigation schemes in all of the provinces (see Table 9.3). The relatively small irrigated area which could be documented is probably an under-estimation. It does not fully capture water use for irrigation by smallholders on tribal communal land, but it also reflects the gradual collapse of farming activities on government-initiated irrigation schemes. There is scope for upgrading 66 700 hectares of under-utilised land and water on existing irrigation schemes where water works have already been constructed (DWAF, 1995).

The most important type of irrigation that is practised in all rainfall regions is permanent irrigation – the crop's full water requirement is applied to avoid plant stress and to achieve optimal yields. The most important methods of irrigation used are flood irrigation on 32.8% of the total area, sprinkler irrigation on 54.4% of the area and micro irrigation on 11.8% of the area (see Table 9.4) (Water Research Commission 1996: 29). Variations of these methods are also used by smallholder farmers with sometimes innovative adaptations, such as short-furrow irrigation (Crosby *et al.*, 2000: 7.1–7.32). In a recent survey it was found that the relative importance of the methods of irrigation has changed somewhat with flood irrigation now 28.5%, sprinkler irrigation (non-mechanised and mechanised) 53% and micro irrigation (including drip irrigation) 18.5% (Water Research Commission, 1999). The efficiency levels that are realistically achievable range from 55–65% for flood irrigation, 70–85% for sprinkler irrigation and 85–95% for micro irrigation, which gives a weighted average efficiency of nearly 75%.

Water is applied for commercial production of field, industrial, horticultural, pasture and forage crops (see Table 9.5). Most of these crops are also grown in smallholder agriculture (see Table 9.3). The stabilisation of variable rainfall through irrigation is conspicuously more important for high-income crops such as potatoes, vegetables, grapes, fruit and tobacco. Although grain and oilseed crops are effectively grown under rain-fed conditions, they will remain part of crop combinations under irrigation for production of seed and as rotational crops to maintain soil productivity. Pastures and forages form an important component of the fodder flow programme for dairying, and sustain extensive livestock production in arid and semi-arid rainfall zones (Water Research Commission, 1996: 37–41).

In all irrigation areas 6–26% of soils have become waterlogged and salt-affected (see Table 9.6). The extent of the problem is cause for concern as, with the exception of the KwaZulu-Natal Province, the area of land with high water tables

Table 9.2: Areas currently under irrigation and land suitable for potential irrigation development in the different provinces (1996).

Province	*Land*								*Water available for further development*
	Very suitable		*Suitable*		*Risky*		*Total*		
	Current ha	*Potential ha*	*Current ha*	*Potential ha*	*Current ha*	*Potential ha*	*Current ha*	*Potential ha*	*ha*
Eastern Cape	34 568	14 524	94 995	19 370	25 367	14 735	154 930	48 629	52 000
Free State	39 104	14 496	41 131	12 580	18 745	2 146	98 980	29 222	10 000
Gauteng	16 925	4 100	6 895	1 600	2 864	0	26 684	5 700	0
KwaZulu-Natal	66 627	52 552	71 762	31 955	31 855	15 056	170 244	99 563	77 000
Mpumalanga	129 225	31 250	25 426	4 070	2 769	490	157 420	35 810	7 000
Northern Cape	80 965	8 707	71 356	5 250	9 553	1 340	161 874	15 297	20 000
Limpopo	61 752	6 100	46 898	3 550	26 496	1 420	135 146	11 070	2 000
North West	83 660	700	16 306	1 800	3 352	100	103 318	2 600	0
Western Cape	100 861	6 480	128 038	14 578	52 637	14 401	281 536	35 459	10 000
Total	613 687	138 909	502 807	94 753	173 638	49 688	1 290 132	283 350	178 000

Source: Water Research Commission (1996)
Note: Based on the Report of the Commission on a Food and Feeding Strategy for South Africa (1990)

Table 9.3: Smallholder farmer irrigation schemes in different provinces (1996).

Province	*No. of schemes*	*Area irrigated ha*	*No. of farmers*	*No. of food plots*	*No. of commercial farmers*	*Main commodities*
Eastern Cape	25	9 460	7 365	3 752	2 613	Maize, vegetables, citrus, lucerne
Western/Northern Cape	5	487	1 004	905	99	Vegetables, deciduous fruit, lucerne
North-West	20	3 874	880	342	538	Wheat, cotton, vegetables, maize, lucerne, fruit
Limpopo	102*	19 895	7 425	310	7 115	Maize, vegetables, groundnuts, wheat, cotton, citrus, fruit
KwaZulu-Natal	33	8 341	18 745**	17 190	1 555	Sugar cane, maize, vegetables
Mpumalanga	17	5 429	1 689	740	949	Sugar cane, vegetables, fruit
Total	202	47 486	37 108	23 239	12 869	

Source: Water Research Commission (1996)

* Includes 22 schemes (5 257 ha) not yet settled
(These figures have been updated by Bembridge (2000) and the Department of Agriculture (2002) in Limpopo Province)

** Details of farmers on 14 schemes not available

Table 9.4: Total areas, type and method of irrigation in different rainfall regions.

			Type of irrigation			*Method of irrigation*		
Region	*Rainfall mm*	*Total ha*	*Permanent %*	*Supplementary %*	*Occasional %*	*Flood %*	*Sprinkler %*	*Micro %*
1	(< = 125)	19 174	92.5	0	7.5	66.6	8.3	25.2
2	(126–250)	161 197	61.1	0.4	38.5	77.1	16.8	6.1
3	(251–500)	399 278	86.7	7.7	5.7	42.8	43.6	13.6
4	(501–750)	488 543	75.2	20.8	4.0	21.0	65.4	10.8
5	(>750)	221 940	81.5	16.6	1.9	5.3	80.9	13.8
Total		1 290 132	78.3	13.1	8.6	32.8	54.4	11.8

Source: Water Research Commission (1996)

and salinity levels exceeds the land area with risky soils (class 3 soils). Taking the total area into account, irrigation schemes in the Northern Cape Province are worst affected (compare Table 9.6 with Table 9.2). This deteriorating water quality is one of the greatest threats to the sustainable use of soil and water resources (Scotney & Van der Merwe, 1995: 51–55). Land reclamation practices on an area of 241 630 hectares, or 18.7% of the total irrigated area, are required if production is to remain sustainable.

Table 9.5: Estimated contribution of irrigation to commercial crop production in South Africa.

Crop	*Area irrigated*		*Production*	
	x 1 000 ha	*% of total area planted to this crop*	*x 1 000 t*	*% of national production*
Maize	110	3	660	10
Wheat	170	12	740	30
Other small grains	52	3	200	6
Potatoes	39	70	1 200	80
Vegetables	108	66	1 330	90
Grapes	103	90	1 300	90
Citrus	35	85	1 100	90
Other fruit	95	80	1 200	90
Oilseeds	54	10	108	15
Sugar cane	60	15	4 000	25
Cotton (lint)	18	17	17	42
Tobacco	12	85	20	90
Lucerne	203	70	1 600	80
Other pastures and forages	104	15	800	25

Source: Water Research Commission (1996)

Table 9.6: Estimated percentages of waterlogged or salt-affected irrigated land in the different provinces.

Province	*Waterlogged or salt-affected*		
	Severely	*Moderately*	*Total*
Eastern Cape	6	13	19
Free State	6	18	24
Gauteng	5	15	20
KwaZulu-Natal	5	9	14
Mpumalanga	1	5	6
Northern Cape	4	20	24
Limpopo	12	14	26
North West	3	5	8
Western Cape	9	15	24

Source: Water Research Commission (1996)

9.3 IRRIGATION POLICY

9.3.1 Background

Major changes have been made to the way in which water resources are managed in South Africa. Riparian ownership of surface water rights and private ownership of groundwater rights have been abolished. The Water Act of 1956 and other related water legislation has been repealed. In doing so, the notion of public and private water, normal flow and surplus water in public rivers and government control areas have fallen away. A distinction is no longer made between public government water schemes and private irrigation board schemes. The previous approach of command

and control for supply management of water resources has also largely been discarded. Subsidies on operation and maintenance cost for the supply of water are being phased out. Due to past political inequality, priority attention is now given to previously disadvantaged individuals and communities, in particular regarding equitable access to water resources and rendering of water services.

Public policy is by definition the decisions and actions of government authorities. Although policy is not binding, it does serve the purpose of convincing members of society that action by government is reasonable (Easton, 1979; Randall, 1987). The key elements of any public policy are its objectives and strategies and the main goal of policy, including irrigation policy, is generally to increase economic and social welfare, with intermediate objectives of improved efficiency and equity. Strategies are the decision-making rules that are followed for the conservation and allocation of water resources (Ciriacy-Wantrup, 1985; Backeberg, 1994).

The current water policy debate concerns which options should be followed for:

(a) use and development; and
(b) apportionment and transfer of water resources for irrigation.

9.3.2 Policy framework

In order to achieve public support for policy reform, the correct process for policy formation involves formulation on a national level, gaining acceptance through several stages of consultation, and implementation on a provincial and local level. There are various activities at different levels of government, but these are linked through feedback, on impacts for example, and therefore involve a continuous process of evaluation (Cameron, 1991). Democratic constitutional changes in South Africa during 1994 have given impetus to a comprehensive review of water policy and legislation. This has taken place through consultation processes with stakeholders, contribution by experts and publication of various Discussion Documents over a period of two years. Thereafter the National Water Policy for South Africa (DWAF, 1997b) was adopted. It contains three objectives for managing water resources:

(a) To achieve equitable access, i.e., equity regarding the use of water resources, benefits from the use of water and access to water services.
(b) To achieve sustainable use of water, i.e., to achieve a balance between water requirements and implementing measures to protect water resources.
(c) To achieve efficient and effective use of water, i.e., to achieve optimum social and economic benefits.

In order to achieve these objectives, important shifts in policy have taken place:

(i) the powers of government will be exercised as a public trust;
(ii) water requirements to meet basic human needs and to maintain ecological balances are reserved as a right and all other water uses will be subject to administrative authorisations; and

(iii) the responsibility and authority for water resource management will be decentralised to enable local participation of organisations and people.

The principal institution to implement the policies for water resource management is the National Water Act (NWA), 36 of 1998, which contains comprehensive provisions for the use, development, protection, management and control of water resources.

In terms of the NWA, a National Water Resources Strategy (NWRS) has been published, which provides information about ways in which water resources will be managed (DWAF, 2002). Water resources are considered to be common to all and all water users, in particular farmers using water for commercial purposes, have had to register their water use. In addition, a number of new water charges have been imposed on raw water users. With reference to the policy objectives and strategies mentioned above, three policy options will be briefly discussed.

9.3.2.1 Water use authorisations

The previous riparian water rights, water quotas and pumping permits are being replaced by water use authorisations. There are three different types of authorisations and, in general, a licence to use water will apply for agricultural water use. As a transitional measure, the NWA permits 'existing lawful water use' to continue until such time as it is formally licenced. A licence to use water specifies that it:

(a) may only be issued by a responsible authority, to which a prospective user must apply;
(b) replaces all previous entitlements, if any, to use water for the purpose specified in the licence;
(c) is specific to the user to whom it is issued, and to a particular property or area;
(d) is specific to the use or uses for which it is issued;
(e) is valid for a specified time period, which may not exceed 40 years;
(f) may have a range of conditions attached to it; and
(g) must be reviewed by the responsible authority at least every five years.

Licence applications give existing or prospective water users authorisation to use water resources for productive or beneficial purposes and applications may be made individually, or generally, as part of compulsory licensing.

The process for compulsory licensing[1] will be as follows:

(a) Existing use and its lawfulness is verified.
(b) The responsible authority issues a notice calling for licence applications. Responsible authorities must also identify other potential or prospective users,

[1] Compulsory licensing can be interpreted as apportionment of water.

especially from marginalised or disadvantaged groups and communities, who have not previously had access to water resources but who may wish to do so, to ensure that available water is allocated fairly.

(c) Users and prospective users prepare and are required to submit licence applications.
(d) The responsible authority evaluates all licence applications.
(e) The responsible authority develops possible solutions to balance – 'reconcile' – water requirements with water availability, whilst meeting requirements for the reserve and water quality.
(f) The responsible authority prepares, publishes and invites public comment on a proposed allocation schedule.
(g) After considering all comments on and objections to the proposed allocation schedule, and making necessary amendments, the responsible authority must prepare and publish a preliminary allocation schedule. Anyone who unsuccessfully objected to the preliminary allocation schedule has the right to appeal to the Water Tribunal.
(h) The responsible authority publishes the final allocation schedule in the *Government Gazette*. As soon as reasonably practicable after the final allocation schedule has been published, the responsible authority must issue licences to water users in accordance with its provisions.

One of the most important attributes of a licence is its period of validity. While the evaluation procedure for licence applications will contain broad guidelines for determining the licence period, each water use is unique, and the guidelines must not be regarded as a rigid prescription. Considerations in determining a licence period include:

(i) For agricultural activities, the nature of the crop, particularly whether it is annual or perennial.
(ii) The physical infrastructure necessary to use the water or in which water will be used, its life span, the capital investments and associated loan redemption periods.
(iii) The extent to which infrastructure has been established to process the products of water use, and whose viability depend on the water use continuing. Examples are timber processing plants, sugar mills, and canning factories.
(iv) The social benefits of the water use, such as the extent to which people depend on the water use for employment.

Transfer of water use authorisations is provided for in Section 25 of the NWA for two distinct circumstances. The first refers to the temporary transfer of water authorised for irrigation either on the same property for a different use, or to another property for the same or a similar use. Although every case will be considered on its own merits and within its local context, in general, temporary transfers will be for one year only, with the option of applying for an extension of a further year. Users must apply to the water management organisation that has jurisdiction in the area for permission to effect the transfer.

The second circumstance refers to permanent water transfers, which may be effected by one user offering to surrender all or part of an allocation to facilitate a licence application by another user. Transfers of this nature constitute trade in water use authorisations, which may be used to achieve equity of access to water, or to increase the efficiency of use by moving water use from lower to higher value uses. In the latter case, the new licence application will be subject to all the relevant requirements of the NWA regarding applications for licences, including the need for a Reserve determination.

Transfers, whether temporary or permanent, will only be permitted where both the original and transferred water use are from the same water resource. Procedures have been developed to deal with transfers of water use authorisations, and Section 26 regulations may be written to provide a nationally-consistent basis for them (DWAF, 2002: 56–62).

9.3.2.2 Water demand management, water charges and water trading

Water conservation and water demand strategies are being developed for three identified water use sectors, namely, domestic, agricultural, and industrial and mining use. The interventions and measures aim to increase and support the efficiency of water use by changing the behaviour of water users and by adopting water-efficient technologies. In addition, there are social, economic and environmental advantages that can be obtained when water users and water organisations:

(a) understand the value of water as a scarce resource;
(b) allocate water to alternative users within a particular sector or between competing uses;
(c) postpone the investment in new infrastructure and delay increases in the cost of supply; and
(d) reduce unaccounted-for water due to leakages and wastage by non-paying water users.

In the case of irrigated agriculture, water losses occur during river or canal conveyance, as well as during on-farm distribution, storage and in-field applications. The water demand management strategy for agriculture seeks to promote efficiency gains by ensuring that:

(a) appropriate measures to influence the reduction in water wastage are implemented;

(b) water user associations and end-users understand and appreciate the need to progressively modernise their water conveyance systems and irrigation equipment;
(c) water allocations promote equitable and efficient utilisation of water;
(d) preventive maintenance programmes are put into place;
(e) sufficient irrigation information is generated and is accessible to all stakeholders;
(f) the concepts of environmental awareness and protection are promoted and accepted by all stakeholders;
(g) water management organisations and service providers implement audits from the water source to end-users and beyond; and
(h) water management and services organisations are encouraged to use the most appropriate technologies in water release and distribution systems (DWAF, 2002: 66–68).

The NWA and the Pricing Strategy for Raw Water Use Charges (DWAF, 1999), makes provision for three types of water use charges:

(a) funding water resource management for activities such as information gathering, monitoring and controlling water resources and their use, water resource protection (including waste discharge and the protection of the Reserve), and water conservation;
(b) funding water resource development and use of waterworks for the costs of investigation, planning, design, construction, operation and maintenance of waterworks, pre-financing of development, a return on assets, and the costs of water distribution; and
(c) achieving the equitable and efficient allocation of water through economic incentives to encourage more efficient use of water, water conservation, and a shift from lower to higher value uses.

In addition, a waste discharge charge system is currently being developed that will be based on the Polluter Pays principle, and will address point and diffuse sources of pollution. It will supplement the more traditional regulatory approach to water quality management, in which standards and objectives are set and enforced, by introducing financial and economic incentives and disincentives to:

(a) ensure that the costs of pollution activities are, as far as possible, borne by the polluter (internalised), and not passed on (externalised) to other water users who could be disadvantaged by the detrimental impacts of waste on water resources;
(b) encourage the minimisation of waste discharge; and
(c) promote efficient and effective water use.

The water resource management charges will be imposed when the majority of water use is registered or licensed and is, therefore, gradually being implemented.

Charges will be based on recovering costs of managing the total volume of water that may be allocated for use in each water management area. The water resource development charges will be phased in over time, with the target being to achieve full cost recovery. For irrigated agriculture on government water schemes this includes depreciation, operation and maintenance costs, although emerging farmers will be decreasingly subsidised over a five-year period.

Charges for achieving equitable and efficient allocation of water may be determined administratively by basing them on the opportunity cost of water. In areas where compulsory licensing has been completed, any remaining water may be allocated for use by public tender or auction and regulations to facilitate this will be introduced in due course.

However, it also stated that administratively determined charges, public tenders or auctions may not be needed in areas where trade in water use is permitted to take place. The NWA allows water users to engage in such transactions in terms of policy developed under Section 25 of the NWA, subject to finding the balance between the general public interest and the interests of water users who wish to trade. To this end, the extent of authorised water use, in terms of quantity and quality, which may be traded across water use sectors will have to be carefully considered and defined before such trading can be permitted. Accordingly, regulations will be introduced in due course, specifying the conditions under which trade will be permitted, after compulsory licensing has been completed. In the interim, applications for trade between user sectors will be considered on an *ad hoc* basis (DWAF, 2002: 71–76).

9.3.2.3 Participation in water management through local water organisations

Nineteen water management areas were delineated after country-wide consultations, for which catchment management agencies have to be formed. These agencies are statutory bodies that will be established by Government Notice, with jurisdiction in a defined water management area. A catchment management agency manages water resources, and co-ordinates the water-related activities of water users and other water management areas within the water management area. The initial functions include the important responsibility for the agency to develop a catchment management strategy which provides the framework for managing the water resources of an area. In particular, it must determine the principles according to which available water will be allocated among competing user groups. Guidelines for the financing of catchment management agencies have recently been published (Pegram & Palmer, 2001).

Water user associations are also defined in the NWA as water management organisations, but with more restricted objectives – both in scope and geographical extent – than those of catchment management agencies. They are, in effect, co-operative associations of individual water users who wish to undertake water-related activities at a local level for their mutual benefit. They operate in terms of a formal constitution as set out in guidelines prepared by the DWAF and are expected to be financially self-supporting from water use charges determined and made in terms of the pricing strategy, and payable by members.

Existing advisory committees on government irrigation schemes, irrigation boards, subterranean water control boards and water boards established for stock-watering purposes, must be transformed to become water user associations (DWAF, 2002: 80–83).

9.3.3 Policy initiatives

A public consultation process for the purpose of irrigation policy development started in 1996. Various Discussion Documents were published to inform participants (Water Research Commission, 1996; DWAF & NDA, 1996), and draft Policy Papers on irrigation and water use in agriculture were circulated for comment by the end of 1997. This culminated in the publication of the Green Paper on Agricultural Policy in South Africa, which includes a section on 'Conserving agricultural natural resources: Enhancing efficiency in irrigation' (NDA, 1998: 66–70).

This section states that irrigation policy is linked to water policy and that provisions in the NWA (1998) have implications for irrigation development. The overall objectives of agricultural policy reform are to create opportunities for smallholder and resource-poor farmers; to improve the efficiency and competitiveness of agriculture; and to utilise resources sustainably. The four principal challenges for irrigation policy are:

(a) rehabilitation of existing irrigation schemes, particularly smallholder schemes;
(b) determination of the development capacity for the expansion of irrigation;
(c) establishment of effective organisational structures for implementing policy; and
(d) increased efficiency of water use.

A study financed by the Water Research Commission showed that despite large investments, the performance of most smallholder farmer irrigation schemes in South Africa falls far short of expectations. Due to mainly financial constraints, provincial governments have withdrawn their support from a number of schemes and this has resulted in considerably reduced efficiency and, in some provinces, almost complete collapse of many irrigation schemes (Bembridge, 2000).

It appears that a lack of progress with policy implementation and/or a lack of co-ordination between government departments, both at national and provincial level, has led to a new policy drive. The first is the Strategic Plan for South African Agriculture (NDA, 2001) which provides a vision, the current realities and the sector challenges for South African agriculture. One of three core strategies proposed is sustainable resource management with regard to water use: mention is made of the enhancement of crop and livestock productivity; that farmer participation is a key success factor; that steps must be taken to overcome the threat of resource degradation; and that water conservation programmes will be focused on areas where there is a reasonable chance of success (NDA, 2001: 17).

The second policy drive is a draft strategy document on National Guidelines for Integrated Management of Agricultural Water Use (NDA, 2002), produced as the result of a number of national and provincial consultation workshops. This document emphasises that:

(a) co-ordination of existing policies rather than development of new policies is the point of departure;
(b) government can provide an enabling environment for entrepreneurial activities in order to reduce the dependency and uncompetitive water use by smallholders;
(c) agricultural water use includes irrigation, rainwater harvesting and conservation, and water for stock-keeping; and
(d) resource limited agriculture means agricultural production by previously disadvantaged individuals.

In view of previous policy failures, three key proposals will be briefly highlighted.

9.3.3.1 Revitalisation of smallholder irrigation schemes

On a large number of farmer settlement schemes, management functions for production and marketing were performed by parastatal corporations or government-appointed private agencies. Now, farmers must be enabled to manage their own holdings and the irrigation scheme through what is known as 'irrigation management transfer' (IMT) which essentially provides a pathway for mobility from smallholder to commercial farming. The requirements for success based on international experience, reformulated national policies, key strategy elements and enabling conditions for successful implementation of IMT have been documented, following a national policy workshop (Svendsen & Merrey, 2000). At least the following requirements must be met:

(a) farmers must be ensured of secure land tenure and water entitlements;
(b) replanning of production systems, capital intensive technology and high-risk cost structures has to be initiated;
(c) the profitability of farming has to increase;
(d) mechanisms to consolidate or write-off debt have to be considered;
(e) farmers have to be encouraged to participate in the establishment of water user associations;
(f) the rehabilitation of infrastructure should be preceded by community involvement and feasibility studies;
(g) support services driven by private enterprise have to be introduced; and

(h) further public investment in irrigation infrastructure has to be done according to accepted guidelines (NDA, 2002: 6–8).

The process of the revitalisation of smallholder irrigation schemes is most advanced in the Limpopo Province and it serves as a good example of the success which can be achieved through concerted efforts (Department of Agriculture, 2002).

9.3.3.2 Mainstreaming the marginalised

The revitalisation of irrigation schemes must be part of a broader rural development strategy to achieve the empowerment of resource-poor farmers. This process of empowerment requires enhanced access to:

(a) land, through land restitution, redistribution and tenure reform;
(b) water, through water use entitlements and infrastructure development;
(c) markets, through improved infrastructure and market information;
(d) credit, through innovative collateral mechanisms;
(e) knowledge, through extension services and on-farm training;
(f) appropriate technology; and
(g) gender sensitive intervention to remove the obstacles that women experience in obtaining land titles, credit, training, etc. (NDA, 2002: 11–14).

9.3.3.3 Improved service delivery

Since the announcement of the Presidential Imperative Programme on Integrated Rural Development in 2000, with its emphasis on poverty reduction, many government departments have launched projects in rural areas. This has often led to fragmentation or duplication of services with proposals made to improve good governance and effective funding on a national, provincial and local government level (NDA, 2002: 15–31).

It is clear that with implementation of the new irrigation policy, preference is given to support services for the upliftment of smallholder subsistence farmers. However, the current reality is that the major share of irrigation water use and the major contribution to agricultural production is by commercial farmers. The implications of these policies are that a carefully managed balance will have to be achieved between revitalising subsistence farming and maintaining the production capacity of commercial farming on existing irrigation schemes. For smallholder subsistence farming, the priorities are enhancement of management capacity and settlement of new farmers, after careful selection based on knowledge of farming. In the case of commercial farming, the priorities are increased production efficiency, which will necessarily have to be combined with land and water resource reclamation (Backeberg & Odendaal, 1998: 53–55).

9.4 KEY ISSUES

According to both the national water policy and the agricultural policy, different strategies are being implemented which vary from the general (for example, national water resource strategy) to the specific (for example, water demand strategy for agriculture). It is important that attention remains focused on the objectives of efficiency, equity and sustainability in relation to:

(a) the policy instruments which are used;
(b) the institutions which are the framework for decisions; and
(c) the people who are affected.

This will make it possible to evaluate whether implementation is successful or not.

9.4.1 Efficient water resource use

The expected growth in South Africa's population, as well as the higher income levels and increased per capita consumption will all lead to increased food demand. At the same time, competition will be experienced with mainly domestic and industrial users for the water that is currently used for irrigation. In South Africa increases in water use in agriculture are unlikely and the challenge is consequently to produce more food with the same or less water. There are technologies available to improve water use efficiencies, but the question is whether existing policy measures can create an enabling environment to increase efficiency (Backeberg & Odendaal, 1998: 55–60).

The success of any type of policy intervention depends on determining the incentives for the behaviour of water users and assessing the prospects for changing that behaviour. Water users will respond to policy measures if their own interests can be improved. These interests include those of households reducing their expenditure on water within the constraints of a budget, or those of firms producing more and earning higher incomes with the same or lower expenditure on water. The precautions that must be taken while achieving these interests involve preventing the deterioration of water quality, decreasing the threat of water-borne diseases, protecting biodiversity, and maintaining the ecological balances of river systems.

Policy measures must therefore enable water users to relate water use to cost and to compare water values with cost. The pre-conditions that must be satisfied are secure water entitlements, the measurement of water use as well as some form of volumetric charging. Essentially, water users must receive the message that there are costs attached to supplying water, that there are alternative uses for water and that water is a valuable resource. A combination of economic and regulatory policy instruments is thus needed to influence water use as well as to combine market and administrative processes to influence water allocation.

The proposed implementation of water use charges and licences as explained in the NWRS mostly meet these requirements, but there are three critical exceptions:

(a) Cost recovery charges are in most cases not volumetric; and accurate, reliable, low-cost water measurement is not presently practically feasible.
(b) The administrative charge based on opportunity cost is a pure tax and will be a disincentive for efficient transfers.
(c) The limited duration of water licences will create insecurity and the practical consequences for efficient water trading will very much depend on how administrative discretion is exercised to determine the licence period (Backeberg, 1997: 358–368).

With regard to water trading, it must be mentioned that empirical studies in irrigation areas along the Orange River (Armitage & Nieuwoudt, 1999) and the Crocodile River (Bate *et al.*, 1999) have shown that efficiency gains have been achieved through trading. The modelling of water values on irrigation schemes along the Great Fish and Sundays rivers (Conradie *et al.*, 2002) and the Berg River (Louw & Van Schalkwyk, 2002) also indicate the potential for efficient market transfers. The extent to which this benefit is realised will again largely depend on how the administrative process is managed so that transaction costs are kept low.

9.4.2 Equitable water resource allocation

A bi-modal or dualistic production structure still exists in South African irrigated agriculture. This consists of commercial farmers, producing for markets, farming alongside smallholders who are mostly subsistence farmers, producing predominantly for household consumption. In general, the majority of subsistence farmers are women, who are marginalised for various social, economic, and political reasons. Social and economic change can be accomplished by, firstly, a process of empowerment, and secondly, a process of integrating subsistence and commercial farming.

The empowerment of subsistence farmers in rural communities involves the following:

(a) training in literacy and numeracy;
(b) obtaining knowledge of farming;
(c) improving skills for management practices;
(d) gaining access to secure land and water rights;
(e) gaining access to finance and markets;
(f) alleviating and reducing poverty levels; and
(g) improving individual leadership capabilities for the local organisation of farming interests (Selener, 1997).

Integrating subsistence and commercial farming requires an understanding of what is common and what is different for each category or farming type. While acknowledging the perils of over-simplification, the main components and relationships can be illustrated in Figure 9.1. Given the available resources, finance, infrastructure and information, food products will be delivered by means of various production systems. Management decisions are taken within an institutional and organisational framework and trust is the central element in the interaction. Institutions include mainly contracts, property rights, conventions and authority (Hubbard, 1997: 239–249). This provides the incentives for entrepreneurs to identify new ideas and opportunities to utilise resources, as determined by objectives and the willingness to accept risk. A range of transactions are made in response to consumer demand in a particular market niche.

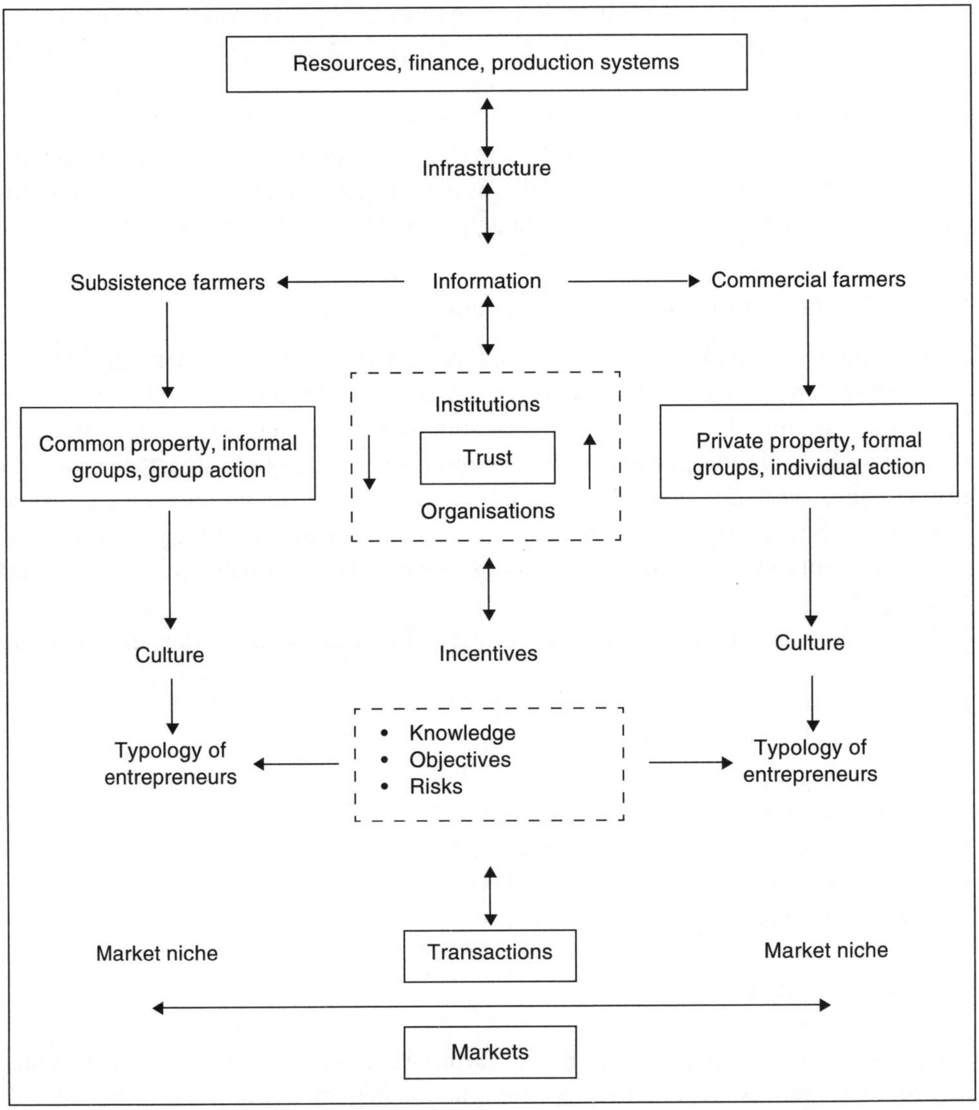

Figure 9.1: Main components and relationships for integrating subsistence and commercial farming.

In broad terms, this value-adding process is common to both subsistence and commercial farming, although the extent and quality of availability of components may differ. It can be argued that the critical differences are: firstly, how are decisions made, as influenced by the form of property rights and the way people organise themselves; and secondly, what is the accepted way of doing things, as influenced by culture and traditions. Following Fukuyama (1995), it is hypothesised that in order to achieve integration between subsistence and commercial farming, trust has to be formed between farmers and their representative organisations and cultural differences have to be respected.

It is clear that improved equitable access to water through secure entitlements, either by apportionment or by transfer, is a necessary but not sufficient condition for the empowerment of water users and the integration of subsistence and commercial farming. If a holistic approach as outlined above is not followed, it will continue to be a cause of policy failure in the sense that real benefits are not experienced by people on the ground.

9.4.3 Sustainable water resource development

Sustainability means that water resource use and development must be undertaken in such a way that benefits are available for present and future generations of South Africans. Analysis of pre-1994 irrigation policy in South Africa shows that emphasis was placed on either soil and water resources or dam and canal technology and development did not succeed when people with management capabilities, capital and markets were not available (Backeberg, 1994). This is an important lesson for smallholder settlement in the new political dispensation (Backeberg & Groenewald, 1995). People and investment in human capital must be placed at the centre of the development of water resources. People make decisions on resource use, the application of technology, marketing and financing within a particular set of institutions and their ability to make correct decisions depends on their knowledge levels and management capabilities (see Figure 9.2). The policy guidelines for the integrated management of agricultural water use succeed in this respect, by following a comprehensive approach to promote development.

For the development to be sustainable, human capital degradation and institutional decay must be prevented (Eicher, 1999). Of course, soil and water reclamation programmes must also be undertaken, but the training and education of subsistence farmers and the protection of secure land and water entitlements must receive priority. Based on the preceding policy review, low training levels and insecure property rights are the weakest links in the policy implementation process.

9.5 CONCLUSION

Since 1995, a consultative process of effective water and irrigation policy reform has taken place in South Africa. Strategies have been formulated and are currently being implemented to achieve efficient, equitable and sustainable conservation and

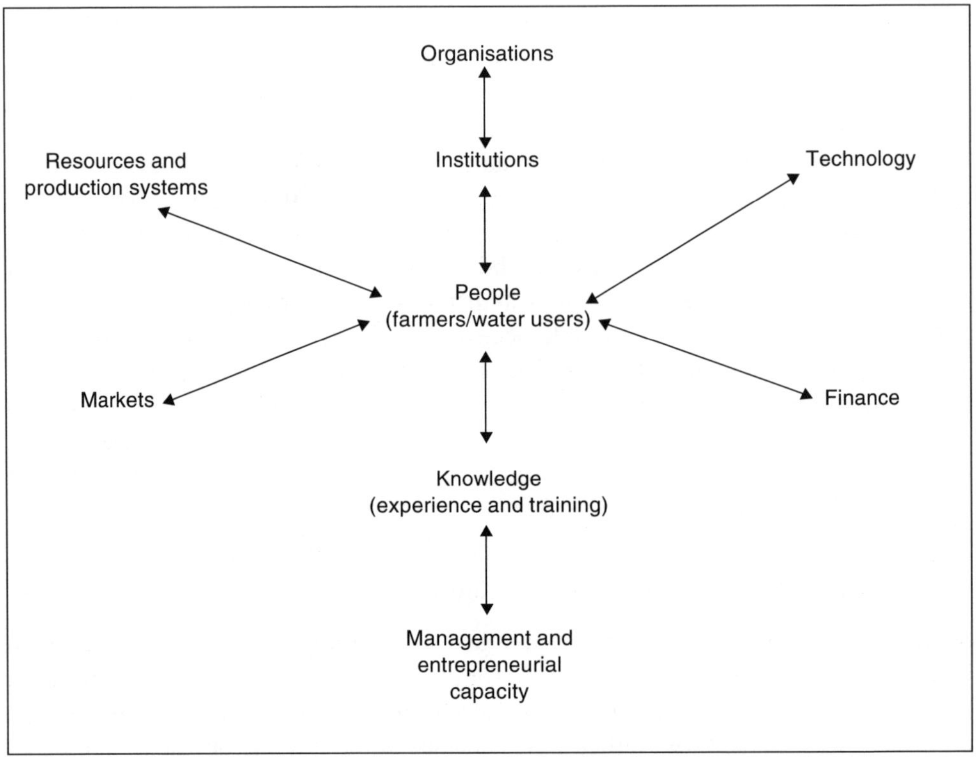

Figure 9.2: The essential requirements for people centred irrigation development.

the allocation of water resources in agriculture. Major tasks still lie ahead to devolve water management responsibilities to local organisations, licence water users and collect water charges. An evaluation of the progress made relative to the stated objectives of efficiency, equity and sustainability show that some achievements have been recorded, but that critical corrections are also required.

Based on the core values of, *inter alia*, individual responsibility and participation by members of society in a constitutional democracy, emphasis must be placed on the people who are using water. In many respects, this will require different approaches to be taken to water management. Although South Africa's water resources are limited, building more dams and depending on the government for a supply of water are no longer appropriate solutions. Nevertheless, government authorities must co-ordinate various implementation actions to ensure that enabling policies serve the interests of water users. The private interests of water users can best be achieved through interaction in the market process, while third party interests are best protected through generally accepted legislation. Achieving economic prosperity will depend on the entrepreneurial initiative of subsistence and commercial farmers to use all the resources available for agricultural production, including water.

REFERENCES

ARMITAGE, R.M. & NIEUWOUDT, W.L. (1999). An economic analysis of surface irrigation water rights transfers in selected areas of South Africa. Report No. 870/1/99, Water Research Commission, Pretoria.

BACKEBERG, G.R. (1994). The political economy of irrigation policy in South Africa. PhD thesis, University of Pretoria.

BACKEBERG, G.R. (1996). Constitutional change and natural resource policy reform in South Africa: conducive or obstructive to sustainable economic growth of agriculture? *Agrekon* 35(4): 160–169.

BACKEBERG, G.R. (1997). Water institutions, markets and decentralized resource management: prospects for innovative policy reforms in irrigated agriculture. *Agrekon* 36(4): 350–384.

BACKEBERG, G.R. & GROENEWALD, J.A. (1995). Lessons from the economic history of irrigation development for smallholder settlement in South Africa. *Agrekon* 34(4): 167–171.

BACKEBERG, G.R. & ODENDAAL, P.E. (1998). Water for agriculture: a future perspective. *FSSA Journal* Pretoria: Fertilizer Society of South Africa.

BATE, R., TREN, R. & MOONEY, L. (1999). An econometric and institutional economic analysis of water use in the Crocodile River Catchment, Mpumalanga Province, South Africa. Report No. 855/1/99, Water Research Commission, Pretoria.

BEMBRIDGE, T.J. (2000). Guidelines for rehabilitation of small-scale farmer irrigation schemes in South Africa. Report No. 891/1/00, Water Research Commission, Pretoria.

BENNIE, A.T.P., STRYDOM, M.G. & VREY, H.S. (1998). Use of computer models for agricultural water management on an ecotope level. Report No. TT 102/98, Water Research Commission, Pretoria.

CAMERON, R. (1991). Public policy. In Venter, A. & Johnston, A. (eds). *Politics: An Introduction for Southern African Students.* Cape Town: Oxford University Press.

CIRIACY-WANTRUP, S.V. (1985). Natural resource economics: selected papers. In Bishop, R.C. & Anderson, S.O. (eds). Boulder, Colorado: Westview Press Inc.

CONRADIE, B., NIEUWOUDT, W.L. & HOAG, D.L. (2002). The value of water in the Fish-Sundays Scheme of the Eastern Cape. Report No. 987/1/02, Water Research Commission, Pretoria.

CROSBY, C.T., DE LANGE, M., STIMIE, C.M. & VAN DER STOEP, I. (2000). A review of planning and design procedures applicable to small-scale farmer irrigation projects. Report No. 578/2/00, Water Research Commission, Pretoria.

DEPARTMENT OF AGRICULTURE. (2002). Revitalisation of smallholder irrigation schemes. Polokwane, Limpopo Province.

DEPARTMENT OF WATER AFFAIRS AND FORESTRY (DWAF). (1995). Future irrigation in South Africa. Discussion document, Pretoria.

DWAF. (1997a). *Overview of Water Resources Availability and Utilisation in South Africa.* Cape Town: CTP Book Printers (Pty) Ltd.

DWAF. (1997b). *White Paper on a National Water Policy for South Africa:* Pretoria: Government Printer.

DWAF. (1999). A pricing strategy for raw water charges. *Government Gazette* No. 1353. Pretoria.

DWAF. (2002). National Water Resource Strategy. (Proposed first edition). Pretoria.

DEPARTMENT OF WATER AFFAIRS AND FORESTRY & NATIONAL DEPARTMENT OF AGRICULTURE. (1996). Towards an irrigation policy for South Africa. Discussion document, Pretoria.

EASTON, D. (1979). *A System Analysis of Political Life.* Chicago, London: University of Chicago Press.

EICHER, C.K. (1999). Institutions and the African farmer. CIMMYT Economics Program. Third Distinguished Economist Lecture, Mexico City.

FUKUYAMA, F. (1995). *Trust: The Social Virtues and the Creation of Prosperity.* London: Hamish Hamilton.

HUBBARD, M. (1997). The 'new institutional economics' in agricultural development: insights and challenges. *Journal of Agricultural Economics* 48(2): 239–249.

LIEBENBERG, K. (1996). *South African Competitiveness Monitor: International Comparisons.* Volume 1. Pretoria: WEFA Group.

LOUW, D.B. & VAN SCHALKWYK, H.D. (2002). The development of a methodology to determine the true value of water and the impact of a potential water market on the efficient utilisation of water in the Berg River Basin. Report No. 943/1/02, Water Research Commission, Pretoria.

NATIONAL DEPARTMENT OF AGRICULTURE (NDA). (1996). Report on the role of agriculture in the South African economy. Pretoria.

NDA. (1998). Agricultural policy in South Africa. Discussion document, Pretoria.

NDA. (2001). *The Strategic Plan for South African Agriculture:* Pretoria: Government Printer.

NDA. (2002). National guidelines for integrated management of agricultural water use. Strategic document (Final Draft), Pretoria.

NELL, J.P. & VAN DEN BERG, H.M. (2001). *The Use of the South African Land-cover Project for National State of Irrigation Reporting.* Warmbad: Congress of the South African Irrigation Institute.

PEGRAM, G. & PALMER, I. (2001). Guidelines for financing catchment management agencies in South Africa. Report No. 1044/1/01, Water Research Commission, Pretoria.

RANDALL, A. (1987). *Resource Economics: An Economic Approach to Natural Resource and Environmental Policy.* (Second edition). New York: John Wiley and Sons Inc.

REPUBLIC OF SOUTH AFRICA. (1998). The National Water Act (Act No. 36 of 1998). *Government Gazette* No 19182. Cape Town.

SCOTNEY, D.M. & VAN DER MERWE, A.J. (1995). Irrigation: long-term viability of soil and water resources in South Africa. Proceedings of the South African Irrigation Symposium. Report No. TT71/95, Water Research Commission, Pretoria.

SELENER, D. (1997). *Participatory Action Research and Social Change.* Ithaca, NY: Cornell University.

SPIES, P.H. (ed.). (1996). *Agrifutura 1995/96.* Stellenbosch: US Printers.

SVENDSEN, M. & MERREY, D. (2000). Bridging the gap between large and small-scale irrigated agriculture: issues and opportunities. National Policy Workshop on Irrigation Management Transfer and Rehabilitation of Smallholder Irrigation Schemes, International Water Management Institute, Pretoria.

WATER RESEARCH COMMISSION. (1996). Policy proposal for irrigated agriculture in South Africa. Discussion document prepared by Backeberg, G.R., Bembridge, T.J., Bennie, A.T.P., Groenewald, J.A., Hammes, P.S., Pullen, R.A. & Thompson. H. Report No. KV96/96, Pretoria.

WATER RESEARCH COMMISSION (1999). *Agrimarket Survey.* Pretoria: Market Surveys and Statistical Analysis Consultants.

CHAPTER 10

Food Security in South Africa

A Case Study of Rural Livelihoods in the Eastern Cape

Gavin Fraser, Nomakhaya Monde and Wim van Averbeke

10.1 INTRODUCTION

According to Maxwell (2001), there are more than 200 definitions of food security. However, the World Bank's (1986) definition of food security has become widely accepted as the one that clearly defines the concept as 'access by all people at all times to enough food for an active, healthy life'. The main elements to consider when defining food security are: sufficiency, access, security and time. The World Bank (1986) describes food sufficiency as enough food to supply the energy needed for all family members to live healthy, active and productive lives. 'Access' refers to whether or not individuals and households have the means to acquire sufficient food, either via production or the earning of income to be exchanged for food. 'Security' encompasses the ability of households and individuals to resist crises that threaten their achieved level of food consumption and lastly, 'time' is concerned with secure access to sufficient food at all times.

For many years the South African government pursued an agricultural policy that emphasised food self-sufficiency (Van Zyl, 1994). Although this objective was largely realised, many people in South Africa are still food insecure. Estimates reveal that in excess of 50% of the population live below the poverty line (DBSA, 1993; Committee, 1992; ECI, 2002), with the largest contributors to this poverty gap being rural blacks domiciled in the former homelands (Simkins, 1991; ECI, 2002). According to Ngwane *et al.* (2001), two-thirds of South Africa's rural population live below the poverty line.

This chapter describes the poverty, rural livelihoods and food security strategies employed by households in two villages located in the Eastern Cape, one of South Africa's poorest provinces. The chapter opens with a discussion of methodology employed in the study. Section 10.3 analyses income sources and their relative contributions to household earnings, the extent and depth of poverty in each village, and the expenditure patterns of households. Section 10.4 focuses on livelihoods, household assets, and strategies adopted by the poor, ultra-poor and non-poor to secure food and to cope with crises. Finally, section 10.5 draws policy conclusions from the various observations made.

10.2 METHODOLOGY

The target areas of the study are the villages of Guquka and Koloni in the Victoria East and Middledrift districts of the former Ciskei homeland which now forms the central part of South Africa's Eastern Cape Province. Guquka is located in the upper sub-humid part of the Tyume River valley. This river forms the western boundary of the village. The annual rainfall in Guquka ranges from 650–800 mm (Van Averbeke *et al.*, 1998). Koloni is situated in the dry southern part of Middledrift District, and the mean annual rainfall is about 450 mm, making rain-fed crop production extremely risky.

Many studies into poverty make use of a poverty line (PL) to define who is poor, and who is not. A similar approach was taken in this study, using a PL developed for rural conditions in South Africa. Although the unit of analysis was a household, the unit of comparison was an adult equivalent which removes the effects of household size and composition on the cost of meeting basic needs. In 1999, the poverty line for rural areas of South Africa was R476.30 per adult equivalent (AE) per month (Carter & May, 1999). Households with an AE income of more than R476.30 per month were considered 'non-poor', those with an AE income less than R238.19 per month (half the poverty line) 'ultra-poor', and those with an AE falling between these two limits as 'poor'. Employing Sen's (1981) entitlement framework for the analysis of food security, income sources were assigned to appropriate entitlement categories that were then compared across the three income groups to reveal relationships between household income and underlying entitlements.

In-depth interviews on strategies employed by households to alleviate poverty and to acquire food were carried out with a subset of households sampled from each village (8 of the 68 households sampled at Guquka, and 8 of the 60 households sampled at Koloni). Purposive sampling was applied to select case studies with different entitlements and depths of poverty in order to reveal relationships between household livelihoods, income and diet, and to investigate ways in which households obtained food. Actual experiences of households in their search for food were recorded and analysed for evidence of food security or insecurity. How households organised the resources at their disposal, who in the household was involved in what economic activities, and how they arrived at the choices they made, were issues that were investigated. The interviews were conducted over a period of one month, during which time the researcher stayed in the villages to enrich the interviews by observing what people did, how they did it, when, and who actually did what, and by seeking explanations for these observations. The interviews were unstructured in that no questionnaire or list of questions was prepared with only the topics to be covered being decided in advance.

10.3 RURAL LIVELIHOODS AND FOOD SECURITY

There is ample evidence that the majority of rural households in the Eastern Cape live in poverty, but little is known as to how deep this poverty is. People are said to be living in poverty when they lack resources or money to satisfy their basic needs. According to Sen (1981), food security arises from entitlements, which fall into any one of four categories, namely:

(a) trade-based (income or food obtained through commodity exchanges);
(b) production-based (on-farm production of food);
(c) own-labour-based (income in cash or kind earned by selling labour); and
(d) inheritance and transfer-based (income in cash or kind given by others, including the state).

Each household tries to achieve food security by employing a network of entitlement relations involving exchanges with nature (on-farm production of food) or with others, including the state. These networks of exchange relations are called 'exchange mappings'. Food insecurity may arise from a shift in exchange mappings, a loss of possessions necessary to conduct exchanges, or a failure to effect exchanges (Omosa, 1998). The particular exchange mappings of a household are the products of social, historical, economical, and political processes that have occurred, and the way the household and its individual members have responded to these processes. Generally, rural areas are thought of as places where a surplus of food is produced for urban centres. However, in the former Ciskei, and in other former homelands of South Africa this is not the case, and it has not been for some time. When local food production fails to meet local demand, access to food depends on supply from other areas, a distribution network, and a range of exchanges.

10.3.1 Household income

Households in Guquka and Koloni obtained their income from multiple sources (see Table 10.1). In both villages, external sources of income made major contributions to household income, amounting to 81.7% in Koloni and 91% in Guquka. State transfers in the form of old age pensions followed by salaries and wages were the main sources of external income in Guquka with salaries and wages followed by pensions the most important sources in Koloni. Local economic activities contributed 18.3% to household income in Koloni, and only 9% in Guquka. Among internal sources of income, agriculture appeared to be an important source as shown in Table 10.1. Agricultural income as a percentage of all income is small. This may understate the importance of agriculture as livestock is their most important asset and the only asset that can be liquidated.

Assigning the various sources of income to entitlement categories as presented in Table 10.1 facilitated an evaluation of the importance of the various entitlements. 'Inheritance and transfers' and 'own labour' were the most important entitlements in both villages. In Koloni, these entitlements made significant contributions (43% and 40% respectively) to household income. By contrast, in Guquka, there were major differences between the two entitlements with 'inheritance and transfers' contributing 66% and 'own labour' 25% to income. The other two entitlements, 'own production' and 'trade' were much less important.

Table 10.1: Household income by source and entitlement category, August to October 1999 (n=128).

Sources of income	*Guquka*		*Koloni*	
	(R)	*%*	*(R)*	*%*
Inheritance and transfers				
Pensions	338.82	45.4	320.67	31.4
Disability	56.47	7.6	26.00	2.5
Cash remittances	59.14	7.9	55.42	5.3
In kind remittances	40.44	5.4	34.00	3.3
Total	494.87	66.3	436.09	42.5
Own labour				
Salaries and wages	185.22	24.8	399.63	39.2
Transporting	1.38	0.2	10.39	1.0
Total	186.60	25.0	410.02	40.2
Own production				
Crops in kind	11.79	1.6	14.39	1.4
Animals in kind	20.54	2.7	35.51	3.5
Total	32.33	4.3	49.90	4.9
Trade				
Hawking food	1.38	0.2	19.35	1.7
Hawking other	0.00	0.0	28.33	2.5
Shop/spaza	7.35	1.0	12.50	1.1
Crops cash	2.58	0.3	14.81	1.3
Animals cash	19.41	3.0	45.93	5.4
Make and sell	2.57	0.3	3.67	0.3
Total	33.29	4.8	124.59	12.3
Grand Total	746.69	100.0	1 020.71	100.0

10.3.2 Degree of poverty in Guquka and Koloni

The poverty status of households is shown in Table 10.2. Households in Guquka were found to be poorer, as 96% were below poverty line compared to 75% in Koloni. Amongst the three poverty classes shown in Table 10.2, the majority of households in both villages belonged to the ultra-poor category; less than 5% of households were non-poor in Guquka, while this group amounted to 25% in Koloni.

Table 10.2: Classification of households in Guquka and Koloni by monthly income per adult equivalent (n=128).

	Guquka		*Koloni*	
	No.	*%*	*No.*	*%*
Above poverty line	3	4	15	25
Below poverty line	65	96	45	75
Total	68	100	60	100
Non-poor (>R476.30/AE/month)	3	4	15	25
Poor	25	37	18	30
Ultra-poor (<R238.18/AE/month)	40	59	27	45
Total	68	100	60	100

Both villages have large poverty gaps as shown in Figure 10.1. The category '<20%' represents households close to the poverty line, i.e., households whose incomes were less than 20% below the poverty line. Likewise, the category '80–99%' represents households far below the poverty line. In both villages, the modal category was 60–79% under the poverty line. The information presented in Figure 10.1 clearly demonstrates that not all households identified as poor or ultra-poor suffered the same degree of deprivation. In 1997, the Eastern Cape Province had the highest share (29.5%) of the national poverty gap, and it was estimated that almost three-quarters of the provincial population was poor (Agricultural Policy Unit, 1997).

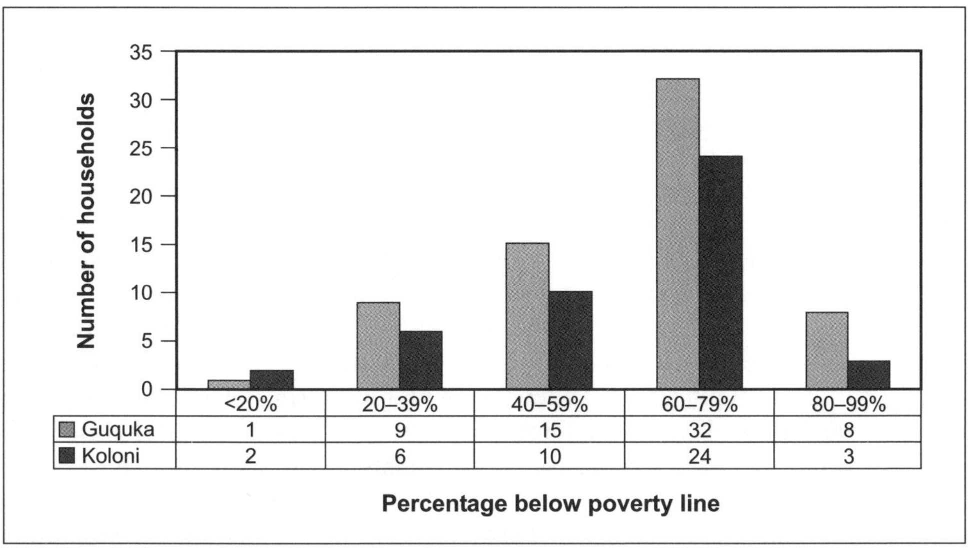

Figure 10.1: Depth of poverty in Guquka and Koloni expressed as the proportion of the poverty line attained by households (n=110).

Table 10.3 shows that the ultra-poor in both villages depend mainly on inheritances and transfers to secure food. In Guquka there was little differentiation in the overall exchange mappings among the three poverty categories and all relied mainly on entitlements based on inheritance and transfers to secure food. The difference in income among categories was essentially determined by the size of this entitlement. In Koloni, there were substantial differences in the exchange mappings of the three poverty categories. The poor differed from the ultra-poor mainly by adding own-labour, trade, and own production-based entitlements to that based on inheritance and transfers. Exchange mappings based mainly on own labour were characteristic of the non-poor in Koloni. These were secured mainly through employment in the civil service as professionals, a status achieved by means of educational qualifications.

Table 10.3: Entitlements underlying the poverty of households in three income groups (R/AE/month) at Guquka and Koloni in 1999.

Entitlement	*Guquka (n=68)*			*Koloni (n=60)*		
	Ultra-poor (n=40) (R)	*Poor (n=25) (R)*	*Non-poor (n=3) (R)*	*Ultra-poor (n=27) (R)*	*Poor (n=18) (R)*	*Non-poor (n=15) (R)*
Inheritance and transfers	101.61	201.34	612.64	134.59	147.12	190.80
Own labour	44.01	96.89	0.00	10.13	59.07	936.45
Trade	8.60	24.51	0.00	13.41	63.05	127.96
Own production	8.90	11.38	0.00	7.52	26.57	38.62
Total	163.12	334.12	612.64	165.65	295.81	1 293.83

10.3.3 Expenditure patterns of sample households

The expenditure patterns presented in Table 10.4 clearly demonstrate how an increase in income enabled households to divert increasingly more of their income to expenditure categories other than food. Apart from the non-poor in Koloni, food was the major item of expenditure for households in both villages. In both villages, saving by ultra-poor households was largely village-based, and had two main objectives: firstly, to ensure that the household could spend its income on non-food expenditure during one month of the year (*mgalelo*); and secondly, to cover at least part of the funeral costs in the event of bereavement (burial clubs). The poor and the non-poor also participated in community-based saving, but in addition they made deposits with formal financial institutions. The non-poor in Koloni invested a large proportion of their savings in insurance policies, which is characteristic for professionals in the civil service of South Africa.

Figure 10.2 shows the relationship between household income and the proportion of total expenditure spent on food. This analysis combined household data from both villages, excluding four outliers that had adult equivalent incomes in excess of

R1 000 per month. The graph shows that households spend proportionally less on food as AE income increases. However, the relationship is not linear, with the proportion of total expenditure spent on food initially dropping significantly as incomes increase, but levelling out at about 25% when AE income approached the poverty line of R476.30.

Table 10.4: Expenditure patterns for households in three income groups (R/AE/month) at Guquka and Koloni in 1999.

	Guquka (n=68)			*Koloni (n=60)*		
Entitlement category	*Ultra-poor (n=40) (R)*	*Poor (n=25) (R)*	*Non-poor (n=3) (R)*	*Ultra-poor (n=27) (R)*	*Poor (n=18) (R)*	*Non-poor (n=15) (R)*
Food	91.74	136.14	135.48	89.08	111.39	213.19
Cleaning	5.49	9.74	17.42	6.02	8.29	17.28
Energy	14.98	22.73	32.52	15.48	29.26	40.83
Medical and educational	12.04	22.07	31.61	12.10	18.81	51.46
Income-related	5.75	11.68	19.35	9.84	41.38	86.56
Semi-durable and durable goods	15.47	49.55	72.58	11.40	24.72	122.68
Leisure	12.93	52.80	41.61	13.95	30.14	94.35
Saving	15.14	22.80	90.97	16.15	35.67	294.95
Total	173.54	327.51	441.54	174.01	299.67	921.31

Note: The fact that expenditure patterns for households in some of the groups exceed the entitlements is a matter of the accuracy of the data collected from the households. Often income is understated and expenditure overstated – there is evidence of this in the socio-economic literature. The results must therefore be seen as an indication of the real picture rather than the exact figures.

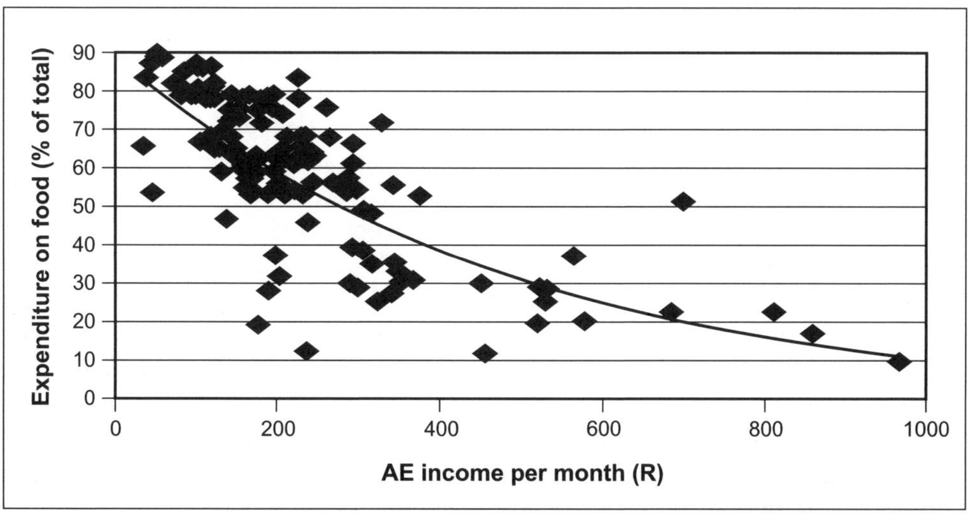

Figure 10.2: Household income (R/AE/month) versus proportion spent on food at Guquka and Koloni in 1999 (n=124).

10.4 COPING WITH POVERTY AND FOOD INSECURITY

Coping strategies, particularly those adopted by rural households in low-income, food insecure areas have become an important area of study (World Bank, 1986). Sijm (1997) defined coping strategies as the bundle of responses by vulnerable households to deal with situations of food insecurity, irrespective of the question of whether these situations are acute, severe, chronic or seasonal. According to Maxwell (2001), food insecurity is no longer seen simply as a failure of agriculture to produce sufficient food at national level, but instead as a failure of livelihoods to guarantee access to sufficient food at the household level. Africa's persistent vulnerability is arguably due as much to a failure of understanding as to a failure of interventions (Devereux, 2001). Despite many efforts, the proportion of population living below the poverty line in South Africa keeps on increasing. A lack of knowledge of how rural people cope with poverty and food insecurity has resulted in the failure of many support measures intended to alleviate poverty. Policy interventions need to support what poor people already do to protect their families and their livelihoods in times of stress.

10.4.1 Household livelihoods

According to Swift & Hamilton (2001), a livelihood comprises the capabilities, assets (including both material and social resources) and activities required for a means of living. A livelihood is sustainable when it can cope with and recover from stresses and shocks, maintain or enhance its capabilities and assets, while not undermining the natural resource base (Swift & Hamilton, 2001). Oshaug (1985), cited by Maxwell (2001), identified three kinds of households, namely, *enduring households*, which maintain household food security on a continuous basis; *resilient households*, which suffer shocks but recover quickly; and *fragile households*, which become increasingly insecure in response to shocks. In Guquka and Koloni, all of the poor and ultra-poor households shared histories characteristic of fragile households. Amongst the non-poor, only one sample household had been able to maintain food security on a continuous basis (an enduring household). The rest of the non-poor suffered shocks but had recovered, i.e., they were resilient households. Most of the ultra-poor households in both villages depended on social welfare grants as a reliable source of income. Those that did not have access to welfare grants survived on piecework, performing small, occasional tasks for other households in the village. This usually involves a daily search for work with uncertain and unreliable reward. Women younger than 50 years of age headed the majority of these households.

The poor also relied on social welfare grants, especially in the form of old age pensions. In addition to these grants, they had other sources of income – such as remittances – that were less reliable. The main livelihood of the non-poor households in Koloni was wage employment, either in a formal job such as teaching or through trading. In Guquka, the non-poor benefited substantially from welfare grants, reflecting a relatively large number of pensioners in these households.

10.4.2 Household assets

Physical assets can be used to construct livelihoods (Swift & Hamilton, 2001), but the ultra-poor households in both villages owned virtually no assets, apart from a garden plot. Garden tools had to be borrowed from neighbours and were not always available when needed. Nombulelo, a pieceworker, explained the problem as follows: 'We rely on my neighbours for the tools. If they are busy with them we have to wait. Sometimes, by the time we get them, it is already dry. We therefore end up not planting anything, or wait for rain again.'

Poor households also owned few assets. Unlike the ultra-poor, some possessed cropland but the majority of their fields lay fallow for reasons including liquidity problems that prevented poor households from buying inputs or hiring labour, tractors and draught power. A few of these households owned cattle and those that did had very small herds. Some of the poor owned garden hoes but very few possessed additional tools.

Most of the non-poor households had cropland, cattle and equipment. These households were actively involved in crop production (especially in Koloni) and many of them owned a variety of implements for use in either field or garden. Cattle were owned in relatively large numbers.

10.4.3 Food security strategies

The principal food security strategies adopted by households that had access to a reliable source of income, whether in the form of grants or salaries and wages, was to purchase food from urban markets (supermarkets in Alice and King William's Town). Respondents claimed that food prices were lower in urban markets than in local markets (village shops), and that urban markets were more appealing because of month-end specials. Ultra-poor households that did not have a reliable source of income relied mainly on piecework for wages or food parcels. If paid in cash, the wage is seldom large enough to warrant a trip to an urban market, and they have little choice but to purchase small quantities of relatively expensive food from local shops. In Nombulelo's words, 'the food in the local shops is expensive, but we do not have a choice, because the money we get at any one time is sometimes only enough for a taxi fare to Alice. It wouldn't be wise to spend all the money we have on transport, whilst we do not have anything to eat.'

10.4.4 Coping strategies adopted by poor and ultra-poor households

The poor and ultra-poor that had access to a reliable source of income were faced with transitory food security. They experienced food shortages towards the end of the month, i.e., a week or so before payday. Those without a reliable source of income were faced with chronic food insecurity and they never had enough food, with the search for food a daily struggle. Even so, some of the food security strategies employed by these groups were similar. The most common strategy was the diversification of income sources, i.e., earning small incomes from various sources including remittances (cash and goods sent home by migrant workers), buying and selling consumables, for example, paraffin, and raising pigs for slaughter. A portion of this meat is sold, usually to pay for other food.

Claiming against family relatives is another popular coping strategy employed by poor and ultra-poor households. Rural households typically belong to several social networks in which membership is often based on kinship and is maintained by exchanging gifts and favours. For example, a household might donate a portion of its garden produce to relatives, friends and neighbours. This 'gift' is a premium for insurance against future shortfalls, as donors are then entitled to seek help from their beneficiaries.

Another strategy is to cultivate a home garden, which has two advantages: reducing household expenditure and providing opportunities to acquire food through exchange: 'When I exchange produce, I'm always in need of the other product immediately. Sometimes I have tea but do not have sugar to put in the tea. So, I exchange a bowl of potatoes for sugar. If I decide to sell the same bowl of potatoes instead, I'd probably not get enough money to buy even the smallest packet of sugar in the shop. Besides people do not always have cash.'

Unfortunately, home gardening is faced with many threats, as during a bad year, for example, when there is drought, poor and ultra-poor households suffer low yields or no yields at all. This raises expenditure on food and limits opportunities to acquire food through exchange. Poor households are often forced to liquidate their limited assets at low prices after a bad season, reduce the number of meals to two or one per day, or gather traditional vegetables (*utyuthu*, *ihlaba*, etc.) from gardens and cropland.

For ultra-poor households the options are far more limited as they do not own marketable assets or crops to consume and exchange for other food products. Members of the poorest households – in some cases including school children – cope by engaging in piecework. Free meals appear to be an important reason for accepting piecework despite its uncertain cash rewards. Without these jobs, some household members would go hungry all day. In essence, ultra-poor households that rely on piecework are trapped in a daily struggle to find adequate food for survival. While poor households could reduce the number of meals per day, some ultra-poor households did not have anything to eat, not even one meal. The question is, how do children cope in such conditions? Nombulelo, a piece-worker with two children at primary school explained: 'Sometimes my children go to school without eating anything. But because they are both still at primary school, they have access to the feeding scheme offered at school.' The feeding scheme is sponsored by the Department of Education and supplies bread for pupils of certain ages. Only one of Nombulelo's children (Mandla) was eligible for the feeding scheme, while the other, Phumzile, relied on a share of his brother's allocation, a quarter of a loaf or a single slice when bread is in short supply. According to the teachers, bread was not always supplied to the school, and for Phumzile and Mandla this often meant going hungry the whole day.

10.4.5 Strategies adopted by non-poor households

Non-poor households were mainly comprised of people who were formally employed as civil servants earning regular salaries. The educational levels of these people were relatively high, and they possessed better skills in relation to the rest of the population. However, amongst the case studies there were two households, a farmer (in Koloni) and a rural entrepreneur (in Guquka), that were classified as non-poor even though their members were neither civil servants nor highly educated. The rural entrepreneur owned various assets including a truck, a light delivery vehicle (bakkie) and a tractor. All three vehicles were used to transport goods (for example, firewood and sand) that were sold to people in local villages, and he also had a garden plot, but did not have cropland. The farmer was endowed with cropland, a substantial herd of cattle, a comprehensive set of implements, and a garden plot. In addition to cattle, he owned a large number of sheep, goats, pigs and chickens. This household survived by producing and selling both crops and livestock. By comparison, a civil servant chosen for the case study owned a relatively well-furnished house with a garden plot, but did not own productive assets such as cropland or livestock.

Although these three households were non-poor, and did not experience hunger (they all had three meals per day), the farmer and the rural entrepreneur were classified as food insecure. A number of factors contributed to this state, including one related to their poor diet. The civil servant's diet was more balanced, and members of this family attained the required levels of energy, protein and fat in all seasons. The other two households also attained the required levels of energy, but their diet was deficient in protein and fat during winter and spring months. In fact, the majority of all non-poor sample households consumed energy in excess of recommended daily allowances, especially in summer and autumn when cereals were relatively abundant. The lack of variety in their diet was attributed to poor knowledge. The civil servant, currently employed as a teacher, had originally trained as a nurse. Her unemployed husband was actively involved in gardening and, acting on her advice, produced a wide range of vegetables in the garden all year round.

Another reason for food insecurity was that livelihoods were subject to many threats. For example, the rural entrepreneur used an old, unreliable truck to transport firewood and sand. In his words, this undermined his family's food security because 'my truck is old, and sometimes it breaks down, and I have to take it for repairs. When this happens, I use my bakkie instead, but then, with the bakkie, it takes two trips to make one load.' The household therefore ends up with less disposable income and a less healthy diet. Farming also faces many constraints, as noted earlier, but the farmer (Mbuzeli) was able to make a living and survive on agriculture alone. He identified drought and labour shortages as the biggest threats. Ploughing is done with animals, and this activity requires at least three people. Since ploughing is a man's task, Mbuzeli has to hire additional labour, reducing his cash income. Other households faced these problems and more, especially shortages of equipment such as yokes, planters, and ploughs. The only way out was to exchange and pool resources, but this was not always an ideal solution. As Mbulezi explained: 'It works but it is not a perfect solution. Sometimes we have to combine labour and equipment owned by five households. The big question is always whose field to start with. If you happen to be the last one, the soil may already be too dry.' To make ends meet, farmers usually concentrated on home gardens. Households in Koloni were at an advantage, because they had access to tap water, which they used to irrigate vegetables grown in gardens. As a result, some households, including Mbuzeli's, cultivated their gardens more than once a year. Apart from home gardening, they always had animals handy to sell to meet other needs, like their children's education.

10.5 CONCLUSION

Households in both villages earned their incomes from various sources with external sources being the most significant contribution to household income. Exchange mappings based on own-labour characterised non-poor households in Koloni, while inheritance and transfers characterised the non-poor in Guquka. For the poor and ultra-poor, exchange mappings were based on inheritance and transfers in both villages.

The findings also showed a high incidence of poverty with relatively more poor households in Guquka than Koloni. The depth of poverty was severe in both villages and the survey data revealed that household food security is under threat. Households are net importers of food and little food is produced in either village, necessitating that a large proportion of household income is spent on food – an important indicator of poverty. Poor households spent more than 50% of their income on food alone.

Improving food security by developing agriculture is a strategy that should be given priority because most rural households have some knowledge of farming and access to some agricultural resources. Although agricultural income is a small proportion of total income, livestock is important as a store of wealth and the only asset that can be liquidated. However, food security is too complex to be dealt with by the Department of Agriculture alone, and it requires better collaboration across sectors, disciplines and organisations. The challenge is to develop policies and programmes capable of supporting what poor people already do, which will require a multi-disciplinary approach with well co-ordinated and targeted interventions. The case studies in Guquka and Koloni portray the nature of rural livelihoods, how rural households are affected by poverty, and how they deal with it. The findings show that the poorest have few livelihood options, and that these are not likely to provide sustainable or secure outcomes. While conditions may appear to be very similar across rural areas, they are in fact quite different. Different types of poverty affect households in Guquka and Koloni, and policy-makers need to recognise these differences. Finally, inadequate purchasing power is not the only cause of food insecurity, as other factors like a lack of education pertaining to nutrition also contribute to the problem. Consequently, there is a need to invest in human capital, targeting women in particular.

REFERENCES

AGRICULTURAL POLICY UNIT. (1997). Food security policy for South Africa. Department of Agriculture and Land Affairs, Pretoria.

CARTER, M. & MAY, J. (1999). Poverty and class in rural South Africa. *World Development* 27(1): 1–20.

COMMITTEE. (1992). Report of the Committee of Inquiry into a Food and Nutrition Strategy for South Africa. Department of Agriculture, Pretoria

DEVELOPMENT BANK OF SOUTHERN AFRICA (DBSA). (1993). A macro economic policy model for human development in South Africa. Development Bank of South Africa, Midrand.

DEVEREUX, S. (2001). Famine in Africa. In Devereux, S. & Maxwell, S. (eds). *Food Security in Sub-Saharan Africa*. London: ITDG Publishers: 117–148.

EBONY CONSULTING INTERNATIONAL (ECI). (2002). The food security effects of the deregulation of agricultural marketing in South Africa. A report by ECI to the National Agricultural Marketing Council, Ebony Consulting International, Woodmead.

MAXWELL, S. (2001). The evolution of thinking about food security. In Devereux, S. & Maxwell, S. (eds). *Food Security in Sub-Saharan Africa*. London: ITDG Publishers: 13–32.

NGWANE, A., YADAVALLI, V. & STEFFENS, F. (2001). Poverty in South Africa. *Development Southern Africa* 18(2): 201–213.

OMOSA, M. (1998). *Reconceptualising Food Security: Interlocking Strategies, Unfolding Choices and Rural Livelihoods in Kisii District, Kenya*. The Netherlands: Market Surveys and Statistical Analysis Consultants.

SEN, A. (1981). *Poverty and Famines: An Essay on Entitlement and Deprivation*. New York: Oxford University Press.

SIJM, J. (1997). *Food Security and Policy Interventions in Sub-Saharan Africa: Lessons from the Past Two Decades*. Amsterdam: Thesis Publishers.

SIMKINS, C. (1991). The social, economic and political content: prospects and options up to 1995. Paper presented at the AEASA Symposium on Agriculture in the 1990s: An Economic-Political Perspective, February, Pretoria.

SWIFT, J. & HAMILTON, K. (2001). Household food and livelihood security. In Devereux, S. & Maxwell, S. (eds). *Food Security in Sub-Saharan Africa*. London: ITDG Publishers: 67–92.

VAN AVERBEKE, W., HARRIS, A.P., MBUTI, C. & BENNETT, J. (1998). Land reform research programme II: an analysis of land, livelihood, governance and infrastructure in two settlements in former Ciskei. ARDRI, University of Fort Hare, Alice.

VAN ZYL, J. (1994). Farm size efficiency, food security and market assisted rural land reform in South Africa. *Agrekon* 33(4): 156–164.

WORLD BANK. (1986). *Poverty and Hunger: Issues and Options for Food Security in Developing Countries*. Washington DC: World Bank.

CHAPTER 11

South African Agriculture and International Trade

André Jooste, Herman van Schalkwyk and Jan Groenewald

11.1 THE NATURE OF MODERN TRADE

World trade is now more interdependent than ever before. This interdependence is illustrated by an example from the cotton and fibre industry: cotton produced in Egypt finds its way to Guatamala where it is processed (spun) and exported to Italy for the manufacture of designer clothes, while advertising campaigns for these same clothes originate in the United States (Cordon, 2000).

Notwithstanding the theoretical advantages of free trade, the world of the late twentieth century has not been ready to acknowledge and implement a fully free market regime (Cordon, 2000). Instead, countries have opted for free trade blocs, of which there are various examples.

Globalisation renders it necessary for enterprises to change their approach to world trade. Production was once considered the cornerstone of the marketing chain, but consumers became more sophisticated and the emphasis moved toward marketing. However, marketing was not enough as many enterprises still lacked the ability to produce and deliver what consumers demanded. Many companies, moreover, were unable to source the products they needed to market successfully (Porter, 1998). This has resulted in emphasis on the organisation of the value chain and multi-national companies in the international trade environment excel with regard to size, profits and market share.

After the Second World War, governments around the world felt the need for a stable trade environment (USDA, 1994). This resulted in the establishment of various international institutions, including the General Agreement on Tariffs and Trade (GATT), which was later transformed into the World Trade Organisation (WTO). It was soon realised that in order to improve welfare through trade, products need to be competitive in markets outside domestic boundaries, and specialisation, technological innovation, structural change, etc. all accompanied the move towards becoming more competitive (Porter, 1998). These changes were, however, largely localised in certain parts of the world and Third World countries, in particular, could not keep up with their more industrialised trading partners.

Growth in the information technology sector changed this situation (Cordon, 2000). Today it is possible for anyone with access to the Internet to source information on any aspect of technology, trade opportunities, trade partners, etc. from anywhere in the world. Communication advanced to such a degree that deals could be clinched without the contracting parties having to meet each other. This, together with the factors already mentioned, resulted in the globalised trade environment that we know today.

South Africa's re-entry into global trade markets after 1994 was fast and uncompromising. Domestic enterprises had to adapt quickly from an inwardly-orientated trade environment to one that is part of the international arena. There is extra pressure on institutions that survived the legacies of a regulated environment and on market structures designed to cater for a controlled trade regime.

11.2 AGRICULTURAL TRADE REFORM IN SOUTH AFRICA AND THE WTO

The South African agricultural sector has emerged from a history of protection and subsidisation at various stages of the marketing chain. With a few exceptions, for example subtropical fruit, vegetables and flowers, the marketing of most agricultural products was regulated. This system, however, became obsolete in the face of both local and international pressure, and the result has been a process of deregulation.

Change came drastically with the repeal of earlier legislation and the promulgation of the Marketing of Agricultural Products Act of 1996. Deregulation was also accompanied by significant changes in the level of direct subsidies paid to farmers. In 1997 interest rate subsidies were discontinued, as were the direct export subsidies which farmers had previously received under the General Export Incentive Scheme (GEIS).

According to Vink *et al.,* (2002), South Africa's trade regime had been characterised by numerous quantitative restrictions, a multitude of tariff lines, a wide dispersion of tariffs, and various other forms of protection such as formulae, specific and *ad valorem* duties and surcharges. These restrictions, a maze of price controls and other regulations, often eliminated any foreign competition, but this state of affairs changed considerably after South Africa became a signatory of the Marrakesh Agreement.

According to Groenewald (2001), the failure of the previous South African government to participate in the agricultural discussions of the Uruguay Round resulted in a lack of negotiating skills and depth in terms of backroom competency to support negotiators. In consequence, the recommendations embedded in the South African WTO modalities pertaining to protection were mainly based on a general guideline that the customs duty should result in domestic production and consumption volumes more or less similar to those produced under import control (Kraamwinkel, 1998).

11.2.1 Main policy instruments

11.2.1.1 Import permits

The Import and Export Control Act of 1963 empowered the Minister of Trade and Industry to limit the importation of goods into South Africa through permits. The Department of Trade and Industry (DTI) is phasing out import permits in favour of tariffs (Cassim *et al.*, 2002).

11.2.1.2 Tariffs

In accordance with the Marrakech Agreement, liberalisation first involved tariff substitution for quantitative restrictions, followed by a reduction in diversity of *ad valorem* tariffs. While the absolute number of tariff lines was well below the 2004 target by the end of 1996, the range of tariffs was still comprehensive (Cassim *et al.*, 2002). Tariffs ranged between 0–131.5% in 1996; and the WTO-bound rates ranged between 0–597%.

Table 11.1 shows progress in tariff liberalisation for the whole economy. Tariff liberalisation proceeded faster prior to 1996, while a modest reduction in the number of tariff lines and in the maximum rates applied have occurred since (Lewis, 2001).

Table 11.1: Changes to the South African tariff structure.

Item	*All rates*		
	1990	*1996*	*1999*
Number of lines	12 500	8 250	7 743
Number of bands	200	49	47
Minimum rate (%)	0	0	0
Maximum rate (%)	1 389	61	55
Unweighted mean rate (%)	27.5	9.5	7.1

Source: Lewis (2001)

Table 11.2 shows average import-weighted tariffs for South Africa. Applied rates were considerably lower in 2000 than in 1996, particularly in the agricultural sector.

Table 11.2: Average import-weighted tariffs for South Africa (1996 vs 2000).

Category	*Applied rates (%)*	
	1996	*2000*
Agricultural products	9.23	1.4
Industrial products	11.40	8.6
Average	11.30	7.3

Source: Van Seventer (2001)

The effective rate of protection for a product is obtained by qualifying its gross protection from tariffs on output by the degree of taxation due to tariffs on its inputs. Agriculture is ranked 53 out of a total of 95 categories in the supply-use tables for South Africa in terms of its nominal rate of protection (the lower the rank, the lower its nominal rate of protection), and 58 for the effective rate of protection. South African agriculture's effective rate of protection is negative (–0.2%) (Cassim *et al.*, 2002), which means that the tariff regime is taxing South African agriculture, something which is confirmed by Jooste & Van Zyl (1999).

Despite reforms, the South African tariff schedule still remains complex and can create uncertainty for businesses that frequently import goods (USTR, 2000; Cassim *et al.*, 2002). Cassim *et al.*, (2002) argue that a highly dispersed and cumbersome tariff structure may mean uneven protection, and limit gains from openness. Trade reform with considerable tariff peaks may be less successful in encouraging exports, especially in sectors that rely on internationally competitive inputs.

11.2.1.3 Export subsidies

South Africa's total export subsidy outlay commitment was R842 million in 1995. The GEIS was terminated in July 1997 and export subsidies are now zero, except for a relatively small producer-financed subsidy on sugar (SAAU, 1999).

11.2.1.4 Domestic support

South Africa fully complies with the Green Box criteria as well as domestic support reduction commitments (the Amber Box) (SAAU, 1999).

11.2.1.5 Standards and regulations

Various government departments and parastatals set and police standards affecting the trade of agricultural products, most notably the National Department of Agriculture (NDA), the Department of Health, the Department of Trade and Industry (DTI), the South African Bureau of Standards (SABS), and the Council for Scientific and Industrial Research (CSIR). Most standards conform, or are in close conformity with, international standards (Jooste *et al.*, 2003), but there are various constraints and areas of inefficiencies that will have to be addressed in the near future.

11.2.2 Institutional framework for trade negotiations

The DTI is responsible for South Africa's multi- and bilateral trade relations and negotiations with the WTO, commissions established to deal with bilateral trade agreements, the United Nations Conference on Trade and Development (UNCTAD), the Organisation for Economic Co-operation and Development (OECD), the

Southern African Development Community (SADC) and the G77 (the latter is a coalition of UN-member Third World countries that aims to provide the means for the developing world to articulate and promote its collective economic interests). There is co-ordination with the Department of Agriculture and the Department of Foreign Affairs is usually involved with the political aspect of agreements.

The relevant government departments liaise with the business sector and organised labour on these matters in the National Economic Development and Labour Council (NEDLAC). A special sub-committee, Teselico, was established by NEDLAC to deal specifically with trade-related matters. Most of Teselico's work involves negotiation and deliberation on the contents of multi- and bilateral trade agreements.

11.3 TRENDS IN AGRICULTURAL TRADE

According to Table 11.3, the total real growth in exports was double the rate in expenditure on gross domestic product over the period 1995–2001, signalling a rise in export intensity. Real imports of goods and services rose equally strongly and the economy became more open.

Table 11.3: Expenditure on gross domestic product: real annual growth (%).

Description	*1990–1994*	*1995–2001*	*1990–2001*
Exports of goods and services	2.64	5.65	4.40
Imports of goods and services	5.11	4.66	4.85
Expenditure on gross domestic product	0.20	2.65	1.63

Source: Jooste *et al.* (2003)

The value of exports as a percentage of the sales of selected domestically produced goods, is an indicator of export intensity and is shown in Table 11.4. All these sectors experienced increases in export intensity.

Table 11.5 shows the import intensities for selected products. In sectors that historically had high import intensities, it did not rise significantly and in some sectors

Table 11.4: Exports/sales (%).

Sector	*1990*	*1995*	*2000*	*Change (1994–2000)*
Agriculture	11	12	14	18
Leather and leather products	17	34	46	37
Footwear	1	4	5	38
Textiles	11	12	18	46
Processed food	9	9	13	49
Wood and wood products	7	15	24	62
Beverages	2	6	13	122
Clothing, excluding footwear	4	6	18	182

Source: Jooste *et al.* (2003)

Table 11.5: Imports/sales (%).

Sector	*1990*	*1995*	*2000*	*Change (1994–2000)*
Leather and leather products	29	32	31	–1.4
Wood and wood products	9	12	13	7.4
Beverages	4	3	4	19.5
Processed food	5	8	10	20.5
Textiles	22	26	32	25.9
Agriculture	4	4	7	52.8
Footwear	10	20	63	219.1

Source: Jooste *et al.* (2003)

the import intensity increased from a low base, hence the high growth rates reported. These developments are nevertheless the logical outcomes of trade liberalisation, especially for an economy that operated in isolation for a protracted period.

Calculations from trade statistics show that the export of primary and processed food rose consistently from 7% of total exports in 1990 to 8.3% in 1995 and 9% in 2001. Increased exports of wine, fish, tobacco and tobacco products and edible fruits and nuts were mainly responsible for the rise in primary and processed foods' share of total exports. Some products with a small share of the total exports also improved their shares while losses were observed in vegetables, cereals, animal and vegetable fats and oils, as well as waste from the food and animal feed industries. The share of total exports for a number of products remained relatively stable over the past decade.

Table 11.6 shows the main export destinations for agricultural exports. The European Union (EU) remains the most important destination for agricultural exports (52.2% in 2001), with gains in export shares to Africa excluding SADC, North and South America and Oceania. The percentage of exports going to the Middle East is shrinking.

Table 11.7 shows the main import origins for agricultural imports. Imports of agricultural products tend to fluctuate severely with no set patterns. The value of imports from the EU remained relatively stable, while imports from the US have decreased considerably since 1995. Import gains from South America, Oceania, SADC and Africa excluding SADC are apparent.

Exports of agricultural products consistently exceed corresponding imports.

11.4 TRADE RELATIONS WITH OTHER AFRICAN COUNTRIES

South Africa is currently a signatory to various trade agreements other than the WTO. Most notable of these are the Southern African Customs Union (SACU), the Southern African Development Community (SADC) and the EU-SA Trade, Development and Co-operation Agreement (EU-SA TDCA). Other less significant trade agreements between South Africa and SADC countries are summarised in Table 11.8.

Table 11.6: Destination of agricultural exports.

Continent	*Country*	*Destination of exports*		
		1990	*1995*	*2001*
Africa (excluding SADC) Total		0.81%	1.59%	2.29%
SADC Total		4.32%	9.19%	10.41%
Eastern Europe Total		0.53%	0.66%	2.25%
Western Europe (excluding EU) Total		0.004%	0.03%	0.03%
	Germany	11.91%	7.48%	4.37%
	UK	21.44%	11.88%	12.78%
EU Total		59.86%	58.39%	52.20%
	China	0.46%	1.26%	1.52%
	Japan	17.75%	4.41%	11.38%
Far East Total		23.37%	14.78%	18.47%
Middle East Total		9.36%	9.48%	6.38%
	USA	0.20%	2.65%	5.09%
North America Total		1.46%	4.24%	6.76%
Oceania Total		0.05%	0.19%	0.57%
South America Total		0.21%	0.12%	0.52%
Other Total		0.03%	1.33%	0.12%
Total (R million)		2 629.00	3 805.60	8 874.50

Source: Jooste *et al.* (2003)

11.4.1 The Southern African Customs Union

The Southern African Customs Union (SACU) is a free trade area, with a common outside border, while recognising each state's sovereignty. Stoneham (1994) and Davies (1994) suggest that intra-SACU trade is of major significance both to South Africa and the other members of the SACU.

The first Customs Union Agreement for southern African countries was signed in 1910. After Botswana, Lesotho and Swaziland gained independence, a new agreement was signed in 1969, establishing the SACU (Cattaneo, 1990; Mayer & Zarenda, 1994. According to Mayer & Zarenda (1994), the SACU agreement was politically inspired at its inception and has remained highly politicised. This view is endorsed by Lundahl & Petersson (1991) who state that the continued involvement of South Africa in SACU was mainly politically and strategically motivated.

The Agreement consists of a Secret Memorandum of Understanding, 22 articles, and a General Memorandum of Understanding. The objectives of the SACU are:

Table 11.7: Origin of agricultural imports.

Continent	*Country*	*Origin of imports*		
		1990	*1995*	*2001*
Africa (excluding SADC) Total		3.18%	9.96%	4.56%
SADC Total		22.37%	13.56%	27.97%
Eastern Europe Total		0.95%	0.35%	1.00%
Western Europe (excluding EU) Total		0.00%	0.04%	0.04%
	Germany	5.37%	3.80%	1.64%
	UK	2.34%	1.50%	1.31%
EU Total		12.56%	10.43%	10.41%
	China	1.63%	4.44%	2.39%
	Japan	0.26%	0.13%	0.22%
Far East Total		15.60%	16.62%	14.11%
Middle East Total		11.13%	6.77%	8.18%
	USA	9.57%	22.04%	6.68%
Northern America Total		28.20%	28.58%	9.19%
Oceania Total		1.64%	2.95%	7.95%
South America Total		3.73%	10.14%	15.90%
Other Total		0.63%	0.59%	0.69%
Total (Rand million)		879.70	2 906.10	3 096.60

Source: Jooste *et al.* (2003)

Table 11.8: Bilateral Trade Agreements (BTAs) between South Africa and SADC countries.

Country	*Objective*	*Date*
Malawi	Duty free access into South Africa for all goods grown, produced or manufactured in Malawi except coffee, tea and sugar which subject to import control	1967
Mauritius	Exemption from surcharge on tea from Mauritius granted by the Republic of South Africa, under Section 48 of the customs and Excise Act of 1964	1964
Mozambique	Tariff concession by South Africa to Mozambique in form of a full rebate of the import surcharge or customs duty under section 75 of Customs and Excise Act of 1964	1964
Zimbabwe	Preferential rates of duty, rebates and quotas on certain products	1964

Source: Poonyth *et al.* (2002)

(a) to maintain 'free interchange of goods' between the member countries;
(b) to enforce 'the same tariffs and commercial regulations regarding goods imported from outside the common tariff area; and

(c) to ensure continual economic development of the common customs area as a whole, to ensure that arrangements encourage the development and economic diversification of less advanced members and afford to all parties equitable benefits from trade among themselves and with other countries.

The SACU currently comprises five member countries: South Africa, Botswana, Lesotho, Namibia, and Swaziland. The acronym BLNS is commonly used when referring to the latter four countries, while BLS is used when referring to the period before Namibia joined the SACU in June 1990.

A formula was established to share revenue obtained from import duties, etc. This formula was revised in 1977 to introduce a stabilisation factor to compensate BLNS countries for the following (Van Dijk, 1994; Stoneham, 1994; Davies, 1994):

(a) price effects of South African control measures on imports from the BLNS countries;
(b) increased prices arising from South Africa's industry protection through customs duties;
(c) time lags in the distribution of shares in the revenue pool;
(d) the negative influence on development in the BLNS countries due to their connection with a relatively developed country (South Africa); and
(e) loss of discretionary authority to pursue an independent national policy (for example, common external tariffs are set by the South African Board on Tariffs and Trade [BTT]).

There has been considerable controversy over the appropriateness of the stabilisation factor and its effects (Van Dijk, 1994; Mayer & Zarenda, 1994; Guma, 1990). Agreement was also reached that each country could maintain its own separate import quota on goods from outside the Common Customs Area (CCA) and that the contracting parties would set up a Customs Union Commission for the purposes of mutual consolidation on the working of the agreement (Hudson, 1981).

The SACU experienced many other problems. South Africa enjoys unquestioned supremacy within the region in terms of its transport infrastructure and, as a result, has strongly influenced trade in the region (Bryant, 1988). South Africa's transport permits policy controls market access, and can apply strict quality control.

The 1969 Agreement also grants South Africa the right to set external tariffs, excise duties and common sales taxes provided that the partners are consulted under certain circumstances (Van Dijk, 1994). In reality, South Africa's partners have lost the tools for guiding their economies (Lundahl & Petersson, 1991). Stoneham (1994) states that intra-customs union trade is conducted at prices well above world levels. Furthermore, the large trade surplus of South Africa with the BLNS members implies that through pricing, South Africa is a major beneficiary of the large transfers of income. Various other problems with the SACU were cited by other authors, for instance McFarland (1983), McCarthy (1985), Margo Commission (1987), Leith (1992), Maasdorp & Whiteside (1993), and Davies *et al.* (1993).

11.4.1.1 Renegotiations of the SACU

The SACU was renegotiated after 1994 and a new agreement was reached in 2002. The main provisions retained from the 1969 agreement are the following:

(a) free trade in locally produced goods;
(b) free movement of goods once cleared through customs;
(c) common external tariffs;
(d) common excise tariffs;
(e) infant industry protection for BLNS;
(f) no intra-SACU restrictions allowed;
(g) similar customs and excise legislation;
(h) import control where each member state has its own regulations; and
(i) freedom of transit and non-discrimination on transit duties.

There was also agreement on various new provisions that include the following:

(a) SACU will be an international juristic person;
(b) six new institutions are to be formed, namely, a Council of Ministers, Customs Union Commission, Secretariat, Tariff Board, Technical Liaison Committees (i.e., Agriculture, Customs Technical, Trade and Industry and Transport), and a Tribunal;
(c) efficient co-operation on customs issues, industrial development, competition issues, agriculture, unfair trade practices and dispute settlement; and
(d) a new revenue-sharing arrangement.

Issues that created problems under the 1969 agreement were thus largely addressed. The accepted provisions provide a proper framework for economic integration and not merely co-operation. The new institutional framework also provides a basis for greater autonomy in respect of economic development and other SACU countries can play a vital role in ensuring that South Africa's political and economic supremacy in the region is used positively to implement mutually beneficial policies. The new Tariff Board effectively removes South Africa's control over tariff-setting for the SACU as a whole, with the consequence that tariffs intended to protect South African manufacturers and primary producers that hold only marginal benefits for partner countries through the tariff revenue sharing formula will now not distort the benefits provided. Greater integration should also entail increased investment in sectors that hold a comparative advantage in BLNS countries.

11.4.2 Southern African Development Community

The attainment of political independence by the majority of southern African states by the 1970s set the stage for economic liberation and regional co-operation. One important initiative was the establishment of the Southern African Development

Co-ordination Conference (SADCC) in 1980 (SADC, 1994; Brown, 1989). The aims of the conference were to reduce dependence on South Africa, increase regional integration, and foster regional development (Mhone, 1991). The liberation of Namibia and democratisation of South Africa led to the establishment of the Southern African Development Community (SADC) which replaced the SADCC in 1992.

The establishment of the SADC meant a movement away from 'conference' to 'community'; a change in orientation. While the 'conference' was an *ad hoc* co-ordinating organisation, the 'community' is a legal entity. Unlike the SADCC, the SADC was to focus not only on co-ordination projects and infrastructure development, but to oversee integration of a number of areas including security, peace, democracy, conflict resolution, energy, trade promotion, investment promotion and information and culture (Tostensen, 1993). The SADC currently consists of Angola, Botswana, the Democratic Republic of the Congo (DRC), Lesotho, Malawi, Mauritius Mozambique, Namibia, Seychelles, South Africa, Swaziland, Tanzania, Zambia, and Zimbabwe.

The objectives of the SADC as stated in its treaty involve development, poverty alleviation, complementarity between national and regional strategies, evolution of common political values, productive employment, as well as sustainable and productive resource utilisation (SADC, 1992).

The SADC also targets the development of policies aimed at the progressive elimination of obstacles to free movement among member states of capital and labour, goods and services, and the people of the region generally (Blumberg & Wentzel, 1994).

The SADC approach is to address national priorities through regional action within the scope of its programme of action. Each member state was allocated a sector to co-ordinate.

11.4.2.1 Implementation of the SADC Trade Protocol

By December 1999 more than two-thirds of member states had ratified the SADC Protocol on Trade, which came into force on 25 January 2000. However, its implementation was delayed until 1 August 2001 to allow member states to deposit instruments of ratification. So far, ten countries have deposited these, while three others are in the process of doing so, and the Seychelles has not acceded to the Protocol (Poonyth *et al.*, 2002).

A very important feature of the SADC Trade Protocol is its intention to stimulate trade between member countries through the reduction of tariffs. The SADC incorporated the principle of asymmetry: the SACU will phase down in eight years (by 2008) while others will do so in 12 years (by 2012). Each non-SACU SADC country prepared two offers: one to South Africa and the other to the rest of the SADC. In order to compensate the less-developed SACU members (BLNS) that would liberalise their imports faster than non-SACU countries, the SACU offer was made conditional upon BLNS being able to maintain all the preferences they had enjoyed in trading with the non-SACU SADC states, for example, enhanced market access for selected products of export significance. Under the principle of asymmetry, there was a general understanding that the 'developing' non-SACU states (Mauritius and Zimbabwe) would mid-load their tariff reductions while the Least Developing Countries (LDCs) would backload.

Products have been classified into four categories for tariff dismantlement. Tariffs on Category A products will immediately be reduced to zero. Category B products will be gradually liberalised over eight years; while Category C products will be liberalised over 12 years. Offers for tariff dismantlement under Categories B and C for almost all countries cover over 85% of their SADC trade. Category E products are considered very sensitive and elimination of duty at the end of the 12 years is not envisaged. Flatters (2002) regards the confusingly differentiated tariff reduction schedules as a major source of concern, since he believes that not enough attention has been given to non-tariff barriers. Effective tariffs are often higher than nominal tariffs because of collection problems at border posts. An African Development Bank study of 1993 identified non-tariff barriers, as opposed to tariffs, as the major impediment to trade in the region (Mayer & Thomas, 1997). The Development Policy Research Unit (DPRU, 2001a, 2001b) mentions other concerns:

(a) poor infrastructure in some of the less developed member nations may divert industrial development and foreign investment to other areas;
(b) the Trade Protocol does not include the supply side measures needed to restructure and diversify industry in the region; and
(c) the Trade Protocol fails to link trade to investment and to industrial development, while there are no measures to compensate countries that may be de-industrialised in response to its implementation.

According to Flatters (2002), SADC rules of origin will hinder regional economic integration and will have no impact on the global competitiveness of regional producers: Maasdorp (2000) mentions six reasons for the SADC's limited progress:

(a) The general political climate has been unfavourable, a situation aggravated by events in Zimbabwe since 2000.
(b) The regional trade union organisation's campaign for unified labour legislation, employment standards and wage rates throughout southern Africa may, if successful, aggravate unemployment in a region where the supply of unskilled labour greatly exceeds its demand as it negates the very basis of competition. If the regional trade union organisation succeeds in its demands, the region will become more uncompetitive in the global trade environment, and any potential for economic growth will be reduced.
(c) The SADC has an inefficient organisational structure. A sectoral portfolio was allocated to each member country, with the portfolio by secondment from the host country, but this system has not been effective. Some allocations could

not be expected to succeed, given the host country's record in that sphere, but an attempt by South Africa and some other countries to reform the system was rejected.

(d) The transport and communication sectors are in a parlous state in spite of considerable funding by donors. The transport policies are badly skewed; the very heavy vehicles have led to the destruction of roads ill-designed to carry such loads, and state-owned railways have deteriorated. In the absence of large improvements in transport and communications, the SADC region is ill-equipped to compete with the outside world. In order to redress this situation, the region requires significant policy changes.

(e) Disease pandemics, particularly of HIV/AIDS and malaria, cause death rates to increase drastically, life expectancy to decrease, health care costs to skyrocket and productivity to plummet.

(f) The negotiation of trade agreements has taken longer than anticipated and some governments sign trade agreements without any serious knowledge on how free or freer trade will affect their economics. 'Free trade doesn't come cheap' (Da Pidada & De Lange, 2000). There are winners and losers (individual firms, subsectors or sectors) in every free trade agreement. Lobbies of vested interests put pressures on governments who have signed agreements, and as a result, there is virtually no trade integration in Africa outside the SACU.

11.4.3 SACU vs SADC vs other trade agreements

The very nature of the SADC and the SACU presents areas of potential conflict. The SACU is an economic entity within the SADC, with the latter representing a broader framework of economic integration in southern Africa. Furthermore, various SADC member states are also members of the Comesa Free Trade Area. Lewis (2001) states that as long as numerous SADC member states remain involved in conflicting regional arrangements or liberalisation commitments, progress may be limited.

A further objective of the SADC is the free movement of the region's capital, labour, goods, services and people. People from poorer areas will most probably resettle to more urban areas where they expect to satisfy their demand for jobs and wealth. South Africa currently has problems controlling illegal migrants and is not likely to agree on the free movement of people in the near future. If and when this objective is achieved, it will conflict directly with the SACU Agreement.

Sectoral co-operation is usually uncontroversial, but the political will to make it effective does not often exist. A lack of regional security and stability may also impede progress. All these factors have been present in the SADC. Trade integration, on the other hand, is a more complex and controversial process, becoming more so with integration envisaged (Maasdorp, 1995).

Overlapping raises a number of issues, namely:

(a) differing rules of origin create a maze of (sometimes conflicting) criteria for the granting of origin and this raises the cost of administering such agreements;

(b) membership of a number of different Regional Trade Agreements (RTAs) for a given country can result in the duplication and maintenance of tariff schedules for each preferential partner with different phase-out periods;
(c) the difficulty in sustaining membership in two or more overlapping plurilateral RTAs should lead to the consolidation of RTAs and rationalisation among participants who may be forced to align themselves with one or another regional grouping; and
(d) the dedication of resources to the negotiation and maintenance of agreements at the regional level may divert attention from negotiations at the multilateral level (WTO, 2002).

The more RTAs a country participates in, the more likely the potential for criss-crossing tariff concessions and rules of origin. In addition, the greater its participation in plurilateral RTAs (which have three or more members), or in RTAs with fewer members, the greater the potential for complexity, especially where each country maintains a distinct preferential regime with each partner. Harmonisation in the granting of tariff concessions and in rules of origin can significantly reduce the administrative burden associated with membership in multiple RTAs covering trade in goods. This scenario becomes even more complex if a country is involved in multiple RTAs covering trade in services, given that the harmonisation of regulations across a number of preferential agreements presents an even greater administrative challenge.

11.5 THE EU-SA TRADE, DEVELOPMENT AND CO-OPERATION AGREEMENT

Only qualified Lomé (now called Cotonou) Convention membership has been granted to South Africa, and consequently the EU has offered the country additional agreements. One of these is the Agreement on Trade, Development and Co-operation Agreement (TDCA), which includes a Free Trade Agreement (FTA). The formal negotiations on the FTA started in June 1995. Detailed proposals were presented in March 1996, the negotiations subsequently started, and the final agreement was accepted at the Berlin European Council in March 1999.

This FTA is the first EU agreement that includes agriculture on this scale and has no predecessor, which is one reason for its long negotiation period (Directorate General VIII, 1999). Normally, agriculture is included in a positive list, which comprises only included products, but this FTA also includes a negative list of excluded products. The aim of the FTA is to liberalise all trade substantially within ten years (a time limit set by the rules of the WTO) of concluding the agreement. In exceptional cases, this time limit could be extended to 12 years – and exception which is used in the case of the South African offer to the EU.

The tariff phase-down started on 1 January 2000 for all products, including wine and spirits, as well as the quotas agreed upon in this sector. The only exception is the 42 million-litre wine quota the EU granted to South Africa as part of the Davos political compromise, as negotiations on wine and spirits could not be completed on time and as a result of the EU's link to the TDCA Wine and Spirits (W&S) Agreement, especially the phasing out of the terms 'port' and 'sherry'.

Several principles form the general basis of the FTA (Directorate General VIII, 1998). Asymmetry is one with the EU to open its markets faster for South African products than *vice versa*, reflecting the developmental approach of this agreement. South Africa is allowed to open its market later to the EU to cater for the FTA within the framework of the SADC (Carim, 1997).

The TDCA is a 'mixed agreement', which has to be ratified by the European parliament, as well as the parliaments of member states. In addition to trade and financial aspects, it contains aspects for which member states have responsibility, such as political, economic and technical co-operation. The ratification process is expected to take up to three years. A transitional agreement was negotiated which will last until the end of 2003 in order to allow the trade and finance provisions to be implemented. Thereafter, if ratification is not complete, the agreement will be extended on an annual basis. Unless the problems regarding the W&S Agreement, as well as those in fisheries are overcome, southern EU countries in particular could delay ratification.

The non-trade benefits of the agreement should not be overlooked or underestimated, especially in the areas of development co-operation, science and technology, and the transfer of skills and knowledge. Another factor often ignored is South Africa's membership, albeit only qualified, of the Cotonou Convention. South Africa has the opportunity of influencing negotiations on so-called Economic Partnership Agreements (EPAs), especially where southern Africa is concerned (Meyer *et al.*, 2002).

11.5.1 Key features applicable to agriculture

The EU will remove duties on about 75% of South African farm exports over 10 years, covering about 1 800 tariff lines, including poultry, eggs, onions, mushrooms and garlic. South Africa will remove tariffs on about 82% of EU farm exports over 12 years, covering about 1 800 tariff lines.

To protect vulnerable and sensitive sectors on both sides and within the southern African region, a limited number of products have been excluded from the FTA's ambit. The EU will not offer tariff liberalisation for about 300 highly sensitive agricultural products, including beef, goat's meat, pork, cereals, sugar and dairy products. Forty other products, including canned fruit and juices, will be given tariff reductions within quotas. Furthermore, South Africa will not remove tariffs on 112 highly sensitive goods, including beef, cereals, sugar and dairy products and tariff liberalisation will begin only in year six of the agreement for products such as fruit and vegetables (South African Journal of International Affairs, 1999).

Some of the main products excluded, which will be periodically reviewed are shown in Table 11.9.

Table 11.9: Products excluded from the EU-SA FTA.

Main products excluded at EU side	*Main products excluded at SA side*
❑ beef	❑ beef
❑ sugar	❑ sugar
❑ some dairy (including milk, butter, whey)	❑ some dairy (including milk, butter)
❑ sweet corn	❑ sweet corn
❑ maize and maize products	❑ maize and maize products
❑ rice and rice products	❑ barley and barley products
❑ starches	❑ wheat and wheat products
❑ some cut flowers	❑ starches
❑ some fresh fruits (certain citrus, apples, pears, grapes, bananas)	❑ chocolate
❑ prepared tomatoes	❑ ice cream
❑ some prepared fruits and fruit juices	
❑ vermouth	
❑ ethyl alcohol	
❑ some fish	

11.5.2 Agricultural trade by member countries

The UK, the Netherlands and Belgium are the SACU's largest trading partners, representing almost 70% of its exports to the EU (see Figure 11.1). The UK remains the largest partner with 31.2% of exports in 2000, slightly down from 35.7% in 1994. Exports to Germany decreased significantly with that country losing its third position to the Netherlands. Finland, Greece, Luxemburg and Ireland were destinations receiving the least exports, with less than 1% being recorded in 2000.

On the import side, major trading partners are the UK, France, Germany and the Netherlands. Trend analysis of the percent change in the distribution of the value of trade between 1994 and 2000 indicates that, except for a small decrease in exports to the SACU from Belgium, France, Germany and Ireland, the situation remains unchanged (see Figure 11.2). The net effect is that the SACU's imports from the EU are decreasing in real terms. As in 1994, UK, France, Germany and the Netherlands retained their positions as the top origins of the SACU imports, accounting for 73% of SACU imports. Less than 1% of imports were sourced from Finland, Greece, Luxembourg and Sweden.

11.5.3 The EU-SA FTA's likely impact on the southern African region

Various studies have attempted to estimate the impact of EU-SA FTA. One such study conducted by the United Nations Conference on Trade and Development (UNCTAD), concluded that EU exports to South Africa would increase faster than South African exports to the EU (SACC, 2000).

Another study, focusing on fiscal policy and labour markets, was carried out by Britain's Institute for Development Studies and the Botswana Institute for

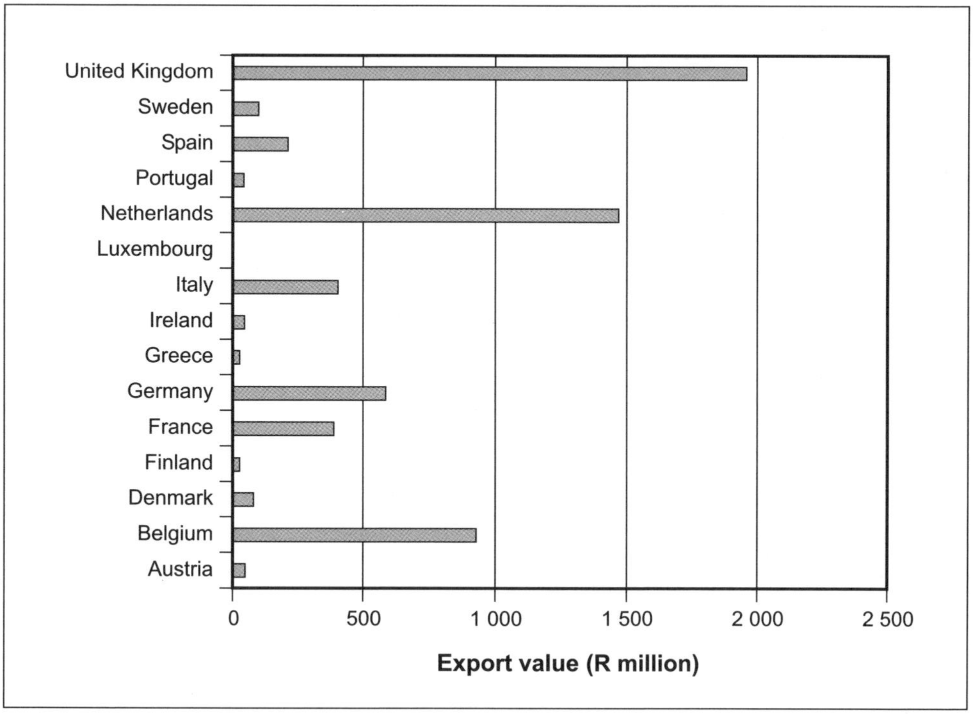

Figure 11.1: SACU exports to EU member states, 2000.
Source: C&E Database

Development Policy Analysis (SACC, 2000). This identified an expected decrease in customs union revenues by one-third or more as a potentially serious matter for Lesotho, Namibia and Swaziland. This may lead to changed SACU revenue-sharing arrangements, but it is certain that BLNS will receive less customs union revenue when the revenue pool declines. This will require significant adjustments in taxation, government expenditure, or both, in Lesotho, Namibia and Swaziland (but probably not in Botswana).

This study furthermore concludes that the overall direct impact of the EU-SA FTA upon BLNS is otherwise likely to be limited. Any significant impact, including the revenue impact, will probably be mainly felt through the South African economy. If the latter is boosted by the introduction of the EU-SA FTA, this could possibly matter as much, or more, to BLNS, than the relatively small impacts revealed by the economic analysis.

Penzhorn & Kirsten (1999) used a general equilibrium analysis to estimate impacts of the FTA on South African agriculture. Their results show that the FTA will yield positive welfare effects for both parties. A sharp increase in South African meat and dairy exports is predicted, although it must be kept in mind that the starting point for these products is very low. Gay & Nieuwoudt (1999) looked at

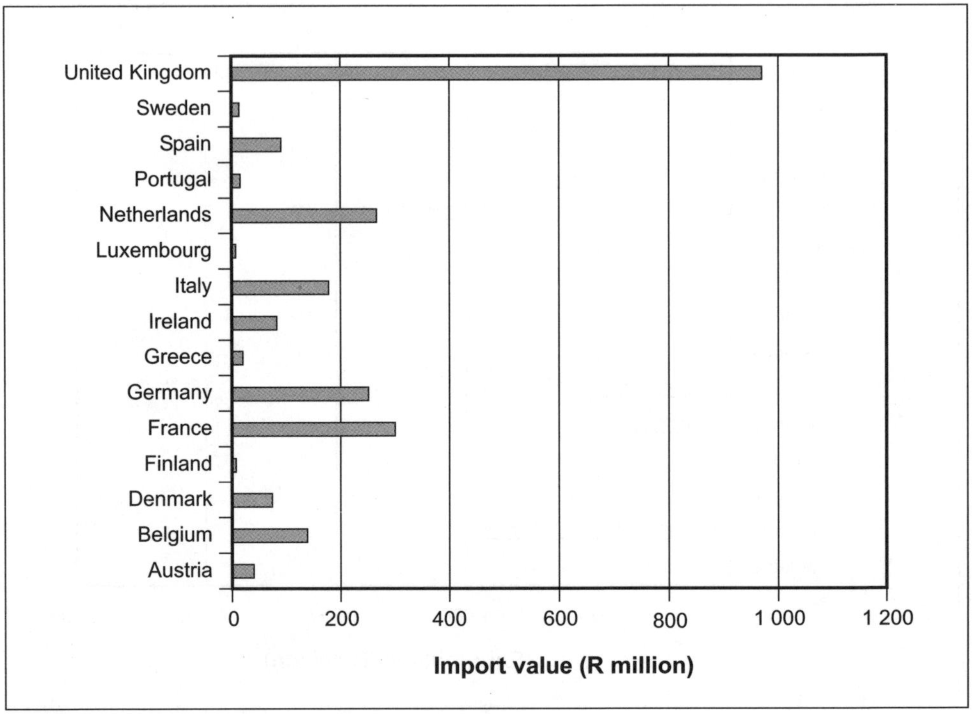

Figure 11.2: SACU imports from EU member countries, 2000.
Source: C&E Database

the effects of changes in tariff for major exports, which in most cases will be very small. This is caused either by their exclusion, or because the EU *ad valorem* tariff is already below 5%. For clementines and table grapes a tariff reduction of almost 20% was observed, which, if realised, will be beneficial for exporters in those industries.

11.5.4 Cotonou and Economic Partnership Agreements

On 9 April 2002, the European Commission (EC) adopted proposed directives for the negotiations of Economic Partnership Agreements (EPAs) with African, Caribbean and Pacific (ACP) countries and regions. The 'negotiating mandate' submitted to the European Council provides the framework within which the Commission will conduct negotiations with the ACP on new trading arrangements, as foreseen in the Cotonou Agreement. EPAs must constitute an instrument for development and will be based on the principles of fully WTO-compatible reciprocal free trade agreements with ACP regions, providing for differential and asymmetric treatment. The agreements should also include trade-related issues (investment, competition policy, environment, sanitary and phyto-sanitary measures, etc.), as well as other appropriate ACP flanking policies and EU support measures. The proposed mandate still raises several questions and concerns, of a general, as well as of a more specific nature, for example, issues pertaining to WTO compatibility, alternatives to EPAs, format of the negotiations, etc. (Bilal, 2002). Another issue that could influence the eventual stance of countries and the impacts of EPAs is the appropriate geographical configurations to be used (Tekere & Ndlela, 2002).

According to Tekere & Ndlela (2002), many ACP countries, including SADC countries, do not really have a choice other than to seriously consider EPAs. If they do not, they risk losing EU aid, preferential market access and the important commodity protocols afforded to them. The present evidence is that a SADC-EU EPA could cause significant revenue losses, resulting in substantial reductions of SADC member governments' revenues since the EU is their major trading partner. These potential losses amount to 7.7%, 10.3% and 9.87% for Botswana, South Africa and Swaziland, respectively.

Tekere & Ndlela (2002) also mention other negative impacts of the EPAs:

(a) eliminating the low tariffs on machinery and basic intermediates may help in kick-starting the region's ailing industrial sector;
(b) increased exports and decreased government revenues in the short to medium term as obsolete equipment is replaced by cheaper technologically advanced equipment; and
(c) the negative revenue reducing consequences of the EPAs will worsen the debt situation within the region, as economies struggle to honour their debt obligations from an ever-dwindling fiscus.

The proposed EPAs also have advantages in that they could provide the basis for maintaining and building on the current economic and trade relations with the EU, especially in respect of current market access preferences. SADC countries could moreover, through EPA negotiations, influence or demand certain concessions from the EU in return for opening up their markets to EU products (Tekere & Ndlela, 2002).

11.6 THE AFRICA GROWTH AND OPPORTUNITY ACT OF THE US

11.6.1 US trade with sub-Saharan Africa

The US has experienced a trade deficit with the sub-Saharan Africa region since 1996. If one excludes the trade in petroleum, the deficit increased from $1.6 billion in 1999 to $3.8 billion in 2000 (US ITC, 2001).

In 2000 US exports to sub-Saharan Africa consisted of the following:

(a) transportation equipment (32.4%);
(b) agricultural products (13.8%);
(c) chemicals and related products (12.8%);

(d) electronic products (12.6%);
(e) machinery (10%);
(f) special provisions (4.8%); and
(g) other (13.6%).

The largest export markets for the US in 2000 in the region were South Africa (50.9%), Nigeria (12.8%), Kenya (4.2%), Angola (4%), Ghana (3.2%) and Ethiopia (3%).

US imports mainly came from Nigeria (43.6%), South Africa (18.9%), Angola (15.1%), Gabon (9.2%), the DRC (2.3%), and the Ivory Coast (1.7%). Merchandise imports amounted to $22.2 billion, an increase of 61.5% from 1999. This increase was mainly due to higher imports of energy-related products and Nigeria was the main beneficiary.

11.6.2 Background of the African Growth and Opportunity Act

On 18 May 2000 President Clinton signed into law the Trade and Development Act of 2000, containing the African Growth and Opportunity Act (AGOA). The Act provides unprecedented opportunities and aims to promote:

(a) increased trade and investment between the US and sub-Saharan African countries by providing eligible African countries with unprecedented liberal access to the US market. Essentially all products of eligible countries will have quota free/duty free access to the almost 10 trillion dollar US market;
(b) economic development and reform in sub-Saharan Africa, moving across a wide range of industries, granting tangible benefits to entrepreneurs, farmers, and families; and
(c) increased access and opportunities for US investors and businesses in sub-Saharan Africa.

AGOA gives sub-Saharan African suppliers a competitive edge over suppliers in other countries with which the United States does not have a free trade agreement. The tariff rates on many of these products are high, and reducing them to zero for exporters in eligible sub-Saharan African countries will enable those suppliers to be more competitive.

11.6.3 Trade under AGOA

The first US imports from AGOA beneficiary countries occurred in January 2001. Imports amounted to $3 billion in the first half of 2001, and to 7.6 billion for the whole year. Overall, US exports to the 35 AGOA-eligible countries were up 19% from 2000, while imports were down 10%. If crude oil as well as precious metals and stones are excluded, imports from AGOA countries increased by 10.7%. Trade in automotive products grew rapidly in both directions, and US apparel imports surged 28% (Feldman, 2002). In respect of year-to-date imports by the US from AGOA-beneficiary countries, the principle suppliers were Nigeria (77%), Gabon (14%), South Africa (5%), Ghana (1%), Cameroon (1%), Kenya (1%) and others (1%) (US ITC, 2001).

AGOA-related imports to the US consisted primarily of energy related products, i.e., 94.2% of total imports. The remaining AGOA-related imports consisted of textiles and apparel, minerals and metals, agricultural products, transportation equipment and footwear. Table 11.10 provides a breakdown of the main imports by the US under the Generalised System of Preferences (GSP) and AGOA.

Table 11.10: US imports under the GSP and the AGOA, by sectors, year-to-date from January–March (AGOA-eligible countries only) (thousand dollars).

Imports	*GSP and AGOA 2001 ytd*	*GSP and AGOA 2002 ytd*	*GSP 2001 ytd*	*GSP 2002 ytd*	*AGOA 2001 ytd*	*AGOA 2002 ytd*
Agricultural products	25 002	39 582	23 678	18 460	1 324	21 122
Forest products	5 749	5 573	5 749	5 477	0	96
Chemicals and related products	29 092	32 769	29 092	30 594	0	2 175
Energy-related products	453 973	1 196 087	2	0	453 970	1 196 087
Textiles and apparel	5 678	186 418	907	783	4 771	185 635
Footwear	34	120	0	0	34	120
Minerals and metals	74 017	69 785	71 934	43 463	2 083	26 322
Machinery	13 872	4 188	13 872	4 182	0	6
Transportation equipment	13 112	120 494	12 354	11 791	758	108 703
Electronic products	3 409	2 250	3 409	2 249	0	1
Miscellaneous manufactures	8 461	6 417	8 461	6 287	0	129
Total	632 399	1 663 683	169 458	123 286	462 940	1 540 396

Source: Compiled by the US International Trade Commission from official statistics of the US Department of Commerce

11.6.4 Potential Free Trade Agreements with sub-Saharan Africa

Section 116 of AGOA calls for the negotiation of FTAs with interested countries in sub-Saharan Africa. In other regions of the world, such agreements have served as catalysts for increasing bilateral trade and investment and for promoting economic growth and development. Preparations are now being made in the US and the SACU countries for such negotiations.

The SACU is a leading destination for US foreign direct investment to the African continent, absorbing over $2.8 billion in 2000. An FTA with the SACU countries would eliminate relatively high tariffs on US exports, reduce regulatory barriers in key service sectors, and strengthen protection of intellectual property rights. It would also level the playing field in areas where US exporters are disadvantaged by South Africa's FTA with the EU and create an environment favouring expanded US commercial presence in the region. As WTO members and leading AGOA beneficiaries that have met the requirements of AGOA's textile and apparel provisions, the SACU countries have already seen the positive role that trade can play in alleviating poverty and creating commercial opportunities. By locking in their AGOA benefits, an FTA would provide the SACU members with the kind of guaranteed access to the US market that supports long-term investment, economic growth and development. An FTA would also support these countries' own economic reform efforts, further the regional integration programme, and lower the perceived risk of investing in southern Africa. The US has relatively few FTAs, and the quality and comprehensive nature of these agreements sends an important signal to the world (USTR, 2002).

11.6.5 Benefits from and problems with AGOA

11.6.5.1 Benefits

According to the USTR (2002) the following benefits have already been derived from the implementation of AGOA:

(a) The Agreement is already helping many countries to expand and diversify their exports and to build a manufacturing base to support long-term economic growth.
(b) The Agreement is encouraging substantial new investments, jobs, and trade in many of the poorest countries in the world. For example, 18 000 new jobs were generated in Namibia, 15 000 in Lesotho, and 11 000 in Swaziland.

11.6.5.2 Problems

AGOA implementation is challenged by a critical lack of institutional and human resource capacity in many sub-Saharan African countries. The number of people living on less than $1 a day in the region has expanded, and the region continues to rank last by nearly every socio-economic indicator.

11.7 CONCLUSION

South Africa's re-entry into the international trade arena was fast and uncompromising compared to many other countries in the world, including the EU and the US. The speed at which domestic producers and agri-business had to adapt received further momentum with the deregulation of the domestic agricultural sector. Producers and agri-business had no choice other than to find ways and means to compete on the international market against products from other countries. As a previous Minister of Trade and Industry stated: 'Producers need to become fit', a comment which underpins the government's direction in terms of international trade. The South African farmer has taken up this challenge and exports have seen remarkable growth during the past decade. Obviously the process has not been painless, as reflected in the growth of imports of agricultural products, although exports still exceed imports.

Many issues still cloud agriculture's performance. The lack of locally based quantitative capacity to measure the impact and consequences of trade integration is reason for concern, as is the general lack of human capacity as opposed to intellectual capacity, which hampers the ability of government to address trade-related issues. For example, only a few key personnel have to resign to create disorder in certain government departments that are involved in trade-related issues, a state of affairs which needs to be addressed. On the positive side, the capacity to research trade issues has grown tremendously outside government and is now mainly situated in parastatals, universities and the private sector. This creates the opportunity for government to interact with such institutions.

In the context of South Africa's future participation in the international trade arena, the following issues are of vital importance for agriculture:

(a) It should improve and strengthen its alliances with those countries or groupings, such as the Cairns Group, that continuously push for lower subsidies and other support measures in, for example, the EU and the US.

(b) South Africa could and should play a leading role within southern Africa in an effort to reduce poverty and enhance the competitiveness of its partners in the SADC. The vehicle in this regard should be the SACU as an efficient customs union could benefit the region tremendously. The SACU should strive for proper integration of economic and institutional activities as mere co-operation on trade-related issues would not be sufficient to reap the benefits of trade creation. A proper functioning SACU could serve as the model for the SADC that ought to be an FTA in 2008 and it could also serve as a model for NEPAD.

(c) South Africa should carefully consider its current drive to engage in too many FTAs. Quantitative backing to properly measure the impact of FTAs presents a real problem and it is also doubtful that the capacity exists to administer and police a multitude of trade agreements. In addition, South Africa needs to consider the likely impact of different FTAs on its neighbouring countries, especially the least-developed partners in the SADC. The pace at which South Africa opens up its market to countries such as the EU, US, Mercosur, Oceania and countries in Asia should keep track of the structural development programmes followed by countries in southern Africa. This would, on the one hand, force government to support and criticise those countries that are hampering progress and, on the other hand, would ensure regional development at a pace that also allows other countries to reap benefits from trade creation.

(d) The simplification of policy instruments that influence trade is needed, especially in the light of further FTAs.

REFERENCES

BILAL, S. (2002). Questions raised by the proposed EC Negotiating Mandate. *Trade Negotiations Insights* 1(2).

BLUMBERG, L. & WENTZEL, W. (1994). Trade relations with Southern Africa: a preliminary legal analysis. Development paper 29, Development Bank of Southern Africa.

BROWN, C.K. (1989). Regional co-operation in post-apartheid South Africa: the need for a uniform information system and data base in the SADCC region. In Odén, B. & Othman, H. (ed.). *Regional Co-operation in Southern Africa: A Post-apartheid Perspective*. Seminar Proceedings No. 22. Nordisha Afrikainstilutet, Uddevella.

BRYANT, C. (1988). *Poverty, Policy and Food Security in Southern Africa*. Boulder: Lynne Rienner.

CARIM, X. (1997). Multilateral trading, regional integration and the SA Development Community. *South African Journal of Economics* 65(3): 334–353.

CASSIM, R., ONYANGO, D. & VAN SEVENTER, D.E.N. (2002). The state of trade policy in South Africa. Unpublished document, Trade and Industrial Policy Strategies (TIPS).

CATTANEO, N. (1990). Piece of paper or paper of peace: the Southern African Customs Union Agreement. *International Affairs Bulletins* 14(1): 44–48.

CHACHOLAIDES, M. (1990). *International Economics*. Singapore: McGraw-Hill Book Co.

CORDON, R. (2000). From market research to national and enterprise export strategies. Seminar on Market Research, ITC, Geneva, 5 October.

DA PIDADA, L. & DE LANGE, R. (2000). Free trade doesn't come cheap. *Traders* 1.

DAVIES, R. (1994). The SACU: background and possible negotiating issues facing a democratic South African Government. In Sisulu, M., Nhosi, M., Setai, B. & Thomas, R.H. (eds). *Reconstituting and Democratising the Southern African Customs Union*. Report of the workshop held in Gaborone, Botswana, 6–8 March.

DAVIES, R., KEET, D. & NKUHLU, M. (1993). Reconstructing economic relations within the southern African region: issues and options for a democratic South Africa. Merg working paper no.1.

DIRECTORATE GENERAL VIII. (1998). The proposed Free Trade Area (FTA). http://europa.eu.int (Accessed 11/98).

DEVELOPMENT POLICY RESEARCH UNIT (DPRU). (2001a). What are the potential benefits and pitfalls of a free trade area in the southern African region. Policy brief No. 01/P8, Development Policy Research Unit, University of Cape Town.

DPRU. (2001b). Free trade areas under Comesa and SADC: what the literature says about the current situation. Policy brief No. 01/P15, Development Policy Research Unit, University of Cape Town.

FELDMAN, G. (2002). *US-African Trade Profile*. Washington DC: United States Department of Commerce.

FLATTERS, F. (2002). SADC rules of origin: impediments to regional integration and global competitiveness. *Southern African Update* 15.

GAY, S.H. & NIEUWOUDT, W.L. (1999). An analysis of the free trade agreement in agricultural products between South Africa and the European Union. *Proceedings of the 12th International Farm Management Congress*. Durban, 18–24 July: 267–278.

GROENEWALD, J.A. (2001). Impediments to delivery of policy for growth with equity in South African agriculture. Unpublished report, Africa Institute for Policy Analysis and Economic Integration (AIPA).

GUMA, X.P. (1990). The revised Southern African Customs Union Agreement: an appraisal. *South African Journal of Economics* 58(2): 63–77.

HUDSON, D.J. (1981). Botswana's membership of Southern African Customs Union. In Horrey, C. (ed.). *Papers on the Economy of Botswana*. London: Heinemann.

JOOSTE, A. & VAN ZYL, J. (1999). Regional agricultural trade and changing comparative advantage in South Africa. Technical Paper No. 94, SD Publication Series, USAID.

JOOSTE, A., KRUGER, E. & KOTZE, F. (2003). Standards and trade in South Africa: paving pathways for increased market access and competitiveness. In Wilson, J.S. & Abiola, V.O. (eds). *Standards and Global Trade: A Voice for Africa*. Washington DC: The World Bank.

JOOSTE, A., VILJOEN, L.J.C., MEYER, N., KASSIER, E. & TALJAARD, P.R. (2001). Competition issues in the agricultural sector. Unpublished research report, Competition Commission, Pretoria.

KRAAMWINKEL, A. (1998). Customs tariff policy. *Poultry Bulletin* October 1998: 459–463.

LEITH, J.C. (1992). The static welfare economics of a small developing country's membership in a customs union: Botswana in the Southern African Customs Union. *World Development* 20(7): 1021–1028.

LEWIS, J. (2001). Reform and opportunity: the changing role and patterns of trade in South Africa and SADC. Africa Region Working Paper Series, #14, World Bank.

LUNDAHL, M. & PETERSSON, L. (1991). *The Dependant Economy of Lesotho and the Southern African Customs Union*. Boulder: Westview Press.

MAASDORP, G. (1995). Southern African Development Community. Briefing document, SADC National Workshop, Helderfontein, 1–2 June.

MAASDORP, G. (2000). Can SADC free trade succeed? *Traders* 1.

MAASDORP, G. & WHITESIDE, A. (1993). Rethinking economic co-operation in southern Africa: trade and investment. Occasional paper, Konrad Adenauer Foundation, Johannesburg.

MARGO COMMISSION. (1987). *Republic of South Africa: Report of the Commission of Inquiry into the Tax Structure of the Republic of South Africa*. Pretoria: Government Printer: WPC-88.

MAYER, M.J. & ZARENDA, J. (1994). The Southern African Customs Union: a review of costs and benefits. Policy Working Paper 19, Development Bank of Southern Africa, Midrand.

MAYER, M.J. & THOMAS, R.H. (1997). Trade integration in the Southern African Development Community: prospects and problems. *Development Southern Africa* 14(3).

MCCARTHY, C.L. (1985). The Southern African Customs Union. A report prepared for the Central Economic Advisory Service, mimeo.

MCFARLAND, E.L. (1983). Benefits to the RSA of her exports to the BLS countries. In: Oomen M.A. (ed.), *Botswana's Economy since Independence*. New Delhi, Tata: McGraw Hill.

MEYER, N., JOOSTE, A. & VAN SCHALKWYK, H.D. (2002). EU-SA TDCA study: agricultural component. Unpublished research report, Fridge.

MHONE, K. (1991). Law as a factor for regional integration: SADCC – problems and prospects. *Comparative and International Law Journal of Southern Africa* 3: 378–385.

PENZHORN, N. & KIRSTEN, J.F. (1999). The impact of the EU FTA on South African agriculture: a general equilibrium analysis. *Agrekon* 38: 4.

POONYTH, D., ESTERHUIZEN, D., NGQANGWENI, S. & KIRSTEN, J.F. (2002). Trade policies and agricultural trade in the SADC region: challenges and implications: South Africa country study. Unpublished research report, University of Pretoria.

PORTER, M.E. (1998). *Competitive Advantage: Creating and Sustaining Superior Performance*. New York: The Free Press.

SOUTH AFRICAN AGRICULTURAL UNION (SAAU). (1999). Negotiations on agriculture during the next round of the World Trade Organisation (WTO). Unpublished document, South African Agricultural Union, Pretoria.

SOUTH AFRICAN COUNCIL OF CHURCHES (SACC). (2000). The EU-SA Free Trade Agreement: implications for Post-Lomé trade relations. South African Council of Churches Public Policy Update, Cape Town.

SOUTH AFRICAN JOURNAL OF INTERNATIONAL AFFAIRS. (1999). *Summary of the SA-EU Trade, Development and Co-operation Agreement in Brief* 6(2): 237–241.

SOUTHERN AFRICAN DEVELOPMENT COMMUNITY (SADC). (1992). Declaration Treaty and Protocol of the Southern African Development Community. A declaration by the heads of state or government of southern African states, Windhoek, Republic of Namibia.

SADC. (1994). Southern African Development Community. SADC secretariat, Gaborone, Botswana.

SOUTHERN AFRICAN CUSTOMS UNION AGREEMENT. (1969). Customs Union Agreement between the Governments of the Republic of South Africa, the Republic of Botswana, the Kingdom of Lesotho and the Kingdom of Swaziland. *Goverment Gazette*, Pretoria.
STONEHAM, J. (1994). Botswana country paper: perspectives on the history of a current situation in SACU. In Sisulu, M., Nhosi, M., Setai, B. & Thomas, R.H. (eds). *Reconstituting and Democratising the Southern African Customs Union.* Report of the workshop held in Gaborone, Botswana, 6–8 March.
TEKERE, M. & NDLELA, D. (2002). Impact assessment of economic partnership agreements on the Southern African Development Community (SADC) and preliminary adjustment scenarios. Draft report, Trade & Development Studies Centre – Trust, Harare.
TOSTENSEN, A. (1993). What role for SADC(C) in the post-apartheid era? In Odén, B. & Othman, H. (eds). *Regional Co-operation in Southern Africa: A Post-apartheid Perspective.* Seminar Proceedings No. 22, Nordisha Afrikainstilutet, Uddevella.
TURNER, B. (1971). A fresh start for the Southern African Customs Union. *African Affairs* 70(280): 269–276.
UNITED STATES DEPARTMENT OF AGRICULTURE (USDA). (1994). *Agricultural Provisions of the Uruguay Round.* Washington DC: US Department of Agriculture.
UNITED STATES TRADE REPRESENTATIVE (USTR). (2000). *Foreign Trade Barriers: South Africa.* Office of the United States Trade Representative.
US ITC. (2001). US trade and investment with sub-Saharan Africa. 2nd Annual Report, Investigation number 332-415, US ITC publication number 3476, United States International Trade Commission, Washington DC.
USTR. (2002). 2002 comprehensive report on US trade and investment policy toward sub-Saharan Africa and implementation of the African Growth and Opportunity Act. A report submitted by the President of the United States to the United States Congress prepared by the Office of the United States Trade Representative, the second of eight annual reports.
VAN DIJK, G. (1994). Customs unions. Internal information document, Department of Agriculture, Pretoria.
VINK, N., KIRSTEN, J.F. & TREGURTHA, N. (2002). South Africa's changing agricultural trade regime. Unpublished article, University of Stellenbosch.
VAN SEVENTER, D.E.N. (2001). Note on the structure of the South African Tariff Schedule. Paper presented at the TIPS Annual Forum, 10–12 September.
WORLD TRADE ORGANISATION (WTO). (2002). Regional trade integration under transformation. Preliminary draft prepared for the seminar on Regionalism and the WTO, WTO Secretariat, Geneva, 26 April.

CHAPTER 12

Status of Agricultural Research and Development Policy in South Africa

Frikkie Liebenberg and Johann Kirsten

12.1 INTRODUCTION

This chapter looks at the history of the agricultural research and development (R&D) system in South Africa against the background of structural, institutional and political changes and highlights the changes in this system during the past 20 years. South Africa's unique history and the successful transition to democratic rule make it necessary to place the current agricultural research dispensation in its historical context.

Public sector finance still remains the most important source of agricultural R&D funding in South Africa, but has come under severe pressure in recent years. The period since 1998 saw an increase in the contributions by producer organisations and international donors to the funding of agricultural research, with universities playing an increasingly important role in providing research. The decrease in core government funding as well as changes in leadership and management styles have led to an exodus of top researchers from the main research provider – the Agricultural Research Council (ARC). The threat of the demise of the agricultural research system has now led to an initiative to co-ordinate the funding and provision of agricultural research in South Africa through a recently established National Agricultural Research Forum (NARF).[1]

Section 12.2 of this chapter outlines the national Science R&D policy within its historical framework. Sections 12.3 and 12.4 provide a detailed account of the agricultural R&D policy by describing the institutional structure, sources of funding and the providers of agricultural R&D. Section 12.5 concludes the chapter by highlighting the major lessons, and summarising the debate about a future and more sustainable national agricultural research system in South Africa.

[1] On 23 May 2002 the first Steering Committee of the NARF was elected by the stakeholders.

12.2 NATIONAL SCIENCE R&D POLICY AND FUNDING TRENDS IN SOUTH AFRICA

As in the rest of the economy, South Africa's science, engineering and technology institutions have experienced momentous policy changes since the early 1990s. Prior to 1994, these institutions were funded under a policy of framework autonomy. The Department of Arts, Culture, Science and Technology (DACST) was established in 1994 from the relevant elements of the Department of National Education. The creation of the DACST has introduced a period of rapid changes in Science and Technology policy in South Africa. Two prominent initiatives taken by the DACST since its inception were the formulation of a White Paper on Science and Technology (completed in 1996), and the establishment of the National Advisory Council on Innovation (Act approved in 1997). The new policy places a strong emphasis on innovation and hence the direction of research resource allocation (Marais, 2000). The Science and Technology branch of DACST took over the administration of an annual budget of about R1.4 billion allocated under the Science Vote. Following the White Paper on Science and Technology, the principle of baseline funding according to a base formula was replaced in 1994, and science, engineering and technology institutions now receive their core funding based on a parliamentary grant allocated on a competitive basis.

Other sources of funding available to science, engineering and technology institutions are the Innovation Fund and the National Research Foundation (NRF) which was established in 1998. The purpose of the Innovation Fund is to encourage and enable 'long-term extensive innovation projects in the higher education sector, science, engineering and technology institutions, civil society and the private sector' (DACST, 1996). The NRF is mandated to ensure the support of research and research capacity-building within the higher education sector and other research institutions. In both cases, funds are allocated on a competitive basis. It is envisaged that the Innovation Fund will grow to about 20% of the Science Vote, forming a strong mechanism to reallocate resources within the national system of innovation. Table 12.1 indicates the Science Vote earmarked to the different science councils for the financial years 1996–97 and 1999–2000. It shows the shift away from funding agricultural and human sciences towards funding the Medical Research Council and the Council for Scientific and Industrial Research (CSIR).

In 1999–2000 the Science Vote also included R78.3 million for national facilities such as the National Accelerator Centre, and a further R146 million for other programmes such as the Innovation Fund (R75 million) and special investigations (R33 million). The growth in the Innovation Fund was paid for from the institutional funding of the science councils, with dire consequences for sustained capacity development. The priority setting criteria of the Innovation Fund are not favourable to primary agricultural research as they have a focus on third generation biotechnology, etc.

In 2002 a new National R&D Strategy was published, accompanied by a change in the structure of DACST (DST, 2002). A new Department of Science and

Table 12.1: Allocation of the Science Vote in South Africa (R'000) in the 1996–97 and 1999–2000 financial years.

Science Council	*1996–97 budget*	*1999–2000 budget*	*Nominal change*
Agriculture (ARC)	319 096	279 237	–12%
Human Sciences (HSRC)	87 632	64 419	–26%
National Research Foundation (FRD/NRF)	138 121	162 000	17%
Medical Sciences (MRC)	57 912	79 566	37%
Minerals (CMT)	82 767	81 773	–1%
GeoScience	63 555	63 794	0%
CSIR	274 355	310 649	13%
Bureau of Standards (SABS)	45 929	73 724	61%
Total	1 069 367	1 115 162	4%

Source: DACST (2001). 'The Science Vote'. http://www.dacst.gov.za (Accessed 26/11/01)

Technology (DST) was established from the relevant elements of DACST. In the current 2003–04 financial year the DST is introducing a number of new competitive funds that focus on the following key technology missions (figures in brackets indicative of the intended programme budget):

(a) technology and innovation for poverty reduction (R150 million);
(b) technology and innovation for advanced manufacturing and logistics (R125 million);
(c) technology and knowledge to leverage resource-based industries (R90 million); and
(d) new technology platforms for South Africa (R300 million):
  (i) information and communication technology; and
  (ii) biotechnology.

12.3 PROVISION AND FINANCING OF AGRICULTURAL R&D IN SOUTH AFRICA

12.3.1 Institutional evolution of the South African National Agricultural Research System

The present structure of the South African National Agricultural Research System (NARS) consists of the following components:

(a) agricultural research institutes operating under the ARC;
(b) departmental research entities;
(c) faculties of agriculture and veterinary sciences;
(d) institutes operating under the Department of Environmental Affairs;
(e) the CSIR; and
(f) some semi-public research agencies supported by the industry.

The ARC is the principal agricultural research entity in the country. It oversees 13 agricultural research institutes that involve a network of experimental farms and modern equipment. Its infrastructure is deployed throughout the country and, except for sugar cane, it supports all the major agricultural commodities in South Africa.

Since its establishment in 1992, ARC institutes introduced a more business-like management style coupled with more aggressive cost recovery from ARC clients on contracted research (the 'users pay' principle). This led to a stronger client orientation than had hitherto been the case. Targets were set to increase external funding rapidly, with the goal of recovering 30% of total expenditures from the commercial agricultural sector (Roseboom *et al.*, 1995). This happened much more rapidly than planned with the contribution of the parliamentary grant to total income declining from 95% in 1994 to 62% in 2000.

The management style of the ARC has changed substantially since the new Science and Technology policy was introduced in 1996. Following various reviews of the agricultural research system and strong criticism of the way the ARC was managed, changes to the governance structures were made and a number of research institutes were merged. One of the important points of criticism was that the research done by the ARC was mainly directed at capital-intensive agriculture and only to the benefit of commercial farmers, while farmers from previously disadvantaged communities did not benefit from the research output. The reviews and recommendations necessitated the introduction of several transformation initiatives by the ARC as its parliamentary grant depreciated in line with the perceived new direction in the Science System, as discussed in section 12.2. Changes in leadership and other factors have seen the ARC moving into a position of increased isolation in the national agricultural research system, since its stakeholders were not effectively informed of the consequences of the new policy directives contained in the White Paper on Science and Technology. Being a relatively young Council, it is possible that it could not have timeously foreseen and communicated the consequences of the changes made. The ARC obviously had to embark on a process to strengthen its relevance and performance if it was to break out of this isolation.

This process has seen the interaction with a number of stakeholders during 1999 where, in a series of meetings with provincial Departments of Agriculture and representative bodies in organised agriculture, stakeholders were asked to criticise the performance of the ARC as a service provider in agriculture. Insights gained from this enabled the ARC to initiate strategic workshops on its research agenda and the funding of agricultural research in South Africa. As a result of the reviews and the perception about the relevance of the ARC research, a system was introduced whereby research on the problems and needs of resource-poor farmers was ring-fenced from the overall parliamentary grant and managed under a separate programme. In addition, greater emphasis was placed on the commercialisation of research outputs.

The new constitutional dispensation in 1994 introduced nine provinces and agriculture became a shared competence between national and provincial governments – legislative authority resides at national level, and development authority at provincial level. The previous agricultural development institutes formed the basis for the establishment of the nine provincial Departments of Agriculture (PDAs). One of the Agricultural Development Institutes, Grootfontein, presented problems in terms of its locality and was transferred to the custodianship of the National Department of Agriculture (NDA), where it still resides.

Funding from the PDAs for agricultural research started deviating fundamentally from the formula guidelines in 1996. Provincial R&D capacity dwindled and in some cases ceased as high costs and poor restructuring plans resulted in the disappearance of agricultural research in some provinces, for example in the Eastern Cape. Only in two provinces, the Western Cape (Elsenburg) and Kwazulu-Natal (Cedara), can one still find a fair degree of research competence, but this is still severely under-funded in some aspects. Most of the provinces have had to rely increasingly on donor funding and the operations of non-governmental organisations (NGOs) as well as producer/commodity organisations. The ARC has recently provided increasing research support to the provinces.

A number of co-ordinating mechanisms were introduced to harmonise the agricultural R&D system – the Minister Members Executive Council (MINMEC) at inter-ministerial level and the Inter-departmental Committee on Agriculture (ITCA) at the Heads of Departments level. The latter had several subsidiary committees, most of which were soon disbanded as they were deemed ineffective. Those mechanisms remaining deal only with natural resource management and veterinary services.

The funding of agricultural R&D changed into two streams: the ARC received its funding from the National Department of Agriculture, and the newly established PDAs were each allocated a portion of the former national agricultural vote according to a formula, although the provincial legislatures were not forced to honour this.

12.3.2 Sources of funding for agricultural R&D in South Africa

System-wide, the funds allocated to agricultural research in South Africa come from four basic sources. At central government level this constitutes the Science Vote allocated by the DST and various national government departments, such as Agriculture, Environmental Affairs and Tourism, Water Affairs and Forestry, Trade and Industry, and the National Education Vote, from which the Higher Education institutions receive their funding. Also at a national level, a number of trusts and levy incomes from producer/commodity organisations, plus research funding from private sector business enterprises, are further sources of income. The latter probably distinguishes the South African NARS from that of other African countries in terms of the prominence that it plays in research funding and the use of research services.

Figure 12.1 shows the flow of funds within the South African NARS for 1999–2000. At central government level, the parliamentary grant from the Science Vote totalled R295.5 million, consisting of R292.9 million (this includes direct transfers in addition to the base funding reported in Table 12.1) allocated to the ARC, and the balance being that assigned by the CSIR to its Division of Food, Biological and Chemical Technologies. The various national departments allocated a further R68.1 million to agricultural research through performance and service contracts and competitive bidding funds – the latter being mainly allocated through the Technology and Human Resources for Industry Programme (THRIP) and the Innovation Fund administered by the NRF, and the lead programmes of the DACST. In addition, some R29.2 million is generated internally by the ARC using its own resources.

Other public sources include R4.5 million allocated to agricultural research by the Water Research Commission. This represents 9.1% of the total research budget of the Water Research Commission, which receives its funding from a levy paid by all Water Use Authorities. Funding from commodity and producer organisations includes research commissioned by the commodity trusts (R26.4 million) and levy income (R39.2 million for non-sugar commodities, and R48.8 million for sugar).

Apart from the structural changes in the agricultural R&D system already mentioned, competitive bidding with other science councils for base line funding/ parliamentary grants was introduced in the 1997–98 financial year. It was furthermore decided that all external contracts would be based on full cost recovery, a principle that was not readily accepted by the various commodity organisations that fund research. As a compromise, a 50:50 cost-sharing arrangement between the relevant institute and commodity organisations was agreed on. As a result, other major funders of agricultural research in the past five years have been the various commodity trusts that were established after marketing deregulation lead to all marketing boards being abolished. The assets of these boards were transferred to trusts such as the Maize Trust, Wool Trust, Red Meat Trust, etc., and the returns from these assets are used to fund the activities of producer organisations and agricultural research. Table 12.2 indicates the extent of research funding provided by commodity trusts and by statutory and voluntary levies managed by certain producer organisations since 1999. Although showing strong growth, it is expected that the level of funds generated through levies will stabilise around the levels reached in 2001.

Funding from business/private enterprises comes mainly from input suppliers and agro-processors that either contract services from research service providers, but also conduct in-house research. Money allocated from these sources amounted to R27 million in 1999–2000.

The ARC and the higher education institutions dominate expenditure by research performers. According to Figure 12.1, the total expenditure in 1999–2000 is estimated at R586 million, of which R139.4 million accrued from non-government income sources. Donor contribution to the NARS is difficult to estimate, but is expected to be relatively small.

Agricultural research at the different faculties of agriculture at South Africa's universities is also funded from a range of sources. Commodity organisations and private companies are often the funders of major and more long-term projects,

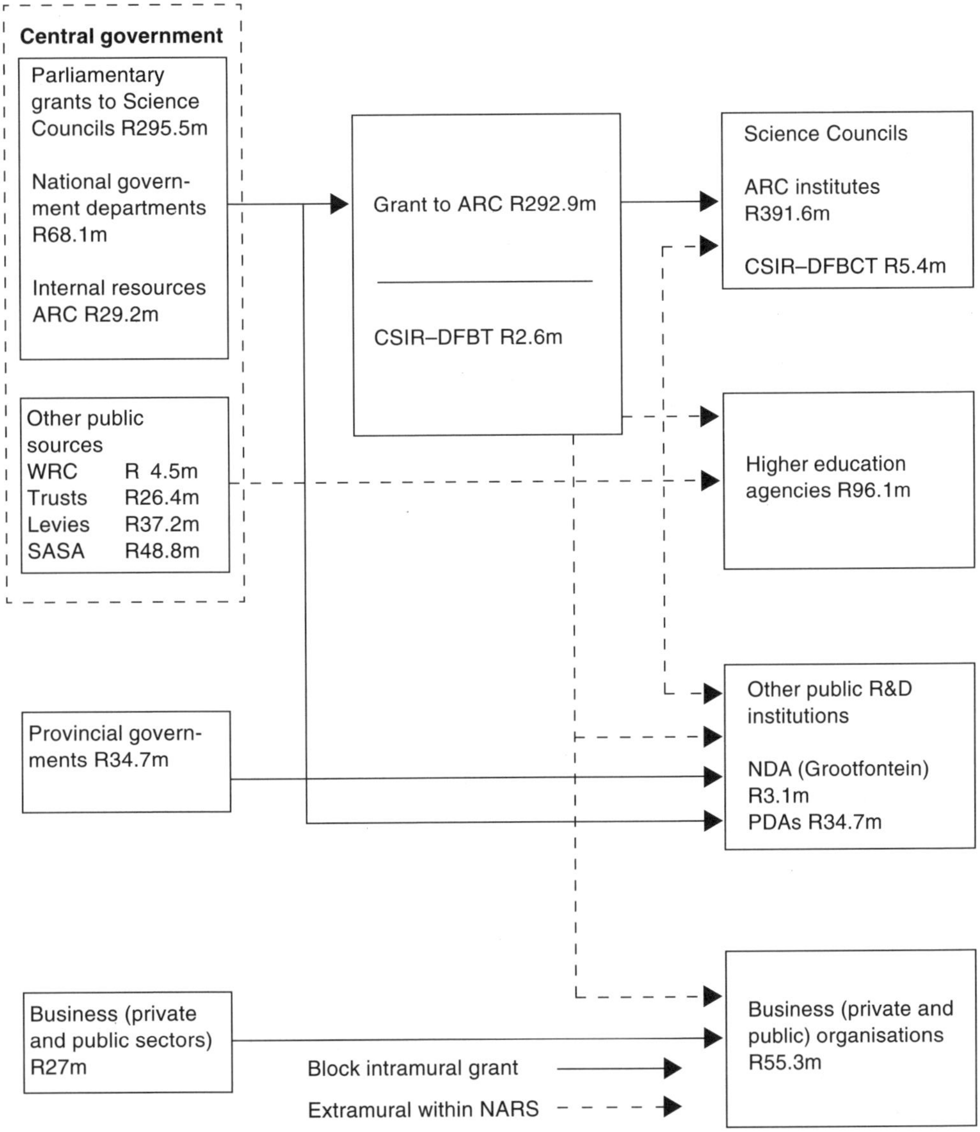

Figure 12.1: Funding channels for South African agricultural R&D, 1999–2000.

Source: Compiled from information adapted by authors from various sources

Note: Data are in nominal South African Rand. Figures may not tally exactly with those reported in other tables/figures because estimates of researcher expenditures were used, and ARC corporate expenditures are included here

Table 12.2: Annual contribution by commodity organisations to agricultural research in South Africa (R million), 1999–2001.

Source	*Year*		
	1999	*2000*	*2001*
Trust contributions			
Animal	3 578 090	3 468 824	7 222 259
Crops	13 060 674	18 732 626	21 338 851
Horticulture	5 280 910	4 200 000	3 684 210
Subtotal	21 919 674	26 401 450	32 245 320
Levy income			
Crops	11 194 274	11 491 691	12 337 120
Horticulture	19 156 306	25 665 738	27 521 309
Subtotal	30 350 580	37 157 429	39 858 429
Total contribution by commodity organisations			
Animal	3 578 090	3 468 824	7 222 259
Crops	24 254 948	30 224 317	33 675 971
Horticulture	24 437 216	29 865 738	31 205 519
Total	52 270 254	63 558 879	72 103 749

Source: Information provided by various trusts and commodity organisations

while funds are also supplied to successful bidders under the Innovation Fund and the NRF. In addition, donor agencies such as the United States Agency for International Development (USAID) provide funding for research and postgraduate teaching initiatives at certain universities.

12.3.3 Agricultural R&D trends

Table 12.3 suggests that the ARC is by far the largest provider of agricultural research in South Africa, being responsible for 57.9% of total agricultural research expenditure in 1999 and housing 59.8% of the researchers in the NARS. Note that the 634.3 full-time equivalent (fte) researchers listed include research technicians with degree qualifications. This shows a significant increase in the relative importance of the ARC since its establishment when it was responsible for 54% of total expenditure on agricultural research (Roseboom *et al.*, 1995). Universities have also shown strong growth in terms of market share since 1992.

The situation may have changed significantly since 2000. The number of research staff at the ARC has dropped from 751 in 1992 to 616 non-full-time equivalent researchers in 2003. The biggest proportional change in terms of qualifications was amongst PhD and MSc qualified researchers. The number of PhDs dropped from 176 in 1998 to 167 in 2000. Of more concern is that the ARC records at the end of April 2003 reflected only 144 PhD qualified research staff employed within all of the institutes, with only 87 being researchers. This meant that 60 key research staff left the ARC since 2000. The corresponding number for

Table 12.3: Composition of agricultural research expenditures, and number of researchers in South Africa, 1999–2000.

Type of agency	*Expenditure (millions)*		*Researchers (fte's)*	*Share (%)*	
	1999 Rand	*1999 PPP Dollars*		*Expenditure*	*Researchers*
Government agencies					
ARC*	431.2	191.5	634.3	57.9	59.8
Other	123.9	55.0	168.4	16.7	12.0
Non-profit institutions	56.3	25.0	56.0	7.6	5.3
Higher education institutions†	107.4	47.7	158.0	14.4	14.9
Business enterprises	25.3	11.2	44.5	3.4	4.2
Total	744.1	330.6	1 061.1	100.0	100.0

Source: Preliminary Agricultural Science and Technology Indicator (ASTI) Survey results

* ARC expenditure reported includes expenditure arising from externally funded sources (i.e. Non-Parliamentary Grant)

† Expenditures for the higher-education agencies are estimates based on average expenditures per researcher for the government agencies

MSc qualified research staff is 69. This rapid drop in qualified researchers is cause for concern and could jeopardise the future agricultural research system in South Africa if the trend continues.

Support staff to scientist ratios also dropped, from as high as 10.7:1 in 1992 to 3.9:1 in 2001. The technician/researcher ratio in the ARC institutes dropped from 1:7 in 1992 to 0.8:1 in 2001. Support available to researchers in the execution of experiments is, therefore, dwindling and the remaining researchers and technicians are now expected to allocate more of their time to support work, instead of concentrating on their research duties. This has severe implications for ARC efforts to maintain past levels of performance and its impact will strongly affect the ability of South Africa's agricultural sector to provide support to regional and local rural development initiatives. The budgetary pressures as a result of the drop in the parliamentary grants could be the main reason behind the change in these ratios.

Baseline funding for agricultural research provided by government through the parliamentary grant system dropped from a high of R337 million in 1997–98 to R262 million in 2001–02. The extent of the decline of funding is emphasised by the rapidly dwindling real value of the grant. In 2001–02 the ARC received only 55% in real terms of the parliamentary grant it received in 1992. As a consequence, external income had to grow significantly to maintain overall spending at R450 million (estimated) for 2001–02. The level of baseline funding of the ARC would have been R481 million if funding were to be maintained at 2% of Agricultural contribution to Gross Domestic Product. Currently it is envisioned under the new Agricultural Sector Strategy (NDA, 2001) that the target for this should be in the order of 3%.

External income for the ARC came from commodity and producer organisations and donor funding. Income from commodity organisations constitutes 13.7% of

ARC expenditure, but within the NARS commodity organisation's fund 19% of total R&D expenditure (see Figure 12.1). Producer levies have been one source of funding for agricultural research. A few statutory levies were introduced by producer organisations as a way to raise funds for amongst other things, agricultural research. Voluntary levies are also used by some commodity groups, but are highly variable sources of income. Donor funding to the ARC is growing, but access to it is severely limited due to policy constraints. On the positive side, it does seem that more donor funding has been flowing to universities to do basic and applied agricultural research. Private funding to the ARC is estimated at 17.6%, while own research in many agricultural input firms has been growing as a result of the high returns on intellectual property/patents. Own income of the ARC from royalties and Intellectual Property Rights has increased markedly over the three years shown, rising from almost 7–12.8%.

Despite the growth in external funding, government contributions (through the parliamentary grant and a range of contracts) remain the biggest single source of funding (around 52%) within the ARC. However, the ARC has now reached a situation where the dual accountability of Institute Directors towards both public and private funders has become a serious issue in resource allocation decisions. The completion of the 2001 ASTI Survey in South Africa will provide complete statistics on the trends in research expenditure according to commodity groups. For the moment, an analysis of the ARC budget for 2000–01 gives a rough idea of the extent of annual research expenditures per commodity group:

(a) horticulture: R103.8 million;
(b) public support institutes: R151.7 million;
(c) grains and industrial crops: R68.7 million; and
(d) livestock: R91.5 million.

12.4 THE PROVISION OF AGRICULTURAL R&D SERVICES IN SOUTH AFRICA

An overview of the relative importance of different research providers in the South African NARS is contained in Table 12.4, and again confirms the dominance of the ARC in the NARS.

12.4.1 Government and local agencies

The difficulties experienced by provincial governments in adapting to the challenges of their new mandate, and pressure on the national government to deliver on its Growth and Development Strategy have seen the national government introduce a number of programmes to direct priority initiatives. This has also impacted on the research systems, especially on the priority setting and activities of the ARC which was continuously urged to adjust in order to operate in line with the

Table 12.4: Agricultural research expenditures by institutional category in South Africa, 1992–2000.

Institution	*Percentage of total*								
	1992	*1993*	*1994*	*1995*	*1996*	*1997*	*1998*	*1999*	*2000*
Government agencies									
ARC	59.5	58.3	59.0	61.2	62.4	62.8	62.4	60.5	57.9
Other	17.9	19.0	18.5	17.6	16.9	16.8	15.7	15.5	16.7
Subtotal	77.4	77.3	77.5	78.8	79.3	79.6	78.1	76.0	74.6
Non-profit institutions	8.5	7.7	7.2	6.6	6.7	6.6	6.5	7.6	7.6
Higher education agencies	10.3	11.1	11.2	10.9	10.5	10.4	12.2	13.6	14.4
Business enterprises	3.8	3.8	4.0	3.7	3.5	3.4	3.2	2.8	3.4
Total, public	96.2	96.2	96.0	96.3	96.5	96.6	96.8	97.2	96.6
Grand Total (R million)	708.3	708.1	682.8	782.4	787.2	767.7	831.5	749.1	744.1

Source: Based on preliminary 2001 ASTI Survey results

seven Presidential Imperative Programmes (PIPs), one of which focuses on rural development. Government departments are 'clustered' around this in terms of their potential ability to deliver on this programme from their own budgets. The Minister of Agriculture and Land Affairs and the Ministers in the provinces have identified five-year priorities for agriculture that are closely related to the rural development PIP. Producer and commodity organisations often form partnerships with public sector R&D service providers such as the ARC. The biggest successes have been in those cases where the ARC took the lead in project management, and the universities, the local provincial department, producer organisations, etc., and farmers were part of the partnership with each contributing an in-kind or monetary share.

12.4.2 Faculties of agriculture at universities in South Africa

Faculties of agriculture at the larger universities in South Africa appear to be in a better position to maintain capacity under the current dispensation. The core funding for universities (salaries/overheads) is obtained via the National Department of Education. Direct costs for research (equipment/running costs/ research technicians) are usually funded through research contracts with producer organisations, private companies and some donors. Researchers at universities also actively compete for research funds such as the Innovation Fund and the annual grant funding for researchers from the NRF. Due to the variety of sources of funding, it is very difficult to estimate the exact level of spending by universities on agricultural R&D. There is a growing concern that universities are venturing into applied research and taking most of the potential projects from the ARC. This is partly due to perceptions held by commodity organisations that the ARC's research capacity is declining and/or that the organisations are attracted by the lower rates charged for this type of research by the universities. Sustainability of this in the light of the cross-subsidisation from the Education Vote is doubtful.

12.4.3 International agencies

The Consultative Group on International Agricultural Research (CGIAR) has provided useful support and information since 1994. There has been little direct involvement over time apart from a few donor-driven projects, but involvement is now increasing, and the recent establishment of a regional office for the International Water Management Institute (IWMI) in Pretoria is an example of this growing trend. The IWMI works in close collaboration with the ARC, government and universities on aspects related to the rehabilitation of irrigation schemes, the development of an irrigation policy, and the development of irrigation scheme management.

Regional R&D organisations, such as the Southern African Centre for Co-operation in Agricultural and Natural Resources Research and Training (SACCAR), are becoming more important. SACCAR, under the auspices of the Southern African Development Community (SADC), used to operate on the basis of allocating the responsibility of certain research initiatives to specific member states of the SADC. It will now take greater direct responsibility for research following restructuring of the SADC. Member states will no longer have the sole responsibility to fund and manage these initiatives with only the review and consultation of the SACCAR Council and its subsidiary technical committees. Initially, the ARC represented South African public sector agricultural research at SACCAR. Whereas universities still have their representation, the NDA has taken over this responsibility from the ARC. In fact, this applies with regard to all foreign representation and liaison of agricultural R&D.

The Special Programme for African Agricultural Research (World Bank) has also been changed to a more permanent initiative with the creation of the Forum for Agricultural Research in Africa (FARA) in Addis Ababa in April 2001. The intent is to provide a forum within which agricultural R&D in Africa can be harmonised through the initiatives of the three regional agricultural research organisations in sub-Saharan Africa. Through this, Africa now takes greater ownership of its own R&D direction. The Mbeki government has taken the lead on this with the development of the Millennium Africa Programme.

12.5 LESSONS AND FUTURE CHALLENGES

Unless public research service providers are willing to reach a mutually acceptable position with private funders on the issues of intellectual property rights and on resource allocation within the institution, it will find itself in a position of conflict with its clients. The result being that it would be unable to ensure a stable fund flow and an inability to retain competent staff. Some will reason that the state should do research of a public goods nature, while the private sector should undertake research of a developmental nature (where user fees can be charged).

The need to harmonise and improve the efficiency of research systems at the international level is also reflected in domestic initiatives. The changes in the

NARS in South Africa over the past 20 years have introduced increased levels of fragmentation characterised by uncompleted processes or efforts to instil greater harmony. Lessons in building a more effective NARS for South Africa will be discussed. Firstly, it is important to maintain continuity in the NARS leadership as the vision for the direction of the NARS and the initiatives to achieve this vision have to be persistent. The team guiding the NARS and its co-ordinating bodies must have direct access to the leadership of key institutions in the NARS. The continued ability to monitor and adjust to change in the operating environment is crucial if the system is to remain dynamic and relevant. Ad hoc, uncoordinated and sporadic response to change in the system is perhaps the most important reason why South Africa experiences fragmentation and duplication within the system.

Secondly, stakeholders as partners need to agree, but always on the basis of proper analytical input into their decision-making. Responsibilities, therefore, need to be clear and agreed upon, something which can be achieved through the use of memoranda of understanding (MOUs) or contracts. Throughout, the focus should be on programme co-ordination and content.

A third important factor is the availability of an enabling policy environment for fund mobilisation, human resource development and communication. This is perhaps one of the most neglected areas of agricultural R&D policy in South Africa. The focus on this is periodic and a consistent effort by a critical mass of policy researchers is needed. It will also be necessary for an effective and world-class agricultural science fraternity in South Africa to encourage increasing numbers of students to train in agricultural science. A new and substantial scholarship programme for students needs to be introduced to counteract the large drop in qualified scientists.

For the effective governance/management of the system, monitoring and evaluation capacity should be developed. Without this, the co-ordination initiatives of the governing entity would at best be frustrating the efforts of the NARS partners throughout the system. The decisions on fund management and allocation should be taken prudently. This chapter was completed just as South Africa is entering into a new phase in the history of agricultural R&D with the establishment of the NARF. This is a critical phase to ensure a positive future for agricultural research in South Africa that will promote the sustained international competitiveness and prosperity of agriculture in South and southern Africa. Sadly, since its establishment, the NARF has failed to become operative as a policy advisory body and to formulate appropriate policy responses to the issues listed here. This is mostly the result of the tensions that arose in the NARS over the issues of ownership of intellectual property generated by public service providers, but partly funded by private means.

REFERENCES

DEPARTMENT OF ARTS, CULTURE, SCIENCE AND TECHNOLOGY (DACST). (1996). *White Paper on Science and Technology: Preparing for the 21st Century.* Pretoria: DACST.

DACST (2001). The science vote. (Accessed 26/11/01).

DEPARTMENT OF SCIENCE AND TECHNOLOGY (DST). (2002). *South Africa's National Research and Development Strategy.* Pretoria: Network Publishers.

MARAIS, H.C. (2000). *Perspectives on Science Policy in South Africa.* Pretoria: Network Publishers.

NATIONAL DEPARTMENT OF AGRICULTURE (NDA). (2001). The strategic plan for South African agriculture: a united and prosperous agricultural sector. (Accessed 15/12/02).

ROSEBOOM, J., PARDEY, P.G., SARTORIUS VON BACH, H.J. & VAN ZYL, J. (1995). Statistical brief on the National Agricultural Research System of South Africa. ISNAR Indicator Series Project. Statistical Brief No.23, September.

CHAPTER 13

Agricultural Finance in South Africa

Gerhard Coetzee

13.1 INTRODUCTION

The agricultural sector echoes the dual nature of the South African economy, with the duality reflected in a commercial and a developing component. Until recently, financial institutions provided exclusive support to the commercial sector, when, for example, the Land Bank expanded its portfolio to cater for small-scale farmers as well as the traditional commercial farming sector.

Table 13.1 presents the outstanding agricultural debt on an annual basis since 1970. It is clear that the main actors in the supply of retail agricultural credit are the Land Bank, commercial banks and agricultural co-operatives. The co-operatives source their funds from the Land Bank and commercial sources, implying that as the original sources of funds the Land Bank and Commercial Banks are the main players in the market. Small-scale farmer agricultural finance will be addressed later in this chapter. It is important to note that no formal assessment is made of the extent of small-scale farmer credit. Table 13.1 represents those institutions that focus primarily on the commercial farming sector, except for the Land Bank which invested in extending their services to incorporate small-scale and developing farmers. This portion of their portfolio is still negligible in terms of the volume lent, but it exceeds the commercial component of their portfolio in terms of the number of clients it has served.

The salient aspects of agricultural finance in South Africa are highlighted in this chapter. Section 13.2 focuses on the financial position of the agricultural sector and is followed by an overview of small-scale farmer finance. Thereafter the focus is on the financing of commercial agriculture and highlights the development of new agricultural finance products. Section 13.4 highlights some agricultural financiers, and is followed by a conclusion on the state and the future of agricultural finance in South Africa. Although this chapter focuses on agricultural finance, it must be seen in the context of a range of support elements for agriculture and within the broader context of rural finance. The approach is therefore that finance is one of many aspects that will support agricultural production, and that finance alone is not enough to ensure production and growth of the agricultural sector. Indeed, this chapter argues that the failure of small-scale farmer agricultural finance in South Africa is largely a function of the failure of other support services to agriculture.

Table 13.1: Total farming debt in South Africa in R million.

As at 31 Dec.	*Land Bank*	*Commercial banks*	*Agricultural co-operatives*	*Department of Agriculture*	*Private persons*	*Other financial institutions**	*Other debt*	*Total*
1970	295.8	281.8	113.4	136.2	242.9	286.9	45.0	1 402.0
1975	429.1	454.5	247.4	151.2	327.3	324.1	70.0	2 003.6
1980	675.6	801.5	866.9	180.0	579.9	612.3	122.4	3 838.6
1985	2 338.4	3 315.3	2 754.0	549.2	792.0	1 128.2	240.5	11 117.6
1990	3 441.1	4 948.6	3 780.3	1 013.1	1 209.4	905.0	675.6	15 973.1
1995	4 679.6	6 385.6	3 950.2	1 414.6	1 641.8	432.1	895.1	19 396.1
2000†	9 606.0	11 172.3	4 500.0	860.4	1 986.7	602.4	1 098.1	29 825.9

Source: Abstract of Agricultural Statistics (2002)

* Includes discount houses, merchant banks, other monetary institutions, insurance companies, pension funds, trust companies, non-monetary banks and trust assets, participation mortgage bonds

† Preliminary

13.2 FINANCIAL POSITION OF THE SOUTH AFRICAN COMMERCIAL FARMING SECTOR[1]

13.2.1 Introduction

Agricultural finance in South Africa stands at a crossroads as an era of direct state involvement is drawing to a close. This is evident in the closure of the Agricultural Credit Board (the National Department of Agriculture's in-house financier) in 1997 and the direction indicated by formal agricultural policy. These things are happening at a time when the government is placing emphasis on ease of entry for those farmers hitherto constrained and prevented from joining mainstream agriculture in South Africa. This creates the perception of conflicting policy objectives. The Land Bank, tasked by government to attend to financing development agriculture while retaining its commercial financing responsibilities, reacted by aggressively expanding its commercial financing, to which commercial banks reacted by taking legal action. Questions are being asked about the future of agricultural finance in the country and the role of the government and it is opportune to gauge the current state of agricultural finance at this turning point.

13.2.2 Main trends impacting on the financial position of the sector

An overview of the general nature of agricultural policies during the years under consideration is necessary in order to understand the economic movement of the agricultural sector as a whole. The discussion on agricultural policies can be divided into two time periods, the 1980s and the period from 1994. Kirsten & Vink (1999) classify the main policy shifts in the 1980s as:

(a) deregulation of the marketing of agricultural products in terms of the Marketing Act and other legislation;
(b) changes in fiscal treatment of agriculture, including the abolition of many tax breaks that favoured the sector and a reduction in direct budgetary expenditure on the sector; and
(c) a start to the processes of land reform, reforms of labour legislation, and trade policy reform.

Policy shifts as from 1994 include:

(a) the promulgation of the Marketing of Agricultural Products Act, 47 of 1996, which represents a radical departure from the marketing regime to which farmers had become accustomed since the 1930s;
(b) trade policy reform aimed at reversing decades of 'inward industrialisation' strategies;

[1] Section 13.2 draws from the work of Meyer *et al.* (2000).

(c) labour market reform, which involves the continuous application of legislation that governs the occupation rights of those who live on farms; and
(d) reform of the state's approach to financial support to the agricultural sector.

The effect of monetary policy on the agricultural sector is important, especially with regard to the depreciation (and recent appreciation) of the Rand exchange rate as well as the farm debt problem and interest rate variability. As Figure 13.1 shows, a negative real interest rate mainly occurred during the 1980s. The reason for this is that the government followed an expansionary monetary policy that implied an increase in money supply (Kirsten & Vink, 1999). This gave rise to the depreciation of the Rand exchange rate, which resulted in the escalation of the costs of imported components of machinery, implements, dips and sprays. The expansionary monetary policy resulted in a high inflation rate (represented by the Consumer Price Index [CPI]) which triggered the negative interest rate. However, in order to curb inflation, the government started to implement a policy of positive real interest rates from 1988 onwards.

Agricultural debt (deflated by the price index of all farming requisites to eliminate the effect of inflation) reached a maximum level in 1985. This phenomenon is mainly due to three factors (Van Schalkwyk & Van Zyl, 1995):

(a) the negative real interest rate brought about higher levels of borrowing than would otherwise have occurred;
(b) subsidised interest rates further increased the availability of cheap credit in the early 1980s, which led to an increased willingness of farmers to borrow more funds; and
(c) the severe drought of 1982–83 caused cash flow problems, which resulted in a decreased repayment ability of farmers.

A steady decline in real total farm debt occurred after 1985. Financial institutions started to acknowledge the increased risk of the agricultural sector (especially when some of the subsidy programmes started to be decreased and withdrawn), and a more strict credit repayment programme was followed. In addition, farmers realised the increased risks and decreased their borrowing of funds.

Because low or even negative real interest rates encouraged borrowing, land prices were stimulated artificially. According to Van Schalkwyk & Van Zyl (1995), the real price of land peaked in 1976. As real interest rates then started to increase, the price of land decreased. They also state that the lag between the 1976 real land price peak and the 1985 debt peak is a direct result of the expectations of agricultural financiers that real farmland prices would increase. After the over-mechanisation in the 1970s and early 1980s (with negative real interest rates), and as interest rates increased and the financial position of the farmers deteriorated, farmers could not afford to replace their machinery frequently. This implies that the life span of these assets had to be increased and this automatically resulted in the

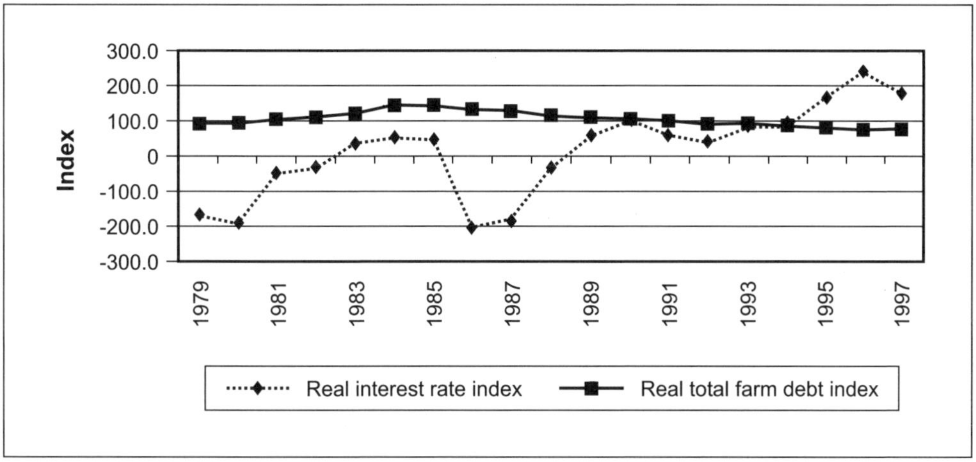

Figure 13.1: Trends in real interest rate index and real total farm debt.
Source: SARB Quarterly Bulletin (March 1999) (constant 1990 prices); Abstract of Agricultural Statistics (2000)

downward trend in the value of machinery. These trends are echoed by Coetzee & Viljoen (1991).

Interest rates have an impact on the inventory behaviour of farmers. Higher interest rates raise the cost of inventory investment, causing stocks to be run down. It is, however, interesting to take a closer look at the trend of real livestock assets where a slight incline has been observed since 1993. This may be explained by the government programmes aimed at converting marginal crop areas to grazing as well as the increasing awareness of farmers in high-risk crop production areas.

It is thus inevitable to draw the conclusion from this data set that the real value of total assets of the agricultural sector has declined over the past one and a half decades. The negative trend between debt and equity in the early 1980s, as shown in Figure 13.2, can be explained by the fact that the severe drought in these years forced farmers to repay their loans by using their equity.

The decrease in the total value of assets does not necessarily imply that the financial position of the farmers has deteriorated drastically if it is measured in financial ratios. The leverage ratio (an indication of the farmers' ability to meet their total liability by using own capital) indicates that except for deterioration in the early 1980s, this ratio has stayed fairly constant since 1990. One reason for this is the steady decline in the real debt value over this period.

The high level of inflation and the deterioration in the exchange rate serve as contributing factors for the cost price squeeze where margins between input costs and producer prices become smaller. Only in recent years was the government able to reduce inflation, although the devaluation of the Rand continued at an increasing rate until 2002 when it started to regain its ground against the main currencies.

All of these factors led to the decrease in real net farm income (NFI) of about 2% since 1980 and up to 2001 (Abstract of Agricultural Statistics, 2000).

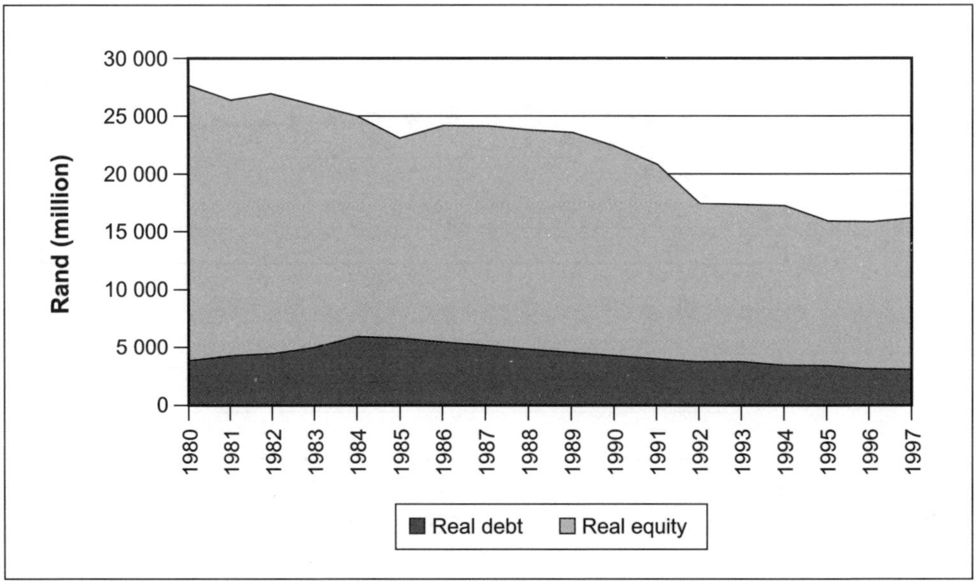

Figure 13.2: Real debt and real equity of South African agriculture.
Source: Abstract of Agricultural Statistics (2000)

Fiscal policies also impact on the financial situation of the farming sector. Fiscal policies are the attempt by the government to meet employment, income, growth and distribution, and other objectives through its powers to tax and spend. Gross domestic fixed investment in agriculture as a percentage of total gross domestic investment shows a decline from 4.38% in 1986 to 3.50% in 1997 (Figure 13.3).

Based on earlier discussion, a few reasons for the deteriorating investment trend from 1988–92 can be identified:

(a) high increases in input costs due to the weakening exchange rate;
(b) rise in the cost of inventory because of the high level of interest rates;
(c) the heightened political uncertainty, which led to a decrease in the confidence level of the agricultural sector; and
(d) the focus of government investment shifted towards social sectors like education, health and housing, which influenced investment in the agricultural sector.

From 1998 these trends continued until the end of 2001, when a severe depreciation of the exchange rate and drought in South Africa's neighbouring countries impacted on the revenues of all export crops. Farmers enjoyed steep increases in revenue and the declining rate of real NFI started to turn around (NDA, 2002). However, the exchange rate recovered in 2002 and a deterioration of farm income is expected after a lag in time.

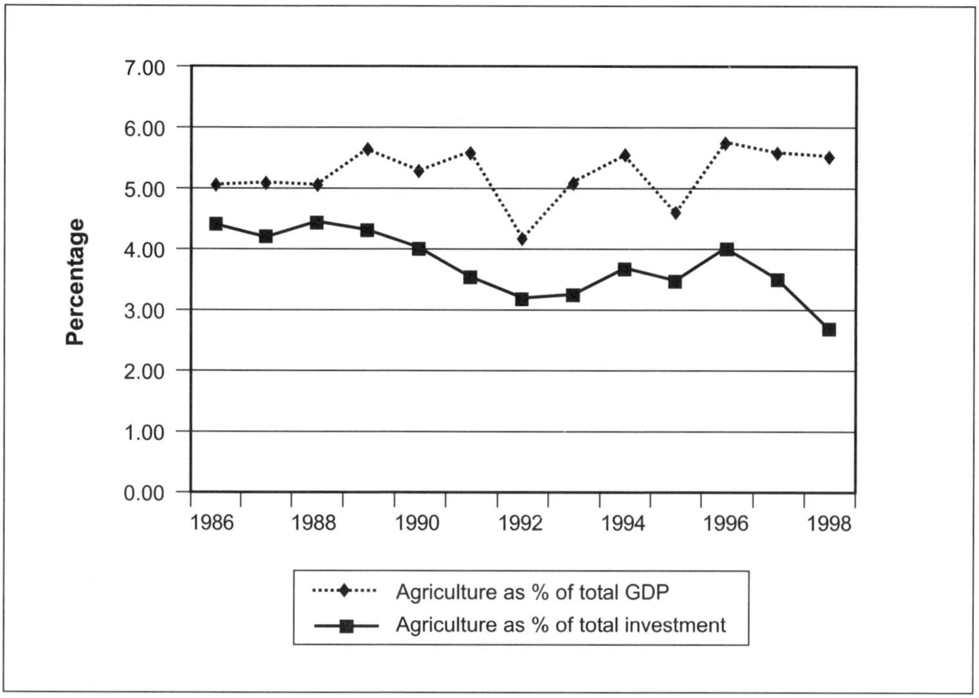

Figure 13.3: Trends in agricultural investment and GDP as a percentage of total GDP.
Source: SARB Quarterly Bulletin (1999) (constant 1990 prices)

13.2.3 Impact on the profitability of farming

Although a negative trend in farming profitability was observed until 2001, it is clear that farmers reacted to severe conditions by changing their business strategies, including their financial strategies. This is shown in their changing credit-use-pattern and the decline in the real liability profile of the farming sector. Agricultural financiers also reacted to these changes, and the commercial banks ensured that their portfolios followed those sub-sectors where export and value-adding processes decreased the risk in the portfolios. The commercial financiers of agriculture are investigating a range of new products to serve farmers which range from structured finance packages to inter-linked products (hedging of price risk linked to production loans).

13.3 RURAL AND SMALL-SCALE FARMER FINANCE

13.3.1 Recent history of rural and small-scale farmer finance in South Africa

Recent history can best be summarised by the work of the Commission of Inquiry into Rural Financial Services, better known as the Strauss Commission (Strauss

Commission, 1996). The Commission was active in 1995–96, and made a range of recommendations of which the most essential are highlighted here.

The Commission considered the major policy objectives of the government and its brief, namely to contribute towards increasing access to financial services for rural people. These services were identified as transmission services, savings products and loan products for consumption smoothing and productive loans (for farm and off-farm activities). The Commission also identified that state grants would be necessary under certain circumstances and it called for a detailed set of guidelines on the management of finite subsidies and grants.

A role was identified for the state to facilitate, as well as co-ordinate, the provision of financial services with special attention to the needs of women. In this regard, the Commission proposed that the Land Bank be tasked to fulfil this role. The way in which the Land Bank, the state and other institutions act, must be in support of the market, but at the same time the inherent weaknesses (not sustainable and low outreach) in the existing institutional fabric must be acknowledged, as these institutions are not able to contribute to the aims of rural reconstruction on a national basis without appropriate and active support.

The Commission Reports emphasised a paradigm shift and a movement away from the supply-led[2] approach, rather than a detailed programme for implementation in order to increase access to financial services in rural areas. This shift is indeed necessary as most of the South African parastatal institutions follow a supply-led approach, as proved by the Commission. Although the Land Bank has been tasked to fulfil certain functions, no point exists where these functions can be co-ordinated with activities in the provinces.

Government accepted the Strauss Commission proposals in August 1997 (Agricultural News, 1997). The Land Bank, which previously reported to the government through the Ministry of Finance, now reports through the Ministry of Agriculture which continues the distinct agricultural emphasis of the Land Bank. This, of course, increases the exposure of the Land Bank to covariant risk, and is a contravention of the Commission's guidelines regarding the minimisation of risk.

The Commission also gave a clear proposal on how the overall development finance system should be co-ordinated, but it did not provide any indication of the relevance of the structure of the development finance system in South Africa. Although this was not included in the brief of the Commission, it did define a role for the Land Bank within this system, without questioning the system itself.

Subsidised interest rates and credit programmes in which the government decides what is best for the client were rejected. The Commission emphasised that a broad range of services should be accessible, namely, savings and transmission facilities, as well as credit products within a demand-driven system. It further emphasised the

[2] Depicting an approach where the demand for types and volumes of financial services are not the driver, but rather a supply of services with the hope that it will initiate economic development.

importance of the retail network in rural areas to achieve access to these services. In this regard, special mention was made of a role for the Post Office in terms of the provision of savings products, transmission services and other agency services. Ultimately, the Commission also stated that although subsidies are necessary, they should be implemented within strict rules and be finite in nature. The Commission supports an integrated approach, in which it is clearly stated that financial services on their own would be inadequate.

13.3.2 The current supply of rural and small-scale farmer finance

Recent estimates (Coetzee & Grant, 2000) provide a good overview of micro-finance in South Africa, although it would be virtually impossible to delineate between rural and urban micro-finance. It is estimated that approximately 30% of 19 500 formal retail outlets are in rural areas (outside major metropolitan areas) where a client could enter into either a loan or a savings transaction. It is also estimated that approximately 30% of the 1 150 000 informal 'outlets' in South Africa are in rural areas.

It is not possible to estimate the overall supply of small-scale farmer finance in the country with any accuracy, as it is suspected that many diversions of loan funds do occur due to the general unavailability of sources of small-scale farmer agricultural credit. Much of small-scale farmer agricultural activity is also cyclical in nature and many 'small-scale farmers' would leave farming if economic conditions improve and wage income opportunities become available. It can also be assumed that a sizeable proportion of small-scale farmer credit gets diverted. Figures 13.3 and 13.4 confirm that rural people have diverse income streams that may include farming. This is not entirely surprising as it represents one of the numerous ways in which poor people manage risk, i.e., by diversifying their income sources.

However, it is assumed that since agriculture plays such a small part in the household income streams of most small-scale farmers, loans for agriculture are quite scarce and targeted towards those who are full-time farmers (ironically the higher-risk farmers as they rely largely on agriculture as an income source). To refer to small-scale farmer finance may therefore be somewhat incorrect, whereas the concept of rural finance is non-sector specific and emphasises rural areas as the locality of financial transactions.

It seems that less than 2% of the portfolio of the Land Bank is for small-scale farmer financing, and it is difficult to estimate the number of outstanding loans to small-scale farmers from provincial parastatal institutions. The number of parastatal financiers is contracting as these entities find it difficult to reach operational and financial sustainability and therefore are dissuaded from financing small-scale farmers.

It is expected that in the portfolios of some institutions without specific farming loan products, a leakage (thus diversion of use) exists to farming activities. These institutions include non-governmental organisations (NGOs), Village Banks, agricultural co-operatives, stokvels (rotating savings and credit associations), and even the micro-lending sector. It is, however, very difficult to assess the magnitude of this leakage. A study in the Limpopo Province indicated that approximately 2.4% of respondents indicated that micro-lender loans were used for agriculture (Kgowedi, 2001).

It is possible to do a rough estimate of credit supply by merely noting the main suppliers of small-scale farmer credit. Private sector financiers appear to provide approximately 25 000 loans to small-scale farmers annually. These institutions include cotton processors, sugar processing firms, agricultural co-operatives and commercial banks. State provision of small-scale farmer loans would probably not exceed 50 000 loans. This information is mostly based on annual reports and information obtained directly from the institutions. It is also difficult to estimate the exact number of small-scale farmer loans as some of these institutions report both small-scale farmers and rural enterprise loans in the same category. The current overall advances based on a range of surveys seem low, ranging from 7–10% of small-scale farmers surveyed in 1997 (Coetzee, 1998).

It was also possible to assess the seasonality of income flow patterns in the 2001 annual survey. It confirms the highly seasonal nature of farm income, and the difficulty in assessing small-scale farmer risk in the loan application process. Further, the low per client loan sizes also increase the transaction costs considerably. Figure 13.4 provides an indication of the seasonal nature of small-scale farmer household income (including all sources) and Figure 13.5 indicates the extreme seasonality of the agricultural portion of the income.

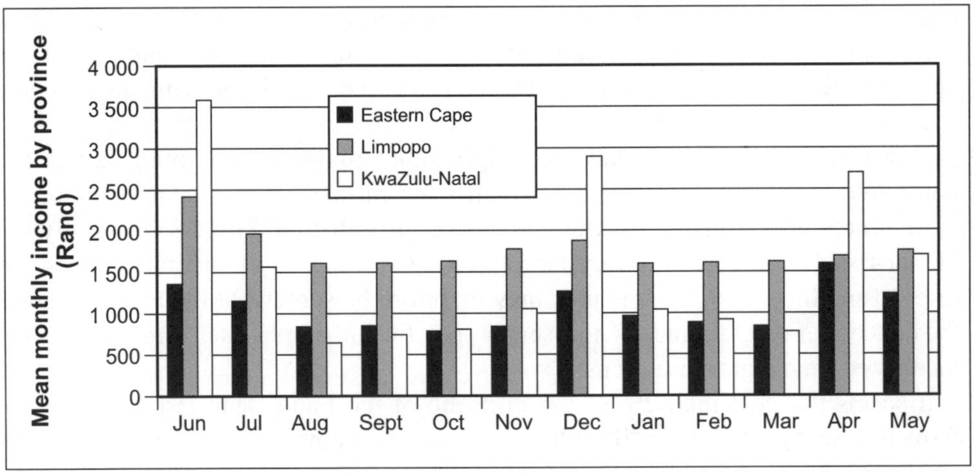

Figure 13.4: Seasonal nature of small-scale farmer monthly household income (all sources) in three provinces (n=521; 2001).[3]

[3] The study refers to work recently completed. This is a joint study by the Imperial College at Wye and the University of Pretoria with research partners being the Land Bank, Uvimba Agricultural Finance and Ithala Development Finance Limited. The study is entitled Diverse Income Sources and Seasonal Agriculture for Smallholder Agriculture: Applying a Livelihoods Approach in South Africa and was financed by the British Department of International Development.

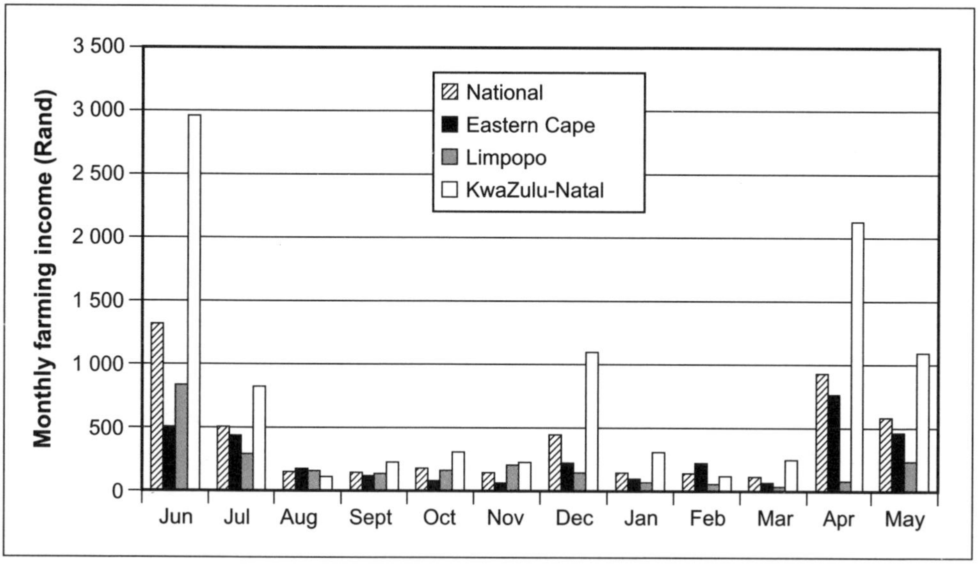

Figure 13.5: Seasonal agricultural income flows in three provinces (n=521; 2001).

A 1998 survey (ESCOM, 1998) indicated that there are approximately 2.1 million small-scale and emerging farmers in South Africa (see Table 13.2). The survey also showed that only 2% of the farmers (thus just over 40 000) used loans for agricultural finance purposes. This is a far lower number than reflected by the estimate based on the supply-side. However, the supply-side estimate includes rural enterprise loans (which may be a major portion). The message is that a small number of small-scale farmers have access to agricultural finance, and that this number has decreased over time. The latter observation is based on comparisons with the results of a number of surveys during 1982–86 (Coetzee, 1998).

Table 13.2: Small-scale farmers in South Africa in 1998.

Province	*Emerging commercial farmers*	*Other small-scale farmers (subsistence)*	*Total*
Western Cape	1 600	18 900	20 500
Eastern Cape	36 300	591 700	628 000
Northern Cape	2 100	6 700	8 800
Free State	5 900	79 700	85 600
KwaZulu-Natal	37 000	533 300	570 300
Gauteng	7 000	85 600	92 600
North West	1 300	65 900	67 200
Mpumalanga	13 600	121 200	134 800
Limpopo	36 200	453 800	490 000
National	141 000	1 956 800	2 097 800

Source: ESCOM (1998)

13.3.3 Access to financial services for rural people

This section presents the expressed demand of rural people for financial services. Demand is equated to expressed demand since no comprehensive demand study with respect to rural finance has yet been conducted in South Africa, barring the analysis of the Strauss Commission (1996) that qualitatively derived demand from a description of economic activities in rural areas. In Tables 13.3 and 13.4 some results of a survey conducted in 1996 are reflected to serve as an indication of the characteristics of rural households and their use of financial services.

It is clear from Tables 13.3 and 13.4 that farming plays a small role in terms of income, although a major proportion of small-scale farmers and small business households (rural households) cultivate the land and produce crops. Only small proportions of crops or livestock are sold and the majority of the households are deficit producers (Van Zyl & Coetzee, 1990). The majority of households in the surveys do not access credit (or are unable to access credit services) while most households are engaged in savings activities.

A rich fabric of informal financial arrangements exists, although not at a frequency comparable with that in urban areas. The low uptake of credit may indicate risk aversion or an inadequate supply of formal credit either in the form of no access or inappropriate financial products on offer. Most people who save money have commercial bank accounts which indicates an effort to deposit savings since the majority of rural branches of commercial banks are in medium to larger towns, normally far away from these clients. The PostBank and NGOs did not feature in the survey results.

13.3.4 Discussion

The present use of banking and financial services by small and micro enterprises (including small-scale farmers) is very limited, with a tiny section of the market – estimated at less than 1% – presently being reached. This is the most unbanked area of micro-finance. No study that provides a comprehensive overview and analysis of overall supply to the lower income and small and micro enterprise market is known.

South Africa has a vibrant micro-finance sector that has reflected immense growth over the last decade. It is, however, focusing on wage and salary earners and mostly targeted as consumer credit. The sector's growth was phenomenal and the current size of the sector on the basis of current portfolio is estimated as R15 billion with the value of loans written per year being estimated at R30 billion. The majority of financial transactions take place in urban areas.

With the withdrawal of the conventional formal private sector institutions from rural areas, and an emphasis on wage and salary earners by the 'new' formal private sector financiers, the financial service provision to rural people is not expected to improve. This is especially true for the self-employed rural entrepreneur and the informally employed in rural areas. The fact remains, if rural people do not turn to

Table 13.3: Illustrative profile of farming households based on recent surveys in two provinces (1997).

		Limpopo Province (n=150)		*KwaZulu-Natal (n=152)*	
Household characteristics					
Male household head (%)		68.0		72.4	
De facto female household head (%)		99.3		66.4	
Household size		–		8.8	
Land tenure					
Cultivate all land (%)		79.3		58.6	
Leased land out (%)		8.0		9.9	
Borrowed farm land (%)		9.3		13.8	
Services and energy use (% of households)					
Use grid electricity		7.3		61.8	
Use flush toilet		1.3		1.3	
Cook with fire wood		95.0		78.9	
Water source unprotected		83.3		66.4	
Public standpipe		44.0		19.1	
Gross annual cash income (R)					
From farming		394		637	
From wages		9 308		14 157	
From micro enterprises		149		1 294	
From remittances		974		4 251	
From pensions		1 489		2 667	
Farm assets – % of households with:					
Car/bakkie		5.3		22.0	
Tractor		5.3		2.0	
Trailer		2.0		1.3	
Plough		4.8		6.0	
Credit used to purchase durables (% of households)	**Main source of credit**				
Car/bakkie	Commercial bank – CB	3.3	CB	9.9	PS
Generator	Private supplier – PS	1.3	–	2.6	–
Refrigerator		4.0	CB	32.9	PS
Television		2.0	–	19.7	–
Savings facilities used (% of households)	**Distance to institution (km)**	%	km	%	km
Banks		30.0	60.7	48.7	45.0
Money keeper		0.0	0.0	3.9	21.0
Informal		80.7	0.0	29.6	19.0
Farm inputs financed on credit (% of households)					
Fertilizers		7.3		9.0	
Seed		6.0		5.0	
Farm equipment hired		2.7		0.0	
Chemicals		0.7		6.0	
Veterinary medicines		0.7		0.0	
Ploughing services		0.0		9.0	
Transportation services		8.0		10.0	
Credit source used to finance farm inputs and implements (% of households)					
Local money lender		0.7		0.7	
Friend or relative		4.7		3.3	
Development corporation		10.0		0.0	
Input supplier		0.0		2.6	

Source: Quattara & Graham (1996) as adapted

Table 13.4: Illustrative profile of small business households based on surveys in two provinces in 1996–97.

	Northern Province (n=270)	*KwaZulu-Natal Province (n=225)*
Average age in years	26	46
Female (%)	21.6	22.9
Women who can sign contract without husband's permission (%)	33.0	70.7
Number of years in operation	8.3	10.0
Number of workers at start	3.2	4.5
Average value of business at start (R)	5 612	4 509
Number of current employees	3.5	4.6
Sector of operation (%)		
Construction	37.8	35.5
Manufacturing	26.3	26.7
Services	35.6	37.8
Multiple sources of financing business at start (%)		
Personal savings	93.3	73.3
Retrenchment package	29.6	12.0
Pension	17.4	0.9
Loan from family or friends	10.0	4.9
Loan from development corporation	5.9	6.2
Income from other business	4.1	13.3
Sole ownership (%)	87.4	93.8
Registered business (%)	33.0	22.7
Place of work (%)		
Home	37.0	34.7
Adjacent to home	7.4	5.3
Market	50.4	33.3
Place of delivery of service	5.2	26.7
Selected income, input and output indicators (R)		
Average wages received per month	1 326	1 102
Average remittances received per month	739	1 360
Average pensions per month	414	438
Average business income for business 1 per month	3 601	2 057
Average business income for business 2 per month	3 796	1 979
Average business income for business 3 per month	7 721	2 800
Average annual agricultural income	1 250	10 886
Average value of physical assets on hand	17 913	12 995
Average value of inputs per annum	1 067	12 703
Use of loans (%)		
Asset finance: Cash loans	10.0	10.0
Asset finance: Hire Purchase	14.0	16.0
Supplier credit	7.0	11.0
Customer advance	16.0	29.0
Customer loans	18.0	16.0
Shopkeeper loans	56.0	–
Savings (%)		
Formal: Mostly commercial banks	92.0	77.0
Informal: Stokvel	26.0	76.0
Informal: Burial society	64.0	22.0
Electronic banking	35.0	18.0
Handling of shocks or misfortunes (%)	**21.0**	**52.0**
Main principle cause: Stolen equipment (% of shocks)	52.0	31.0
Main fallback mechanism: Personal savings (% of shocks)	67.0	24.0

Source: Quattara & Graham (1996) as adapted

self-help organisations, no real volume of financial services will be forthcoming in rural areas. This emphasis on self-help schemes is evident from the rich fabric of informal financial groups in rural areas. These groups or collective action endeavours come in many guises and take place in many different locations.

When we consider different financial products, it is clear that different supply situations exist. For short-term financing of agricultural production, a range of institutions (from private sector to public sector) is involved. However, this represents a problematic area, as no major expansion or innovation in the financing of small-scale farmer agricultural production is evident. The picture is even more bleak when term financing of agricultural activities is considered. This is an international phenomenon and not unique to South Africa.

13.3.5 Innovations in small-scale farmer finance

Recent research (see the note accompanying Figure 13.4) in South Africa focuses on the profile of small-scale farmer households in an attempt to improve small-scale farmer finance products based on a better understanding of the households. The Land Bank, Uvimba Rural Finance and Ithala Development Finance Limited[4] took part in a study that used both qualitative and quantitative research techniques to understand client profiles better. The study took place in three provinces and the results indicate different livelihood strategies followed by small-scale farmer households. They further reflect the impact of seasonality on both income and expenditure of rural households. The results from these surveys will be applied to refine existing products of the financial institutions, or to design new products. It is clear that new products should be more flexible and less focused on agriculture (good examples exist in other parts of Africa, Coetzee *et al.*, 2002), and should be in support of all the economic livelihood strategies of poor rural households. A narrow focus on agriculture ignores the range of livelihood strategies, as well as the inherent risk of the agricultural activities of households.

13.4 SELECTED SUPPLIERS OF AGRICULTURAL FINANCE

In this section important stakeholders in the provision of financial services to farmers are discussed. The National Department of Agriculture (NDA) influences the agricultural sector with its policies, while the Land Bank, commercial banks and co-operatives represent important sources of agricultural finance.

The NDA has phased out almost all past subsidies and is investigating several new interventions that seek to remove constraints on rural lending by reducing the risks involved in lending to emerging black farmers. Among other potential interventions, the NDA seeks to launch a new approach to the support and

[4] Uvimba is a parastatal financier in the Eastern Cape Province. Ithala is also a state-supported parastatal operating in KwaZulu-Natal. These two institutions, the Land Bank and the University of Pretoria and Imperial College at Wye collaborated on this study in 2001 and 2002.

structuring of a co-operative based rural finance programme, integrated with the rest of the financial sector. The purpose of this approach is not only to encourage retail institutions to finance a new segment of small-scale farmers, but also to assist rural people to initiate self-help financial institutions in the form of savings and credit co-operatives.

13.4.1 The Land Bank

The Land and Agricultural Development Bank of South Africa has been financing agriculture since its inception in 1912. The Bank's Head Office is situated in Pretoria, with 27 branches and 45 satellite offices spread across the country.

At present the Land Bank finances approximately 25 000 commercial farmers and over 90 000 small-scale and emerging farmers, and rural dwellers. The 'development' portfolio of the Land Bank is reaching the R450 million mark (Land Bank, 2001) and the Bank is challenged with the design of more appropriate products and improving its knowledge of development and rural finance amongst its front-line staff.

13.4.2 Commercial banks

The current activities of the private sector in the provision of rural financial services can be seen as limited in terms of the lack of a full range of services in all areas. However, the commercial banking sector has the highest number of branches in the rural areas (Strauss Commission, 1996), and together with the Post Office they represent the best distribution of branches. South Africa also has a higher ratio of branches per population than anywhere in Africa with the average number of people per commercial bank branch in Africa estimated at 450 000:1 (Coetzee, 1992), whereas South African data indicate a countrywide average of 15 000:1 (estimated from information in the final report of the Strauss Commission, 1996). However, even with this reach they are not interested in financing the self-employed and informally employed. They cannot access the risk in small businesses and small-scale farm businesses, and the small loan units result in high transaction costs, a scenario not palatable for commercial banks with performance and return requirements from shareholders. Commercial banks finance commercial farmers, mostly by using traditional agricultural finance instruments, but lately also by means of innovative financial products, with one such product range emphasising hedging against price changes and access to loan finance for crop production.

13.5 THE FUTURE OF AGRICULTURAL FINANCE IN SOUTH AFRICA

The greatest developments in agricultural financing in South Africa have to do with the changes that have taken place in the agricultural sector and the normalisation of the political situation in the country.

The change with regards to domestic marketing have also been accompanied by extensive factors relating to international marketing. As international markets have been opened to South African farmers, new requirements have been set for the financing of exports. Three financing needs can be identified in this regard. Firstly, the financing of domestic markets is important, but in many cases, the financing service provided by the marketing boards now has to be taken over by the individual farmer. The financing of marketing also requires the necessary financial instruments needed for participation in futures and options markets. Secondly, exports to neighbouring countries are on the increase and the financing of this aspect is a challenge, both for the farmer and for financial institutions. South African banks have begun to establish themselves in neighbouring countries, which goes a long way to facilitating financing services. Thirdly, international marketing makes its own demands on both the farmer and the financier. Over the past 15 years, agricultural foreign trade has increased from 7.1– 7.5% of total foreign trade (Abstract of Agricultural Statistics, 2002).

The broadening of the agricultural sector to include all part-time and small-scale farmers increases the demand for innovative agricultural financing products. The development of new products for this sector requires a fresh look at small-scale farmers and their households. The phenomenon that the majority of small-scale producers farm on a part-time basis and that the farm business contributes a small portion of total household income creates problems and challenges for financiers and there is not yet a widespread example of the successful financing of small-scale farmers in South Africa. The sugar industry probably has the largest number of small-scale farmers that were financed without state help.

The small-scale farmer sector also offers a lesson in diversification for most farmers. Most small-scale farmer families have a variety of income sources, and in most cases agriculture is not the most important source of income. Many larger-scale farmers focus on a small number of income sources, and in most cases only on a single source of income. A diversified approach to income sources is one of the most effective methods of addressing the issue of risk at farm level. Value-adding at farm level is another way in which to diversify income sources.

Certain farmers in South Africa have the opportunity of entering the tourism sector (and some have done so already) as a result of the diverse income possibilities offered by agricultural assets. This places new demands on agricultural financing and many farmers are involved in sophisticated financing schemes to finance these non-farming investments. In this manner, financiers that are not traditionally associated with agriculture are coming into contact with the farming sector. As a result, the number of different financiers involved is diversified; allowing for more competition in the rural financing market and for more innovative financing products and approaches.

The recent trend of broadening the ownership of farm businesses (usually in order to give the farm workers the ability to build up capital reserves and thus be more closely associated with the success of the farm business) is a positive step towards lowering the risk of the individual farmer. Just as the profits are shared in these situations, so too are the risks (and possible losses) also shared.

It is clear that the sector offers many opportunities for innovation. The current model for deferred payment financing packages for the purchase of agricultural land is a good example of an innovation created by broad co-operation between practice, academics and the state. Agricultural financing for commercial and small-scale farmers is confronted with great challenges, and challenges give rise to innovation and opportunities. In general, this should be the basis for a positive approach to agricultural financing.

REFERENCES

ABSTRACT OF AGRICULTURAL STATISTICS. (2000). National Department of Agriculture, Pretoria.

ABSTRACT OF AGRICULTURAL STATISTICS. (2002). National Department of Agriculture, Pretoria.

AGRICULTURAL NEWS. (1997). Newsletter of the National Department of Agriculture, 8 September. National Department of Agriculture, Pretoria.

COETZEE, G.K. (1992). Conference on 'rural development and finance', University of Ouagadougou (Burkina Faso, 21 to 25 October 1991). *Development Southern Africa* 9: 89–104.

COETZEE, G.K. (1998). Retail rural finance in South Africa: from policies to practice. *Agrekon* 37: 517–526.

COETZEE G.K. & GRANT W.J. (2000). Department of Trade and Industry: interest rates study of small loan sector. Unpublished report, Department of Trade and Industry.

COETZEE, K. & VILJOEN, M.F. (1991). Assessing economic values for inclusion in machinery replacement decisions. *Agrekon* 30: 230–236.

COETZEE, G.K., KABBUCHO, K. & MNJAMA, A. (2002). The rebirth of Equity Building Society in Kenya. Unpublished report for MicroSave-Africa, Nairobi, Kenya.

ESCOM. (1998). *Consumers in South Africa*. ESCOM, Sunninghill, South Africa.

KGOWEDI, M.J. (2001). Providers of informal finance in the Malatji District of Limpopo Province. MSc thesis, University of Pretoria.

KIRSTEN, J.F. & VINK, N. (1999). Agricultural policy in South Africa in the 1990s. Unpublished mimeo, Universities of Pretoria and Stellenbosch.

LAND BANK. (2001). *Annual Report*. Pretoria: Land Bank.

MEYER F., COETZEE G.K. & ADAM H. (2000). The financial position of South African agriculture. Unpublished mimeo, Rural and Agricultural Finance Programme, University of Pretoria.

NATIONAL DEPARTMENT OF AGRICULTURE (NDA). (2002). *Economic Overview of South African Agriculture*. Pretoria: National Department of Agriculture.

QUATTARA, K. & GRAHAM D.H. (1996). Rural finance in the Republic of South Africa: selected farm household case studies. Report for the Development Bank of Southern Africa, Rural Finance Programme, Ohio State University, Colombus, Ohio.

SARB QUARTERLY BULLETIN. (1999). March, South African Reserve Bank, Pretoria.

STRAUSS COMMISSION. (1996). *The Interim and Final Reports of the Commission of Inquiry into the Provision of Rural Financial Services*. Pretoria: Government Printer.

VAN SCHALKWYK, H.D. & VAN ZYL, J. (1995). The South African land market: an analysis of land prices. Unpublished mimeo, University of Pretoria.

VAN ZYL, J. & COETZEE, G.K. (1990). Food security and structural adjustment: empirical evidence on the food price dilemma in Southern Africa. *Development Southern Africa* 7: 105–116.

CHAPTER 14

Agricultural Trade Patterns in the Southern African Development Community

Howard K. Sigwele

14.1 INTRODUCTION

This chapter examines the role of agriculture, agricultural trade, and linkages of the sector to the economies of the Southern African Development Community (SADC) countries. It then analyses trade patterns in the SADC with special reference to tariff structures, factors influencing trade flows, the main traded farm products, product destinations (both regionally and globally), and provides an economic rationale for trade. It also explores opportunities for and threats to increased intra-regional trade consistent with the aims of the SADC Trade Protocol implemented in late 2000. Some policy considerations to improve intra-SADC trade and enhance economic integration are also discussed.

14.2 THE ROLE OF AGRICULTURE IN SADC ECONOMIES

People in rural areas still account for at least 70% of the SADC's population which is estimated at 200 million (IFAD, 2001; Lewis, 2001), and depend on agriculture, particularly rain-fed farming, livestock, forestry and fishing. At least 40% of the SADC population lives below the poverty datum line (daily per capita food consumption of about one US dollar) and most of the poor are, as expected, in the rural areas (World Bank, 2001). A high proportion of these people live in female-headed households whose resource base is limited given the historical and cultural/legal practices found in the sub-region (World Bank, 2001; IFAD, 2001). Poverty or income insecurity is also increasing in the urban areas of the SADC economies, as sustainable alternative employment opportunities in the formal sector are limited. Unemployment is relatively high with some countries facing unemployment rates of 30% and more.

Despite the risk and generally low productivity of agriculture in the SADC, the sector accounts for at least 10% of the Gross Domestic Product (GDP) in Lesotho,

Malawi, Mozambique, Namibia, Swaziland, Tanzania, Zambia and Zimbabwe, and is the single largest source of employment and foreign exchange earnings (World Bank, 2001; IFAD, 2001). Table 14.1, providing a list of SADC countries, shows generally low and fluctuating average growth rates of real GDP for the SADC, especially before 1994. After 1994, the SADC generally witnessed positive average annual and real GDP growth rates, although the rates were below 2% for countries like Zambia and the Democratic Republic of the Congo (DRC). In the DRC, the internal civil war is partly responsible for low GDP growth, while inappropriate macro-economic and sectoral policies coupled with low private and public investment could be possible reasons in other countries.

South Africa, the most technologically advanced country in the SADC and in Sub-Saharan Africa (SSA), accounts for 76% of the SADC's GDP and about 43% of SSA's GDP. The aggregate GDP for the SADC is about US$184 billion (Lewis *et al.*, 2002). South Africa is also a major trade market for the SADC's products (Lewis, 2001; Yeats, 2000), while Botswana, Mauritius, Namibia, Seychelles and South Africa have higher per capita incomes than the rest of the SADC. Given the high income inequalities existing even among the richer SADC members, the GNP per capita income data should be treated with caution (World Bank, 2001).

While Table 14.1 shows that agriculture's contribution to GDP has declined in many countries since independence, the sector has very strong forward and backward linkages in their respective economies. For Botswana, Malawi, Zambia, Zimbabwe, South Africa, Mozambique and Tanzania, the sector has strong intra-sectoral linkages and income and expenditure relations with households and other

Table 14.1: Basic population and economic indicators of SADC economies.

Country	*Population (millions)*[a]	*Agriculture % of GDP*[b]	*GNP per capita at current US dollars*[c]	*Agriculture labour share*[b]	*Average annual real GDP growth % p.a*[a]		
					1980–89	*1990–94*	*1995–99*
Angola	12.4	7.8	220	7.5	2.6	–5.9	6.8
Botswana	1.6	4.2	3 240	46	10.6	4.6	4.8
DRC	49.8	50.0	130	62	1.8	–8.6	0.9
Lesotho	2.1	11.0	530	40	3.6	4.4	3.9
Malawi	10.8	44.2	190	87	1.7	1.0	7.3
Mauritius	1.2	9.4	3 590	17	4.2	5.4	5.2
Mozambique	17.3	39.4	230	83	0.1	2.6	8.7
Namibia	1.7	13.6	1 890	49	1.1	4.3	2.8
Seychelles	0.1	4.0	6 540	11	2.1	4.8	2.4
South Africa	42.1	4.4	3 160	13	2.2	0.2	2.3
Swaziland	1.0	12.5	1 360	49	6.7	3.8	2.9
Tanzania	32.9	46.2	240	84	3.9	2.7	3.7
Zambia	9.9	18.7	320	75	1.4	0.2	1.3
Zimbabwe	11.9	17.1	520	68	5.1	2.1	3.1
Total	194.7		931		3.4	1.5	4.0

Source: [a] Lewis (2001); [b] IFAD (2001); [c] World Bank (1999)

financial and non-financial institutions (McDonald & Townsend, 1998; Sigwele, 1999; McDonald & Walmsley, 2001; Lewis *et al.*, 2002).

This circular flow of income and expenditure is captured in a Social Accounting Matrix (SAM) that also provides information about the structure of the economy and the distribution of income among households. Information about the SAM is important especially for agriculture, since its GDP contribution underestimates the income and expenditure inter-relationships influenced by farming in most SADC countries. Policy studies using SAM-based and computable general equilibrium (CGE) models confirm the sector's strong income multiplier effects, especially among low-income countries, like those in the SADC region (Sadoulet & De Janvry, 1995; McDonald & Townsend, 1998; Townsend & Sigwele, 1998).

Given exogenous shocks like changes in farm income caused by productivity improvements or agricultural export earnings (due to increased market access to industrialised economies), the sector demonstrates large income multiplier effects to several economic agents (factors, household, productive activities, government) (Ehui & Delgado, 1999). These effects underscore the importance of agriculture in enhancing social welfare. Dynamic economy-wide models like CGE show how improvements in factor productivity through technological and human capital developments can increase agricultural incomes, stimulate agro-processing, enhance economic diversification and generate growth in trade (Ehui & Delgado, 1999; Lewis, 2001).

Similarly, unlimited market access of agricultural exports (both primary and processed) from low-income countries like some SADC states to industrialised countries can generate additional foreign earnings estimated at US$60 billion per year (World Bank, 2002). Unlimited access requires total elimination of tariffs and farm subsidies estimated at US$400 billion per year by mostly the EU, US and Japan, and the removal of discriminatory non-tariff barriers like technical quality standards, disease control, quotas, etc.

14.3 AGRICULTURAL TRADE PATTERNS IN THE SADC

The SADC states as a regional economic grouping began as an organisation promoting co-operation through development projects in physical infrastructure, industry, agriculture and natural resources (water, tourism, energy, etc). In 1992, the SADC Treaty was changed to promote regional integration led by among others, trade and development. This strategy necessitated the harmonising of macro-economic and sectoral policies to promote trade free of barriers, cross border investment, and liberalised financial and capital markets, including the foreign exchange market. This has been a daunting task – some countries still pursue inward agricultural and industrial policies to replace imports that, in turn, require high import duties and other non-tariff barriers to protect local industries and raise government tax revenue. Conversely, other countries have opened up their economies mainly due to donor conditions (International Monetary Fund [IMF] and World Bank) or multilateral trade obligations like fulfilling World Trade Organisation (WTO) requirements.

While acknowledging the challenges of regional integration partly led by intra-trade, the SADC adopted a 'Trade Protocol' also known as the Maseru Protocol in 1996. The Protocol, crafted along the WTO framework, advocated freer trade within the region and the promotion of the free movement of capital. It further obligated each member state to trade liberalisation by 2008, when up to 85% of the SADC intra-trade should be free and open to intra-regional competition. Trade liberalisation means unilateral reduction of tariffs and non-tariff barriers (quotas, licences, etc.) by all member states in a transparent and consistent manner.

Parallel institutional configurations complicate the realisation of the SADC Trade Protocol. The SADC region has several trade and economic groupings, some of which pre-date the creation of the SADC. These configurations have different and sometimes conflicting legal and technical requirements among the same SADC member countries. Table 14.2 shows the various trade organisational formations and overlaps among SADC member states, with notable overlaps between the SADC, the Common Market for Eastern and Southern Africa (COMESA) and Cross-Border Initiative (CBI), and to some extent, the Southern African Customs Union (SACU). Except for the SACU, none of the organisations have harmonised common tariffs and technical quality standards.

The SACU was formed in 1910 by the then Union of South Africa and British-protected states of Botswana, Lesotho and Swaziland (the BLS countries). At that time Namibia was also still colonised. After independence, BLS re-negotiated the SACU agreement in 1969 but left the political and legal powers to vary tariffs in the Agreement to South Africa. The frequent changes to tariffs and other trade barriers in the SACU, and lack of adequate consultation, led to frequent concerns by the BLS countries. When Namibia joined the SACU after independence in 1990, there were already concerns about the inequitable and undemocratic nature of the SACU

Table 14.2: Membership of regional organisations in southern Africa.

Country	*SACU*	*CMA*	*SADC*	*COMESA*	*CBI*
Angola			X	X	
Botswana	X		X		
DRC			X	X	
Lesotho	X	X	X		
Malawi			X	X	X
Mauritius			X	X	X
Mozambique			X		X
Namibia	X	X	X	X	
Seychelles			X	X	X
South Africa	X	X	X		
Swaziland	X	X	X	X	X
Tanzania			X		X
Zambia			X	X	X
Zimbabwe			X	X	X

Source: Lewis (2001); Yeats (2000)

structures. In 1994 the process to renegotiate the new SACU gathered momentum, partly due to the democratisation of South Africa.

Since 1994, negotiations have taken place on the democratisation of the SACU and tariff reductions by South Africa in accordance with WTO obligations. A new SACU agreement with powers vested in the Council of Ministers representing member states was adopted, and a secretariat to be based in Namibia will soon be set up to administer the Agreement. A common external tariff will still be maintained, while freer movement of goods will be enhanced within the SACU territory. A tariff tribunal will advise on equitable tariff structures. Given the possible establishment of a common market by the SADC, it remains to be seen whether the SACU will still remain relevant.

Within the SACU, a Free Trade Agreement (FTA) between South Africa and the EU was recently established, but the BLNS (BLS plus Namibia) countries still have to decide on whether to join the Agreement. The SACU countries have a common external tariff, so it would be almost impossible to identify and separate goods destined only for South Africa under the EU-SA FTA. Lewis *et al.* (2002) indicate that the BLNS countries can benefit if they join the EU-SA FTA, while also having unlimited access to the EU market. They predict an increase in GDP and export growth, but Botswana and South Africa will benefit relatively more from the FTA arrangement. There will also be government revenue losses due to lower tariffs, etc. The study, however, advocates a SADC-wide FTA with the EU to remove concerns about rules of origin. As the BLNS countries have not yet joined the FTA, there is an urgent need to resolve their status.

The SADC was formed in 1980 as a regional economic grouping to reduce dependence on South Africa during the apartheid era, and to co-operate in development projects like infrastructure, early warning, animal disease control, and human resource/technology development. Regional integration through tariff reductions and other barriers will be complicated further by the presence of *strategic and sensitive industries* in both the SADC and the SACU such as the red meat, grain, dairy, textile and motor industries. This has led to the imposition of very high duties to protect these industries in the two trading blocks. The differential treatment and tariff structures of industries may delay trade liberalisation in the SADC.

Agricultural intra-regional trade is also hampered by sanitary and phyto-sanitary (SPS) measures used to protect against the spread of livestock and plant diseases and pests. The outbreak of diseases like Foot and Mouth, and New Castle, and the spread of insects, fungal and bacterial diseases in plants, can adversely affect trade-led economic integration. SPS measures have already been applied to suspend intra-regional trade among SADC countries. SPS measures based on transparent, consistent and scientific evidence are admittedly necessary, but sometimes they may be used to protect inefficiency and rent-seeking groups who do not want competition.

It also remains to be seen whether by 2008 most of the anticipated SADC intra-regional trade will materialise. Political and fiscal considerations by member states, coupled with ensuring economic and commercial interests of rent-seeking groups like trade unions, farmers and industrialists, are mainly responsible for the slow pace of trade liberalisation in most economic groupings (Krueger, 1974; Jammeh, 1988; Sodersten & Reed, 1994; Lewis, 2001). The SADC countries face major political and economic challenges in trying to achieve regional integration partly through trade. Strong political will is needed to achieve the SADC's vision. Economic and trade integration is probably the most rational route, but there are costs at both regional and member state levels in terms of loss of political power and reduction in tariff revenue.

The Common Market for Eastern and Southern Africa (COMESA) was established in 1981 as a Preferential Trade Area (PTA) to liberalise trade among member states through tariff reduction and the free movement of goods. It aims to establish an economic union with intra-regional co-operation in fiscal and monetary matters (almost like the present EU). To date this objective has not materialised, nor is there a common external tariff. Except for Botswana, Lesotho, Mozambique, South Africa and Tanzania, the remaining SADC countries are also COMESA members. COMESA also includes the Comoros, Djibouti, Ethiopia, Kenya, Madagascar, Rwanda, Sudan and Uganda. With overlapping membership between SADC and COMESA, it is doubtful whether these organisations can work together without conflicts in tariff structures, rules of origin, SPS standards, etc. Some rationalisation will be required to sustain the two organisations that currently depend heavily on the same international donors for capital projects and part of operating expenses.

The Cross-Border Initiative (CBI) is another economic grouping, covering both SADC and COMESA members, aimed at further trade liberalisation and the promotion of cross-border investment among members. Sponsored by the EU, the World Bank, the IMF, and the African Development Bank, CBI – like the SADC and COMESA – is expected to expedite the removal of investment and trade barriers among members, but major political, economic and financial costs are at stake for each country. CBI barriers across members are still in place despite the deadline to remove them by 1996. Among SADC members Angola, Botswana, the DRC, Lesotho, Mozambique and South Africa are not CBI members. Overlapping membership and duplication may again seriously prejudice efforts towards intra-regional integration in southern and eastern Africa.

The Common Monetary Area (CMA) essentially covers South Africa, Lesotho, Namibia and Swaziland where the South African Rand is legally accepted as a currency to conduct trade among the four countries. Botswana pulled out in 1976 when it established its own Pula currency. While Lesotho, Namibia and Swaziland have their own currencies, the currencies are linked to the Rand and the latter circulates freely in these countries.

Finally, some SADC countries have several bilateral agreements that are in conflict with WTO provisions (Maasdorp, 1998). South Africa has bilateral agreements with Malawi, Mozambique and Zimbabwe, while Botswana, Malawi and Namibia have trade agreements with Zimbabwe. Bilateral trade agreements discriminate essentially against third parties, whereas according to the WTO, all members should enjoy the most favoured nation (MFN) status. The MFN status in trade relations is non-discriminatory as it advocates equal treatment among countries. It remains to be seen whether these multiple memberships by SADC countries in several trade organisations augur well for integration.

Intra-SADC tariff protection could also influence trade flows within the region. Yeats (2000) found that while intra-SACU import protection rates were very low and standard, tariff rates among other SADC countries outside the SACU were too high to promote trade integration. Membership of several economic groupings as shown in Table 14.2 complicates the setting of equitable tariff rates, and customs administration. Section 14.4 summarises the tariffs used to protect sectors and products within the SADC. These differential tariff rates may be responsible for the current trade patterns and low intra-regional trade. Further, historical links such as belonging to the same trade/custom block like the SACU, geographical proximity, logistical support and convenience including the use of easily convertible currencies, could also partly explain existing trade patterns (Maasdorp, 1998; Yeats, 2000).

14.4 TARIFF STRUCTURES IN THE SADC

Import tariff protection rates, to protect local industries, are higher for processed products such as food, textiles and apparel as shown in Table 14.3. For instance, South Africa has a 100% levy on processed food from Malawi, 71% against other non-SACU SADC countries, 66% against Zimbabwe, 50% against Zambia and 49% against Tanzania. Botswana imposes import duties on processed food from as high as 70% against Zimbabwe to zero for most SADC countries (Lewis *et al.*, 2002).

Mozambique has a levy of 50% for processed food from Tanzania and very low rates under 10% against other SADC countries. Zimbabwe's import protection rates on processed food range from 50% against Tanzania to 24% against Botswana, Malawi, Mozambique and Zambia, while the rest of the SADC nations are levied at about 27%.

Table 14.3 shows tariffs by agricultural product in the SADC, and import duties against similar goods from the EU. For instance, South Africa imposes duties of 37.5, 37.3 and 38.8% on grains imported from Tanzania, Zimbabwe and the EU, respectively. No import duty on this product is imposed on Botswana because of the SACU Agreement.

In comparison with processed food commodities, import protection rates on primary agricultural products like grains, vegetables, livestock, cotton, etc. are generally low, partly supporting the initial post-independence policies on industrial import substitution. The only countries that impose duties on grains against some SADC countries are the SACU countries. Botswana and South Africa have import protection rates of about 40% each against food grains from Tanzania, Zimbabwe and Zambia. Other SADC countries have zero import duties on grains. Tanzania,

Table 14.3: Sectoral bilateral import tariffs and non-tariff barriers (*ad valorem* %).

Importing country/product	*Exporting country/product*								
	South Africa	*Botswana*	*Malawi*	*Mozambique*	*Tanzania*	*Zambia*	*Zimbabwe*	*Rest of SADC*	*EU*
South Africa									
Grain	0.0	0.0	0.0	0.0	37.5	0.0	37.3	33.3	38.8
Fruit and vegetables	0.0	0.0	0.0	0.0	20.0	0.0	25.7	0.0	25.5
Other agriculture	0.0	0.0	0.0	0.0	13.4	0.0	12.9	0.0	9.3
Livestock	0.0	0.0	0.0	0.0	7.7	0.0	7.2	0.0	7.3
Forestry and fisheries	0.0	0.0	0.0	0.0	0.0	0.0	0.0	0.0	0.3
Food processing	0.0	0.0	100.0	0.0	48.7	50.0	65.3	71.4	71.4
Botswana									
Grain	0.0	0.0	0.0	0.0	0.0	37.0	40.7	0.0	0.0
Fruit and vegetables	0.0	0.0	0.0	0.0	0.0	0.0	0.0	0.0	50.0
Other agriculture	0.0	0.0	8.6	0.0	0.0	34.1	9.3	0.0	0.0
Livestock	0.0	0.0	0.0	0.0	0.0	0.0	0.0	0.0	0.0
Forestry and fisheries	0.0	0.0	0.0	0.0	22.2	44.4	43.4	0.0	50.0
Food processing	0.0	0.0	0.0	0.0	0.0	0.0	69.8	0.0	67.2
Malawi									
Grain	0.0	0.0	0.0	0.0	0.0	0.0	0.0	0.0	0.0
Fruit and vegetables	0.0	0.0	0.0	0.0	0.0	0.0	0.0	0.0	0.0
Other agriculture	0.0	0.0	0.0	0.0	0.0	0.0	35.4	0.0	100.0
Livestock	0.0	0.0	0.0	0.0	0.0	0.0	0.0	0.0	12.5
Forestry and fisheries	0.0	0.0	0.0	0.0	0.0	0.0	0.0	0.0	0.0
Food processing	0.0	0.0	0.0	0.0	0.0	0.0	20.3	0.0	32.0
Mozambique									
Grain	0.0	0.0	0.0	0.0	0.0	0.0	1.2	0.0	2.4
Fruit and vegetables	0.0	0.0	0.0	0.0	0.0	0.0	8.7	0.0	25.0
Other agriculture	0.0	0.0	0.0	0.0	0.0	0.0	11.2	0.0	4.1
Livestock	0.0	0.0	0.0	0.0	0.0	0.0	0.0	0.0	0.0
Forestry and fisheries	0.0	0.0	0.0	0.0	0.0	0.0	0.0	0.0	0.0
Food processing	0.0	0.0	0.0	0.0	50.0	0.0	5.1	0.0	32.2

Table 14.3: *Continued.*

Importing country/product	*Exporting country/product*								
	South Africa	*Botswana*	*Malawi*	*Mozambique*	*Tanzania*	*Zambia*	*Zimbabwe*	*Rest of SADC*	*EU*
Tanzania									
Grain	17.9	0.0	16.7	29.5	0.0	0.0	0.0	0.0	17.6
Fruit and vegetables	50.0	0.0	21.4	0.0	0.0	0.0	0.0	0.0	8.0
Other agriculture	25.0	0.0	0.0	0.0	0.0	0.0	39.1	0.0	30.0
Livestock	0.0	0.0	0.0	27.3	0.0	0.0	18.2	0.0	28.1
Forestry and fisheries	50.0	0.0	0.0	0.0	0.0	0.0	0.0	0.0	10.3
Food processing	28.1	0.0	27.8	27.8	0.0	33.3	30.2	33.3	22.2
Zambia									
Grain	0.0	0.0	0.0	0.0	0.0	0.0	3.1	0.0	0.0
Fruit and vegetables	0.0	0.0	0.0	0.0	0.0	0.0	25.0	0.0	13.0
Other agriculture	0.0	0.0	0.0	0.0	0.0	0.0	8.5	0.0	5.0
Livestock	0.0	0.0	0.0	0.0	0.0	0.0	0.0	0.0	8.4
Forestry and fisheries	0.0	0.0	0.0	0.0	0.0	0.0	0.0	0.0	0.0
Food processing	0.0	11.1	0.0	0.0	25.0	0.0	10.9	0.0	16.5
Zimbabwe									
Grain	7.4	0.0	0.3	0.0	0.0	0.9	0.0	0.0	6.3
Fruit and vegetables	35.0	0.0	25.0	33.3	0.0	0.0	0.0	0.0	14.3
Other agriculture	55.1	0.0	11.9	18.8	95.5	11.8	0.0	0.0	8.0
Livestock	13.7	0.0	0.0	0.0	0.0	0.0	0.0	0.0	4.2
Forestry and fisheries	10.8	0.0	0.0	0.0	0.0	0.0	0.0	0.0	0.0
Food processing	30.4	24.3	24.5	24.3	50.0	24.1	0.0	27.6	43.0
Rest of SADC									
Grain	0.2	0.0	0.0	0.0	0.0	0.0	2.1	0.0	0.5
Fruit and vegetables	22.7	0.0	0.0	0.0	0.0	0.0	26.5	0.0	5.5
Other agriculture	15.4	0.0	0.0	0.0	0.0	0.0	28.8	0.0	13.1
Livestock	1.8	0.0	0.0	0.0	0.0	0.0	0.0	0.0	2.5
Forestry and fisheries	9.1	0.0	0.0	0.0	0.0	0.0	0.0	0.0	9.1
Food processing	44.9	0.0	0.0	0.0	22.5	0.0	7.0	0.0	29.6

Source: Southern African Model database derived from GTAP 5.0, final version

Note: Tariffs are for imports *from* column country *to* row country

however, has a 17% import levy on grains from South Africa, while fruit and vegetables from South Africa are levied at about 50% by Tanzania.

Most SADC countries embarked on import substitution industries immediately after independence as a strategy to replace imported manufactured agricultural and industrial goods. A combination of fiscal instruments, such as high import duties on competing goods, and discriminatory non-tariff barriers like quotas, monopoly licencing, differential technical quality standards, and rules of origin, were applied to protect infant domestic industries. Added to this objective was the need to raise government revenue, which is very important in several SADC countries (Lewis, 2001).

Besides primary and processed food products, the SADC also has high import duties (not shown in Table 14.3) on many commodities consumed within the region. Botswana has about a 25% duty on textile imports from Malawi, Zambia and Zimbabwe, while the rest of the SADC outside the SACU pays a duty of about 26% on textiles. Malawi imposes an import levy of 36% and 39% for textiles from South Africa and Zimbabwe, respectively. Zimbabwe levies duties of at least 27% on textiles imported from all SADC countries while South Africa, although with lower duties ranging from about 10–20%, imposes levies on imported textiles from all non-SACU SADC countries (Lewis *et al.*, 2002). Apparel/clothing also attracts import duties as do equipment, wood and paper.

Further, capital goods such as machinery and equipment attract very low import duties in the SADC, unlike agricultural, consumer and intermediate products. Presumably, the scarcity of capital goods influenced countries to levy low duties to attract investment. Also, given the high level of unemployment among the young and unskilled people in the SADC, preferential tariffs on capital goods might prejudice employment creation in sectors with high potential such as farming. Given the intra- and inter-sectoral linkages in an economy, discriminatory import tariff policy could harm the growth and viability of farming activities and SADC agricultural trade.

While SADC countries export the bulk of their agricultural and non-agricultural products to the EU under favourable tariff concessions, the EU still imposes import duties (not shown in Table 14.3) on SADC products. Processed food products from the SADC attract the highest duties, while primary agricultural goods enjoy preferential rates. Most import duties are, however, claimed back by the SADC countries under rebated products. For processed food commodities, the EU tariff rates range from about 29% against Mozambique, 36% against Tanzania, 44% against South Africa, 71% against Malawi and Zimbabwe, 74% against Zambia to 86% against Botswana. The rest of the SADC countries attract an import duty of about 66% on processed food commodities. Grains, fruit and vegetables also attract higher duties than other primary products like livestock, cotton, forestry, etc. Textiles and apparel attract lower import duties from all SADC countries. Similarly machinery, equipment, basic intermediate goods, wood and paper enjoy preferential tariff concessions from the EU.

Notwithstanding the current preferential African, Caribbean and Pacific (ACP)/ EU trade agreement, there are some inherent trade challenges or threats to SADC-EU trade in particular. Developing countries such as SADC countries are allowed to levy duties on goods imported from the EU, while the latter accords preferential tariff concessions to the former (developing countries). The agreement is non-reciprocal, as the developing countries are not expected to grant trade concessions to EU goods. Within the SADC group, South Africa is classified as a developed country and therefore its trade relations with the EU require reciprocity.

Note, however, that in its present form the current ACP-EU trade agreement is discriminatory and in conflict with the MFN arrangement advocated by the WTO. Other developing countries outside the ACP grouping specifically do not enjoy preferential tariff concessions. It is partly due to this trade discrimination that the banana exporting countries in Latin America challenged the existing ACP-EU agreement through the WTO as they also wanted to access the relatively lucrative and large EU market. The banana exporting countries won their case and arrangements are underway to improve their market access to the EU.

Further, the current ACP-EU trade agreement is expected to remain in force in its present form until 2007 when the agreement should comply with WTO provisions. This poses a major challenge for the SADC as the preferential treatment will most likely be reduced or removed while competition for the EU market will intensify as other trading countries join. The EU producer prices rank among the highest in the world because of large subsidies under the Common Agricultural Policy (CAP). The CAP favours producers both in the EU and those that freely access this market like the ACP/SADC countries.

14.5 TRADE STRUCTURE IN THE SADC

Before examining agricultural trade flows within the SADC, concerns that hamper the growth of trade in the sub-region and the structure of the region's trade will be identified. Trade flows are influenced by a complex set of factors of which supply, regional and international demand, trade policy, the macro-economic environment, logistical/infrastructural support, political stability and competition policy take precedence. Trade facilitation through efficient and transparent customs and immigration services is also critical to promote intra-SADC trade (Maasdorp, 1998; Lewis, 2001; Lewis *et al.*, 2002). Inordinate delays at custom border points in the SADC are a serious weakness in promoting intra-regional trade.

14.5.1 Factors affecting trade in the SADC

With regard to logistical/infrastructural support services, the SADC adopted a Protocol on Transport, Communications and Meteorology in 1996 to try to expedite trade by establishing standards and guidelines. Despite this Protocol, the movement of goods within the SADC is hampered by poor road and telecommunications infrastructure, some of which require urgent rehabilitation while certain transport/road user charges are too high to encourage growth in intra-SADC trade. Some countries levy road users in foreign currencies like US dollars, a practice that could divert trade to other countries. Railways and ports also need upgrading and regular maintenance and efforts are being made by the SADC in this regard, but a high dependency on donors to fund infrastructure may not be sustainable in future.

Turning to political stability within the region, prospects for peace have improved following the end to civil war in Angola while progress in the DRC to end the internecine war is encouraging. These wars paralyse the flow of goods and reduce incentives to increase the domestic production of tradables in these high-potential agricultural countries. Political instability makes domestic and intra-regional trade difficult, while the production of and investment in traded commodities becomes risky.

While the SADC Trade Protocol of 1996 covers competition law/policy to promote fair and transparent commercial practices, few member countries have formulated legislation to prevent uncompetitive business behaviour which can in some cases lead to price and market collusion to the detriment of poor households. The formulation of the competition law in the SADC is also complicated by the existence of public monopoly organisations that provide services such as agricultural marketing, farm input supply, utilities, post and transport at uncompetitive and inefficient rates due to public subsidies. The promotion of private sector investment to achieve competition is, in most cases, resisted by rent-seeking groups while the majority suffer.

Prior to the economic and trade reforms by the IMF and World Bank in the 1980s and the democratisation of South Africa in 1994, most SADC countries pursued inward-looking or import substitution policies which discouraged the expansion of exports through overvalued currencies, state exporting monopolies and poor incentives to producers. Lack of developed infrastructure and financial services discouraged the development of exports in agriculture and exports declined primarily due to supply constraints. Farm inputs, like fertilizers and equipment, were either not easily accessible or beyond the reach of most producers due to a lack of credit facilities. In those countries that administered subsidies, chronic budget deficits followed while high inflation reduced the real purchasing power of farm incomes.

While economic and trade liberalisation reforms have been undertaken in many SSA countries including the SADC, there is concern over welfare effects and the provision of public goods by respective governments (Lipumba, 1996; Stiglitz, 1998). After fixed and controlled exchange rates were phased out in many countries, some exporters benefited while importers of food and farm inputs faced high prices due to currency devaluation. Government subsidies in farming were either curtailed or totally removed, which in turn could have encouraged private sector participation. However, the lack of short-term credit to purchase farm inputs, previously supplied by the government, is a constraint for small farmers. Almost all IMF/World Bank-sponsored structural adjustment programmes advocate reduced public expenditure, and the active participation of the private sector. Some services such as marketing of agricultural products, supply of inputs, etc. can in most cases be undertaken by the private sector.

The provision of public goods, such as roads, health, water, education, basic technology development, animal and crop disease control, conservation and the maintenance of law and security in many countries, including industrialised states, still remains the responsibility of government (Stiglitz, 1998). In many SADC countries the private sector is still small and weak and reluctant to locate to remote rural areas where infrastructure is underdeveloped. Consequently, the production and marketing of traded farm products from these areas are adversely affected.

The SADC produces a diversity of commodities such as grains, fruit, vegetables, cotton, livestock, fisheries and forestry (Lewis *et al.*, 2002; Lewis, 2001). Agricultural production ranges from high input and capital-intensive systems by large-scale farmers to low input, risk-averse small-scale producers. The former specialise in export and high value commodities while the latter – the majority in SADC countries – generally produces domestically consumed products where they face problems accessing inputs like credit. Minerals and energy are important production activities in Angola, Botswana and South Africa, while processed foods are important for Malawi, Mozambique, Tanzania, Zambia and Zimbabwe. Basic intermediate inputs, machinery and equipment are vital to the South African economy. Other productive activities like trade and services are important for all SADC countries.

Differences in technical quality standards heavily influence trade within the region. Animal and plant disease and pest control requirements specifically have a bearing on trade. Certain SADC countries require high hygiene and safety standards in livestock products, while some enforce equally tough plant quarantine measures. The efficiency and effectiveness by member states to enforce technical standards vary from country to country, while some states might use these measures to protect inefficient sensitive industries.

The use of genetically modified organisms (GMOs) is a technical requirement that is likely to influence the growth in cereal trade. While the Food and Agricultural Organisation (FAO) observes that GM grains or crops can increase agricultural productivity by using less agro-chemicals and varieties that withstand unfavourable weather and soil conditions, there is concern over human safety, contamination of traditional crop varieties, the development of more aggressive weeds, and possible destruction of bio-diversity (FAO, 2002). In the SADC, despite the current hunger and food insecurity affecting about 15 million people, some countries such as Zambia have already declined donations of GM maize.

Zimbabwe, Malawi and Mozambique have imposed stringent conditions on the consumption of GM maize, including the transportation of the grain through their territories. These countries have strict surveillance on the transportation, milling and distribution of GM grain. The SADC plans to set up a GMO technical team to advise member states and consumers about the use of grains from genetic engineering. Information about GM grains is critical for SADC trade since maize is the single largest food source in the region.

14.5.2 Structure of agricultural exports and imports

The SADC is highly dependent on trade, with exports/GDP ratios of least 23% for most countries except Mozambique and Tanzania whose ratios are below 20%. The region's imports/GDP ratios are similar to export ratios for almost all countries except Malawi (Lewis *et al.*, 2002). It can be concluded that SADC economies are relatively open despite high tariffs (Table 14.3). With regard to agricultural exports, primary products such as grains, fruit, vegetables, groundnuts, soybeans, cotton, tea, coffee, sugar, tobacco, livestock, forestry, fisheries, etc., make up at least 18% of total exports for SADC countries except Botswana, South Africa and Zambia. Exports of primary agricultural products, namely grains, fruit, vegetables, livestock, etc., expressed as a percentage of total agricultural exports (primary commodities plus processed foods), show high dependencies on primary agricultural exports by almost all SADC countries. The exception is Botswana where primary agricultural exports contribute 8% to total agricultural exports. In Malawi, Tanzania, Zimbabwe and Zambia the ratios are over 65%, partly due to the relatively high contribution of the agricultural sector to GDP, as well as the high agricultural labour share in the sub-region's economy (see Table 14.1). Exports from Namibia, Swaziland, and Lesotho are also likely to be highly dependent on primary agricultural products, while exports from Angola and the DRC are dominated by energy and minerals. Botswana is highly dependent on primary mineral exports. Basic intermediate products, machinery and equipment are important exports for South Africa while Zambia exports intermediate products for external processing/assembling.

Unlike SADC countries, the EU exports mainly manufactured goods and services, including machinery and equipment, basic intermediates, trade and private services. However, if exports of primary agriculture are expressed as a ratio of total agricultural exports, the EU farm sector depends significantly on primary exports (18%). Export subsidies under the Common Agricultural Policy (CAP), coupled with high domestic support in the farming sector, are largely responsible for this relatively high export dependency. The high agricultural subsidies by the EU constitute major trade distortions in global trade and caused considerable debate, especially in WTO multilateral negotiations. Annually, the EU and other industrialised countries spend at least US$400 billion on agricultural subsidies, an amount which is twice the SADC's aggregate GDP (see Table 14.1).

Few SADC countries import primary agricultural goods to a significant extent, as they depend on own production to meet domestic requirements. Imports of primary products constitute less than 5% of total imports for almost all countries (Lewis *et al.*, 2002). As in the case with primary agricultural exports, when the imports of primary agriculture are expressed as a ratio of total agricultural imports, the import share of primary products is significantly higher. Except for Tanzania with about 17%, other SADC countries show higher import ratios, ranging from 27% for South Africa, 29% for Botswana, 40% for Malawi to 58% for Zimbabwe.

The aggregate value of SADC agricultural exports exceeds the aggregate value of agricultural imports by at least threefold, implying that the SADC has an agricultural trade surplus. However, in Botswana and the rest of the SADC, agricultural exports cannot pay for agricultural imports, hence the dependence on other exports such as minerals, oil, textiles, or sometimes food aid.

14.5.3 SADC traded goods: a Heckscher-Ohlin model

Based on the type of traded goods by the SADC and the EU, as a developed region, it is evident that almost all SADC countries except South Africa specialise in exports of primary products both in agriculture, livestock, forestry and minerals. Specialisation in these products reflects different factor endowments by each country to produce and export, such as natural resources (land, water, minerals and soils) and relatively abundant unskilled labour. These differential factor endowments lead to comparative cost advantages to produce particular primary commodities. Mozambique, Malawi, Zimbabwe, Tanzania, the DRC, Angola and certain parts of South Africa, unlike other SADC countries, have comparative cost advantages to produce primary agricultural goods (maize, fruit, vegetables, cotton, groundnuts, tea, sunflower, sugar, tobacco, livestock, etc.). This means countries with relatively abundant natural resources and unskilled labour, as in several SADC countries, will produce and export goods that are intensive in the use of these resources. SADC exports employ relatively more unskilled labour and use more land, while those from the EU are technology and capital intensive.

At least 65% of the total labour force in the SADC is unskilled, and most of them reside in the agricultural sector (see Table 14.1; Lewis, 2001; Lewis *et al.*, 2002). Given this high level of unskilled labour, investments that prejudice employment creation against this group of potential workers could exacerbate poverty in most SADC countries. Unlike SADC countries – that are mostly low-income countries – the EU specialises in manufactured exports that do not require natural resource endowments (land, fertile soils, unskilled labour, etc.) but rather technology, skill and capital-intensive production inputs, etc.

Given this background of trade based on factor endowment or comparative cost advantages for SADC countries, SADC trade, both intra-regionally and internationally, seems to conform to the Heckscher-Ohlin theorem. According to this model, countries trade in primary commodities based on different factor endowment or natural resources. Differential factor endowments in labour, land, water, etc. lead to different relative costs to produce goods, leading to specialisation in traded commodities. Primary products like fruit, vegetables, cotton, tea, coffee, livestock, tobacco, etc. generally require more unskilled labour and land rather than machinery and equipment. Mozambique, Tanzania, Zimbabwe and Malawi export primary products to other countries because of relatively cheaper labour and land, and favourable climate and fertile soils.

Inter-industry trade between the SADC and industrialised areas such as the EU differentiates between exports/imports of primary goods, on the one hand, and machinery and equipment, on the other. SADC countries mainly export primary agricultural products to the EU, while the EU exports skill-intensive, capital-intensive and technologically advanced goods like computers, machinery, television sets, etc. to the SADC. Intra-industry trade is normally associated with trade between industrialised countries as the countries produce and trade in similar goods such as motor cars, machinery, equipment, processed food, etc. There is growing evidence of intra-industry within SADC countries, as some countries export similar products to others like fruit, vegetables, grains, livestock, processed food, etc. For instance, South Africa exports fruit and vegetables to other SADC countries and yet these countries also export the same products to South Africa.

14.6 SADC AGRICULTURAL TRADE FLOWS BY COUNTRY

Lewis *et al.* (2002) estimated total exports for SADC at about US$51.2 billion and total imports at US$45,8 billion. SADC exports are generally dominated by primary agriculture, while few countries depend on energy and mineral exports. For South Africa, however, basic intermediate goods and machinery and equipment are a major source of export revenue, while processed food items are also significant exports for Mozambique, and to some extent for Tanzania. SADC imports are concentrated on capital goods like machinery, equipment and intermediate items, and, to a lesser extent, processed food items.

Intra-SADC trade has generally increased since 1985 but the rate of growth has been slow due to factors identified earlier including differential and high tariff rates among members outside the SACU. Intra-SADC trade is low, compared to other regional trade and economic groupings worldwide (Yeats, 2000; Lewis, 2001). At least 28% of SADC total exports go to the EU under the preferential ACP-EU agreement (Lewis *et al.*, 2002). SADC primary agricultural exports include fruit, vegetables, tea, coffee, tobacco, cotton, timber and fish, while processed foods cover meat, juices, sugar, etc. Other SADC agri-business exports to the EU include textiles, clothing, furniture, etc. The dependency on the EU as a destination for SADC exports ranges from about 28% for Zambia, 35% for South Africa, 37–39% for Zimbabwe, Tanzania, Mozambique and Malawi, to about 76% cent for Botswana (mainly diamonds and beef). The rest of the SADC countries depend on the EU for up to 30% of their exports.

At least 50% of SADC exports of primary agriculture go the EU under the ACP-EU agreement, which attract low duties because of preferential treatment. If exports of processed food commodities are included, the reliance on the EU is even higher. For instance, South Africa exports about 70% of fruit, vegetables and livestock to the EU, while over 70% of Botswana beef goes to this market. The EU is the largest export-trading partner for the SADC's farming sector, accounting for about 50% of the SADC's total agricultural exports.

14.7 TRENDS IN SADC AGRICULTURAL TRADE

There is a paucity of reliable, long-term trade statistics in the region, but it is apparent that since 1994 when South Africa became a democratic state, trade opened to the world, including the SADC region. South Africa's agricultural exports to non-SACU SADC states increased by at least 10% in real value per year during 1990–2000. The main importing countries are Zimbabwe, Zambia, Malawi, Mozambique and Mauritius. South Africa is the largest SADC exporter and importer of agricultural products (Lewis, 2001; Lewis *et al.*, 2002). Trade with other non-SACU SADC countries also increased and South Africa enjoys a trade surplus. Agricultural exports by South Africa to other non-SACU SADC countries mainly cover processed food products such as dairy, milled cereals and sugar.

During 1990–2000, there was also an upward trend in South Africa/SACU imports from the rest of the SADC. As South Africa has a common external tariff with Botswana, Lesotho, Namibia and Swaziland under the SACU Agreement, agricultural imports are simply classified as SACU imports. Goods entering the SACU territory from any custom border point attract the same duty. The duties are pooled and divided among members according to an agreed formula, based on imports. In general, Zimbabwe, Zambia, Malawi and Mozambique have increased their imports from the SADC, but there have been marked yearly fluctuations which are possibly due to supply constraints caused by weather, domestic economic policies and infrastructural problems. The harmonisation of tariffs, technical standards, sectoral and macro-economic policies (exchange rate, taxation, etc.), the prevalence of peace and stability and access to improved infrastructure should contribute towards the upward trend in intra-SADC trade.

14.8 OPPORTUNITIES FOR SADC TRADE

Opportunities for increasing intra-regional trade have improved following global trade liberalisation under the WTO, the removal of exchange and price controls, alignment of currencies to market-determined exchange rates and the phasing out of monopoly state agencies in trade and marketing. The democratisation of South Africa, and efforts to open up SADC economies, has further promoted regional integration. Table 14.3 shows relatively low tariffs for primary agricultural commodities, while tariffs for other goods have also declined due to economic and trade reforms mainly orchestrated by the IMF and the World Bank.

Reduced tariffs and the gradual removal of other trade barriers within SADC have provided scope for improved food security, consumer choice and resource efficiency as countries specialise according to comparative cost advantage. The increase in intra-regional trade observed in section 14.5.3 underscores agricultural trade mainly based on factor endowment. There is scope for intra-industry trade in the SADC if skills, technology and capital are available.

As the SADC is exposed to both regional and global trade liberalisation, opportunities arise to access cheaper food, farm inputs, technology, and investment that could increase the region's GDP and possibly welfare/food security (see Table 14.1). Lewis *et al.* (2002) show that with the creation of a SADC free trade area, harmonisation of tariffs and unlimited access to industrialised markets, especially the EU, GDP will increase in several SADC countries, and import penetration will improve efficiency and possibly competitiveness. Exports of agricultural products like fruit, vegetables, tea, cotton, etc. will also increase in the region. However, not all countries will benefit equally: some sectors will lose due to external competition; unemployment of unskilled labour will increase in the short to medium term due to structural changes; and governments will lose tariff-based revenue.

With the possible return to political stability and peace in the sub-region, it should soon be possible to promote cross-border investment and trade with less business risks. The internal civil wars in Angola and the DRC have had a negative impact on cross-border business ventures. With peace, the movement of human capital across borders to improve agricultural productivity and trade competitiveness will hopefully be enhanced. The SADC's agriculture is characterised by low productivity partly due to limited skills and investment, especially for infrastructure and technology.

14.9 THREATS TO SADC AGRICULTURAL TRADE

While trade liberalisation can be seen as an opportunity, some countries might lose customs revenue. Agricultural products, particularly those that are normally processed, attract higher duties (see Table 14.3). If the SADC pursues its goal of trade liberalisation under the Trade Protocol, there could be very serious fiscal and political consequences for several SADC countries. If trade is liberalised through an FTA and tariffs are reduced and harmonised, countries such as Zambia, Zimbabwe, Malawi, Tanzania and Mauritius could suffer serious tariff revenue losses (Lewis, 2001).

The outbreak and spread of livestock diseases and plant pests have been common in the SADC. Political instability has contributed to the failure to control diseases and pests, despite intra-agricultural trade in the SADC being dominated by primary products. It is imperative to guarantee consumer's high levels of food safety and hygienic standards. The use of GM products has also created a threat to SADC agricultural trade. Scientific evidence and safety assurance are critical for allaying fears about GMO technology.

The SADC region has been plagued by the HIV/AIDS pandemic for the last two decades. Within this region, Botswana, Lesotho, Swaziland, South Africa, Zambia and Zimbabwe have prevalence rates above 20% for the 15–49 age group that constitutes the most productive labour force in the economy, including farming. Agricultural production, exports, investment and the overall region's economy are affected by this disease. Resources to provide public goods (infrastructure, education, environmental protection, animal and pest control, technology development, etc.) are diverted to the control of the disease. Measures to control HIV/AIDS in the region should be intensified.

One of the threats facing SADC trade, especially agricultural trade, is the current ACP-EU agreement, which in its present form is discriminatory against non-members as it does not offer them the MFN status. Cognizant of this inherent discrimination, it is expected that by the year 2007 the agreement will comply with the WTO provisions. The SADC, like the rest of Sub-Saharan Africa, therefore faces a serious threat to their current preferential treatment and tariff concessions. SADC agricultural trade depends heavily on the EU market, and transitional agreements will be required to enable the SADC to acclimatise to a global economy with reduced EU preferences.

14.10 CONCLUSION

This chapter has examined the role of the agricultural sector in the SADC economy, described the existing trade tariff structure, the value of agricultural trade, agricultural trade flows and trends in the sub-region's farm trade. Trade opportunities and threats posed by economic liberalisation to the achievement of the goals of the SADC Trade Protocol have also been identified. The main objective of the Protocol is to liberalise trade in the SADC in order to help improve human welfare and reduce poverty via higher per capita incomes.

Empirical evidence indicates gains from SADC trade liberalisation through the formation of FTAs or similar structures and unlimited access to markets such as the EU, low and harmonised tariffs, custom procedures, technical standards and sectoral and macro-economic policies. Agricultural trade can thus contribute towards the realisation of the objectives of the SADC Trade Protocol. However, political and economic costs must be noted, as studies of trade liberalisation in the SADC indicate differential benefits and costs, with some countries likely to face serious tariff revenue losses. Transitional support arrangements need to be explored for the net losers from trade liberalisation under the SADC Trade Protocol.

The SADC community needs to improve its analytical capacity to understand the costs and benefits of trade reforms. Lewis accurately observes 'there must also be recognition of the constraints on policy capacity that exist in many SADC member governments, to ensure that focus is not spread thinly across too many initiatives, and prevent any of them from being successful' (2001: 41). Enhanced capacity to analyse policy options in trade and economic reforms, and membership to/partnership with certain trade groupings is critical for SADC integration and long-term sustainability. Currently, this capacity is lacking in the region and has contributed to the failure to comply with several international trade provisions (WTO, 2001).

As the SADC strives to develop skills, infrastructure, and a competitive economic environment, and improves access to information, technology and capital, it is hoped that opportunities will exist for intra-industry trade (Yeats,

2000). The SADC, like most developing regions in Africa, has been exporting primary products with very little value adding or processing and this has reduced the growth of intra-industry trade as industrialised countries monopolise the trade in processed agricultural products and other manufactures.

REFERENCES

EHUI, S. & DELGADO, C. (1999). *Economy Wide Impacts of Technological Change in Agro-food Production and Processing Sectors in Sub-Saharan Africa.* Washington DC: IFPRI.

FOOD AND AGRICULTURAL ORGANISATION (FAO). (2002). Statement on biotechnology. Rome.

INTERNATIONAL FUND FOR AGRICULTURAL DEVELOPMENT (IFAD). (2001). Poverty report. Rome.

JAMMEH, S.C. (1998). State intervention in agricultural pricing and marketing in Senegal. Unpublished paper, University of Michigan.

KRUEGER, A. (1974). The political economy of the rent society. *American Economic Review* 64: 291–303.

LEWIS, J.D. (2001). *Reform and Opportunity: The Changing Role and Patterns of Trade in South Africa and SADC.* Washington DC: World Bank.

LEWIS, J.D., ROBINSON, S. & THIERFELDER, K. (2002). *Free Trade Agreements and the SADC Economies.* Washington DC: World Bank.

LIPUMBA, N.H.I. (1996). *Liberalization in Africa.* Helsinki: UNU/WIDER Publications.

MAASDORP, G. (1998). Trade integration and economic development: some Southern African issues. In Petersson, T. (ed.). *Post-Apartheid Southern Africa: Economic Challenges and Policies for the Future.* London: Routledge Publishers.

MCDONALD, S. & TOWNSEND, R.F. (1998). Biased policies, agriculture and income distribution in South Africa: a Social Accounting Matrix approach. *Journal of Studies in Economics and Econometrics* 22: 18–35.

MCDONALD, S. & WALMSLEY, T. (2001). *Bilateral Free Trade Agreements and Customs Unions: The Impact of the EU-South Africa Free Trade Agreement on Botswana.* Purdue University: Global Trade Analysis Project.

SADOULET, E. & DE JANVRY, A. (1995). *Quantitative Development Policy Analysis.* London: John Hopkins University Press.

SIGWELE, H.K. (1999). Keynote Address: 'Botswana's livestock industry: the way forward'. Ministry of Agriculture, Botswana.

SODERSTEN, B. & REED, G. (1994). *International Economics.* London: Macmillan Press.

STIGLITZ, J. (1998). More instruments and broader goals: moving towards the post – Washington Consensus. Wider Annual Lecture, Helsinki, Finland.

TOWNSEND, R.F. & SIGWELE, H.K. (1998). Socio-economic cost benefit analysis of action and alternatives for the control of Contagious Bovine Pleuropneumonia (CBPP) in Ngamiland, Botswana. University of Pretoria.

WORLD BANK (2001). *World Development Report.* Washington DC: Oxford University Press.

WORLD BANK (2002). Speech by Chief Economist of the World Bank to the Center of Economic Studies, Munich.

WORLD TRADE ORGANISATION (WTO). (2001). The Doha Declaration, Doha, Qatar.

YEATS, A.J. (2000). *What can be Expected from African Regional Trade Arrangements?* Washington DC: World Bank.

PART 3

CONCLUSION

CHAPTER 15

Demands on and Challenges for South African Agriculture

Jan Groenewald and Lieb Nieuwoudt

15.1 INTRODUCTION

At the outset of the third millennium, South Africa is faced with several challenges. Inequality, largely a legacy of past thought, attitudes, actions and mistakes has to be replaced with equitability, and there is an urgent need to eliminate, or at least to substantially reduce, poverty. Living standards have to be improved by attending to all facets of socio-economic welfare, including per capita income; the distribution of income in the population; the supply of basic needs such as clothing, food, health, shelter and education; decent working conditions; and decent living conditions. Being relatively developed in an African sense, South Africa also has a duty to contribute to improved welfare elsewhere in the continent. Experience worldwide clearly shows that economic growth is a prerequisite for the ideals mentioned above. In short, South Africa needs growth with equity.

The agricultural and rural complex has a crucial stake in this process. It has to contribute to growth and to equitability, but it should also share in the fruits of development. Equitability in itself implies that the agricultural and rural sector must not, as in other parts of the world, be left behind in the ability to share in the benefits of development. Not only does a major portion of the South African population reside in rural areas, but agriculture must contribute to the development of the economy through the production of food at reasonable prices, the earning of foreign exchange, and through forward and backward linkages.

Agriculture can be viewed as having four functional components (Mosher, 1971):

(a) Farm-based inputs and inputs originating in the rest of the economy are used to produce products that flow to the rest of the economy. Only in a completely subsistence system is farming synonymous with agriculture, although it is always a central component;
(b) Commercial agri-support services, which deliver for direct remuneration those goods and services needed for agricultural production, sustenance and product marketing. Supply chains on both sides of the farm gate are included;

(c) Non-commercial agri-support activities supply other necessary goods and services, but are not usually paid for directly. These include government services such as the provision of fixed transport infrastructure and information; and
(d) The agricultural milieu, which embraces the social, political, institutional and legal environment within which the other functional components operate.

The real and urgent challenge is to modernise all facets of South African agriculture. According to Mosher (1971), a modern agriculture is one in which:

(a) there is continuous improvement in the technology and efficiency of farming;
(b) the types of commodities produced change continuously (at least in proportion to each other) in response to changes in market conditions and costs of production;
(c) there is steady and continuous improvement in the quality of land, the competence of labour and the forms and quality of capital goods used in agriculture;
(d) the proportions in which land, labour and capital are combined keep changing in response to changes in population growth rates, alternative employment opportunities and technology which is reflected in relative price changes for the factors of production; and
(e) the agricultural sector is served by private and public agencies that are continuously adapted to serve new functions in new ways.

Section 15.2 discusses the changing environment in which the functional components of South African agriculture are operating, and how they are adapting to the challenge facing the country.

15.2 THE CHANGING ENVIRONMENT FACING SOUTH AFRICAN AGRICULTURE

15.2.1 South African agriculture in perspective

South African agriculture is decidedly dualistic, with a commercial modern sector and a subsistence and emerging farmer sector. There are currently about 50 000 large-scale commercial farmers in South Africa (Department of Agriculture, 2001), with this commercial sector being predominantly, but not exclusively, drawn from the white population. This sector exported about R16 billion (US$1=R8) in 2000, or nearly 10% of South Africa's total exports, and employs about one million workers, which is equivalent to some 11% of the total formal employment of South Africa. Another 240 000 small farmers provide a livelihood to more than one million of their family members and occasional employment to another 500 000. In addition, an estimated three million farmers, mostly communal in the former homelands, produce food primarily for subsistence (Department of Agriculture, 2001).

Primary agriculture accounts for about 4.5% of the GDP of South Africa, while the larger agro-food complex accounts for another 9% of GDP. The relatively small contribution to GDP, and the relatively large number of people residing in the sector (70% according to Department of Agriculture, 2001) indicate that the sector will keep shedding people in future. These people often have low skills which means that rural education is important (Schuh, 2002) and deserves the priority it currently receives. It also implies that demands on urban housing will continue. High urban unemployment and squatter conditions near some of South Africa's main cities are not making it attractive for people to leave rural areas. The rural/urban interface needs to be understood better and innovative solutions need to be found to cope with the influx of people. High unemployment is a major concern in both the rural and urban sectors and rural living standards cannot be raised in the absence of a growing non-farm economy to which farm workers have access (Gardner, 2000).

The most important challenge in the large-scale commercial farming sector is the redistribution of land to previously disadvantaged communities. In the small-scale subsistence sector, low productivity is a major challenge. Low productivity can be attributed to past policies that discriminated against people on the basis of their race, coupled with other problems such as tenure insecurity, and very small land holdings. Tenure insecurity leads to insufficient incentives to invest in and/or improve productivity in this sector. Given that most people in South Africa are poor, there are challenges for agriculture not only on the production side but also on the demand side as outlined in section 15.2.2.

15.2.2 Future food demand and food security in South Africa

For the agricultural sector to expeditiously adapt to future challenges, it is important to study the structural changes in demand for food in South Africa. McGuigan & Nieuwoudt (2002) have developed an interactive model to predict the demand for food in South Africa in 2010 and 2020. The model allows for expectations regarding population growth rates, urbanisation rates, a narrowing in income per capita growth, and declining income elasticities of demand for food over time. Statistics for the different population groups were included as they experience different demand shifts. Results indicate that the growth in food demand is largely explained by population growth, as expected income growth is matched by expected population growth, with per capita income remaining almost constant. The demand for food will increase markedly under high-income growth scenarios, as the income elasticities for food are higher for the poor than for the relatively more affluent. Although income elasticities are generally high for all foods, the income elasticity of demand appears negative for maize and bread, predicting a switch away from these foods with an improvement in living standards. Population growth rates are also higher among poor people. These forces have caused a decline in per capita consumption of some foods during periods of increasing consumption per capita for both poor and affluent; for example, dairy products where consumption per capita for the more affluent is significantly more than for the poor.

Urbanisation causes the composition of the demand for food to change, as the opportunity cost of time is higher in an urban environment. Some switching from maize (corn) meal to bread, a more convenient food, occurs. The increase in the opportunity cost of time (for women) associated with urbanisation is also expected to have a negative impact on population growth (Fairlamb & Nieuwoudt, 1991). This is consistent with the theoretical arguments of Friedman (1962: 208–211) and Becker (1965). Under a worst-case AIDS disease scenario, the annual population growth rate was projected to fall to 0.39%. This is likely to slow the rate of growth in demand for most foods as population growth declines, while disposable incomes fall (due to decline in productivity).

A major concern on the demand side for food is the distribution of food. Although food security can be promoted if food is available and abundant, many people in South Africa do not have entitlement to a nutritional diet. Entitlement can stem from income, other possessions, or own production (Sen, 1981). According to Wilson & Ramphele (1989), at least two million urban workers' wages were below supplementary living standards in South Africa. The positive relationship between nutrition and productivity has been repeatedly proven internationally (Fogel, 1998). A school lunch programme should again be considered in South Africa, as its administration cost may be low (no verification of claims required), while it can be targeted at the needy. State pensions that are not much below the minimum wage in South Africa provide some food security to a vulnerable section of society. Basic income grants could also be considered, and more attention could be given to promoting job creation and training to improve skills and access to income.

In relatively high-income countries, food prices are influenced less by prices of agricultural commodities than in low-income countries (Antle, 1999), because in high-income countries a large proportion of food is consumed in a processed form. The high incidence of poverty in South Africa makes many consumers vulnerable to food price increases. This effect is magnified by the larger share of income poorer consumers tend to spend on food.

It is thus not surprising that there was great concern when the prices of basic foods such as maize escalated in South Africa towards the end of 2001 and beginning of 2002. The increases could partly be attributed to (a) the marked depreciation of the Rand; (b) the change in maize from an excess to a deficit position in the southern Africa region; and (c) higher world prices due to drought in the US, Australia and Argentina. The difference between the export realisation price for maize and the import landed price is significant due to the long distances between South Africa and its markets or alternative supply areas.

Some analysts called for a return to marketing control boards in South Africa while others called for food price ceilings as in Zimbabwe. This pressure must be resisted as it can be expected that grain production will increase significantly with more attractive prices given the surplus capacity in grain production. In contrast to the short-run elasticity of supply that in agriculture is invariably low, the long-run supply in South Africa is estimated to be close to unity or above (Nieuwoudt, 1970; Van Schalkwyk & Groenewald, 1993), which is an international phenomenon (Chhibber, 1982; Peterson, 1979, 1988). Price ceilings on food as adopted in Zimbabwe will, through lower supply in the longer run, put upward pressure on prices.

15.2.3 Supply forces in South Africa

Binswanger & Townsend (2000) attributed poor growth and under-capitalisation in sub-Saharan Africa over the last century to adverse resource endowment, poor policies, remoteness, low population density, increasing transport and transaction costs, and institutional failures. Poor policies include an urban bias (Lipton, 1993) and policies that tend to tax agriculture, with farmers receiving a producer price lower than the world price equivalent. These conclusions are important and South Africa should not repeat the policy mistakes of other countries by under-evaluating agriculture. Such policy mistakes have seriously retarded economic and agricultural development in many countries, including Nigeria (Ilorah, 2000), China (Lin, 1994) and India (Srinivasan, 1994).

South Africa's natural resource base is on average not favourable for farming, due to unfavourable weather, uneven topography and varying soil conditions. In his international comparison, Peterson (1987) estimated South Africa's land quality index at 56, given a world average of 100. There is, however, wide variation within South Africa. Using a methodology similar to that of Peterson (1987), Van Schalkwyk & Groenewald (1993) classified South African regions. The extensive grazing areas' indices varied between 50 and 90 (with the country average being 100), while some others, such as the fruit areas of the Western Cape and parts of KwaZulu-Natal, scored over 210. The quality in these areas is comparable to that in countries with above-average land quality such as Benin, Ivory Coast, Denmark, the Netherlands, Norway and New Zealand.

Many adverse climatic effects apply to agriculture in South Africa, namely low rainfall in the main cropping areas, combined with a high risk of drought; high transport costs due to distances from global markets; and the absence of internal waterways. At present, South African farmers receive no state subsidies. In Chapter 1, Vink & Kirsten show that the aggregate level of support (mainly via tariff protection) to South African farmers is currently low, although sugar farmers receive relatively more protection. Poor road transport in the small-scale sector is a constraint in the marketing of produce and in the acquisition of inputs. Transport costs in the commercial sector is increased by the lack of internal competition in rail and port facilities, which has resulted in unreliable and expensive services (Department of Agriculture, 2001). According to engineers, many roads are severely damaged by heavy vehicles, and significant cost savings to producers, traders and the state would be possible if the rail service in South Africa could be upgraded (or privatised).

Past investment in research has been relatively high, notably in crop and horticultural production, and in eradicating livestock and human diseases. There is, however, concern that funding for agricultural research has declined rapidly and that this will affect research, and ultimately competetiveness, negatively. The AIDS pandemic may still have a devastating impact on society with southern Africa experiencing the highest incidence in the world (World Bank, 2002). Life expectancy at birth in South Africa has fallen from 62.8 years (over 60 years for all population groups) in 1994 to 47.8 years in 1999 (World Bank, 2002). The latter figure is significantly below the concurrent world figure of 66.5 years. The cost of the disease in terms of foregone income, productivity losses, health care and suffering is enormous.

Commercial farms in more developed countries and in South Africa have increased in size in recent decades. Unfortunately in South Africa, this has a racial dimension (lack of ownership by previously disadvantaged people) that has to be removed. In a globalised economy the South African farming sector must compete with other food exporters such as the US. It must be noted that the number of farms in the US fell from roughly six million in 1940 to two million in 2000 (Gardner, 2000), with concomitant increases in farm sizes. Previously disadvantaged farmers in South Africa will have to be accommodated on both small and large farms. A criticism in the past was that commercial farming in South Africa was overcapitalised as a result of subsidised interest rates and favourable tax allowances on machinery purchases. This criticism is less valid now as farmers are exposed to high real interest rates, while machinery prices have escalated due to a falling Rand exchange rate.

15.2.4 Changing agricultural policy environment in South Africa

After 1994 the new government in South Africa adopted new policies to move from a more closed economy where the focus was on self-sufficiency (for example, synthetic oil from coal) to a more open economy with the scrapping of currency control measures and the abolishment of import quotas. A more market-related approach was also adopted in agriculture. Policy shifts that impacted on agriculture include:

(a) deregulation of the marketing of agricultural products;
(b) abolishing certain tax concessions favouring the sector;
(c) reductions in budgetary expenditure on the sector;
(d) land reform (restitution, redistribution and tenure reform);
(e) trade reform, including the tariffication of farm commodities; and
(f) the broadening/introduction of new labour legislation.

The net effect of these changes is that the South African agricultural sector has become less dependent on state support and internationally more competitive, although many sectors had to go through a painfall period of adaptation. The ability of South African farmers to compete internationally is still hampered by tariff and non-tariff (for example, phytosanitary regulations applying to citrus exports) barriers imposed by, and subsidies paid to farmers by, countries that South Africa competes with – particularly the US and the European Union (EU).

The impacts of market forces are clearly evident in the South African grain sector. The area under maize declined from over five million hectares in 1986–87 to 2.7 million hectares at present, while the area under wheat declined from over two million hectares (during 1971–73, 1982 and 1988) to about 0.7 million hectares at present (The Directorate: Agricultural Statistics, 2002). The decline may be attributed to declines in real grain prices (aided by the removal of marketing boards) and expectations that the present government will not provide additional drought assistance to grain farmers. The previous government gave summer grain farmers significant drought assistance during the 1980s and once again in 1992. The aid could, in any event, only provide temporary relief; farmers with poor management and very unfavourable debt:asset ratios would still go bankrupt within the next decade (Van der Watt & Groenewald, 1996; Swanepoel *et al.*, 1998). The South African crop insurance programme, whereby the state subsidised 25% of the premiums, also failed in 1989. These forces caused farmers to move out of grains in marginal cropping areas, where many farmers have successfully converted maize and wheat land to grazing. The conversion of cropland to grazing also caused the livestock factor to increase. This improved cash flow and introduced more stability for farmers in relatively higher risk production areas. It is significant that the summer grain area has not had a major crop failure as a result of drought in recent years. Most of the drought disaster payments in the past were allocated to farmers in this region.

Although agriculture in these areas has now adapted better to its environment, droughts will recur and it is imperative to have risk management strategies. The South African government has been considering such strategies, particularly for small-scale farmers. Lessons may be learned from the US Federal Crop Insurance Program, which is well developed but heavily subsidised. During 1998, US growers paid $900 million in premiums while the US government spent $1.2 billion per year on subsidies during 1995–98 (Dismukes, 1999). Risk is systemic in the South African summer rainfall area (large areas are affected by drought at the same time), and an area insurance programme where farmers are insured *as a group* may be more appropriate. This programme overcomes problems inherent in individual crop insurance such as adverse selection, moral hazard and the cost of farm visits (verification of claims).

Since administration costs account for a large part of the costs of insurance programmes, an alternative Income Equalisation Deposit (IED) scheme should be considered. Under this programme, farmers can invest in a bank in good years and pay tax on the income only once it is withdrawn. A similar type of programme receives strong support from the Australian government and requires no state contribution, not even in lost taxes. Farmers can misuse the programme for tax purposes given the current South African In/Out tax provision (they could deposit funds into the IED to deliberately understate average income and so pay lower tax rates), but this can easily be avoided by imposing a rule that no funds can be deposited if current taxable income is below a farm's five-year moving average taxable income (Nieuwoudt & Howell, 2000).

A tax on the 'improved' value of farmland is currently being considered in South Africa, payable to local municipalities. Although land taxes are common in other countries such as the US and in the EU, farmers in the EU typically receive five times more in subsidies from the state than they pay in land taxes. Agricultural sectors are commonly subsidised in high-income countries and taxed in low-income countries (Antle, 1999). The current debt:asset ratio in South African agriculture is about 30% (Directorate Agricultural Statistics, 2002). A land tax of 2% on the value of land can increase this debt:asset ratio to between 50–60% because of an expected fall in property prices, assuming that the Ricardian rent return on farmland is 5% (Nieuwoudt, 1980). This expected change in debt:asset ratios will negatively affect the solvencies of financial institutions (Van Schalkwyk, 1995), and the borrowing capacity of many farmers. Most lenders prefer that borrowers have at least as much investment in their own business as the lender does, implying a standard rule of thumb for the maximum debt:asset ratio of about 0.5 (Barry *et al.*, 1995).

A land tax also reduces net returns to land and so will increase the size necessary for a farm or holding to sustain any living standard, rendering land reform more difficult. A land tax rate of over 6% will destroy property values, thus also farmers' equity (and collateral for further borrowing) in most parts of the country (Van Schalkwyk, 1995; Darroch 2003). High taxes on land are likely to cause bankruptcies and increased cash flow problems, especially in drought years. A land tax not only reduces the incentive to invest in new land improvements, but also reduces the ability to invest as investment in agriculture is largely financed out of retained earnings.

A unique feature of South African commercial agriculture is its labour intensity. Agriculture, Forestry and Fishing generate 24.2 jobs per R1 million production (indirect, direct and induced effects included), which is significantly higher than for any other sector in the economy (Conningarth Economists, 2001). Labour numbers on commercial farms have, however, declined in recent years. According to the Directorate of Agricultural Statistics (2002), farm employees and domestic servants on farms declined from 1 638 800 in 1971 to 914 500 in 1996 or by 45%. In Chapter 1, Vink & Kirsten report a decline of 50% over this period. This decline was accompanied by an increase in real wages on farms of 46% between 1970–98 – suggesting that the input demand elasticity for labour is about –1. It is interesting that Latt & Nieuwoudt (1985), using a simultaneous equation modelling system, estimated the demand elasticity of regular farm labour as approximately –1 for KwaZulu-Natal. This estimate is similar to the –1.1 figure reported by Lianos (1972) for the wage elasticity of hired farm labour in the US.

Since 1996, the South African government has instituted legislation regulating working conditions, and outlawing child labour, on farms. New minimum wage legislation implemented from March 2003, however, threatens to have severe negative effects on agricultural profitability and employment in a situation of large-scale unemployment given the estimated elasticity of labour demand of about –1. The farm labour market in South Africa has not been adequately researched, and research on the effects at a regional level is urgently needed before large-scale application of minimum wages. In the US, it was found that minimum wages penalised marginal farm workers (Lianos, 1972), and depressed wages in other non-affected sectors and areas as displaced farm workers sought alternative employment (Gardner, 1972).

15.2.5 Macroeconomic policy changes in South Africa

The South African government has followed fairly sound monetary and fiscal policies in recent years with the annual inflation rate relatively low at around 7%, and positive real interest rates. These macroeconomic policies have important impacts on agriculture. The South African Rand has operated under reasonably free market conditions that have generally benefited the farming sector. The South African Reserve Bank has raised nominal interest rates in recent years to achieve inflation targets and to achieve positive real rates. Although nominal interest rates are relatively low for a developing country, they are still high enough to create cash flow problems for the commercial farm sector and the agro-business sector. Agricultural debt is about R30 billion for 2000 (Directorate: Agricultural Statistics, 2002), which translates into R5.1 billion worth of interest payments at 17% nominal interest rates. The shift towards higher real interest rates in recent years has depressed property prices and reduced farmers' collateral for borrowing.

15.3 FUTURE DEMANDS AND CHALLENGES

15.3.1 Land redistribution and social upliftment in rural areas

A land reform programme was initiated by the government during 1994 with the aim of redistributing 30% of the total agricultural land to previously disadvantaged people. Under the initial programme that operated between 1994–99, applicants were given grants of R15 000 per household. Because this amount was far too small for land purchase and development, farmers pooled their grants and purchased farms as groups (of up to 500 households). Research indicates that insecure tenure of these projects impacted negatively on the ability of the beneficiaries to access credit, and on production (Lyne, 2001). This programme was replaced by the LRAD (Land Redistribution for Agricultural Development) programme in 2001 that allows disadvantaged farmers to acquire more land if they have an own contribution (financially or in own labour) according to a sliding scale. Beneficiaries do not have to be poor to qualify for the minimum grant of R20 000.

There is concern by many analysts that the pace of land redistribution is too slow. Role players and bottlenecks need to be identified and the process needs to be expedited. The government has in recent years been investing in the rural poor through improved provision of water, electrification of rural areas, education and better pension schemes. Transport infrastructure in the communal areas is also inadequate. Concern has also been expressed that land redistribution may not benefit the target group. For instance, farmers can purchase a grazing farm as a group but only those who have cattle benefit from grazing in terms of an income stream (Shinns & Lyne, 2003). Clearly, the very poor (those without cattle) are not better of. The management of communally bought properties is often problematic, with negative effects on production and on the survival of the proposed enterprise.

Over the period 1997–2000, market-driven land reform initiatives by sugar millers transferred more land in KwaZulu-Natal to disadvantaged farmers than did state initiatives (Lyne, 2001). The sugar millers gave 18% of the purchase price of farms they sold (to previously disadvantaged buyers) to Ithala Bank to reduce the then interest rate from 16.5% to 10% in the first year on mortgage loans the buyers obtained from that bank. This enabled Ithala Bank to grant them an interest rate subsidy in order to overcome the cash flow problem associated with the early years after buying land. Given a rate of return of 5% in agriculture and nominal interest rates then of 16.5%, the cash flow problem arises for a resource poor farmer if the mortgage bond exceeds 30.30% of the purchase price of land (5% of R100 = 16.5% of R30.30). Most of these farmers could only contribute 5–10% of the purchase price of the farms that they bought (Mashatola & Darroch, 2003). The interest subsidy was phased out over six years as the farmers' nominal returns were expected to increase in line with anticipated inflation. The cash flow problem arises from relatively high inflation rates in South Africa that increase the immediate cost of capital through high nominal interest rates while deferring returns (profits increase over time). This financing approach was partly based on a theoretical paper discussing the cash flow problem under inflationary conditions (Nieuwoudt *et al.*, 1993).

The approach by the private sector is laudable, but it is somewhat ambitious to require farmers to repay 10% of the purchase price in the first year. Given some studies showing that the rental return in agriculture is around 5% (Nieuwoudt, 1980; Van Schalkwyk, 1995), it is recommended that in order to avoid a cash flow problem, farmers should not be required to pay more than this in the first year. It simply means that a greater subsidy is required to facilitate the transfer of land. It is not surprising that about 20% of clients are in arrears on their graduated payments, although this figure would have risen to 30% if some loans had not been rescheduled in 2001 (Van der Heever, 2002). Perhaps farmers should not be required to repay (as at present) the principal in the early years, but only interest. After about 10 years the subsidy on the interest payment can fall away; inflation in farm profits will have eliminated the cash flow problem associated with the repayment of principal.

In the Western Cape export fruit growing areas of South Africa, and more recently in Gauteng and KwaZulu-Natal, land reform has in some cases been achieved by equity sharing agreements between commercial producers and workers, who through this process became part owners. Workers have the incentive to make the scheme work by sharing in enterprise profits, and they can benefit from the specialist managerial skills provided by the former sole owner.

Black farmers were historically excluded from the commercial farming sector, while freehold farming was impossible in the former homeland areas. The latter areas are communal with very small farms which are used for subsistence. Policy suggestions have focused on improving farmers' incentives to make investments, as the areas are underdeveloped and overgrazed. The most promising suggestions involved development and improvement of a rental market of land-use rights and the upgrading of infrastructure (roads). The establishment of co-operatives in these areas could facilitate the flow of information and technology. The fact that early efforts have failed does not mean that problems cannot be solved; many reasons for the failure of earlier co-operatives have been identified (Machethe, 1990).

15.3.2 Market access and market improvement for all participants

The demise of controlled agricultural marketing in South Africa since 1996 did, in a sense, create a vacuum in the marketing of all the products that were subject to either one-channel marketing or strict access control. Farmers, farmers' co-operatives and traders operating close to the farm gate did not have real commercial marketing functions to fulfil; they assembled and stored commodities and awaited instructions from control boards to dispatch the product to a given destination. Neither did those in the next step(s) in the supply chain – such as millers and bakers – have to do procurement management; they simply informed the control board of their needs, and the required products were delivered in due course, or almost so. It was an inefficient, monopolistic and costly system, but convenient for many. Under this system, important marketing and procurement skills were hardly used and became underdeveloped. The scrapping of the control board system suddenly made these skills vital. Although the agricultural marketing industry has been remarkably successful in adapting to the change in a short time, certain serious bottlenecks, such as a lack of adequate market information, remain.

If it is accepted that an efficient food and fibre marketing system will be characterised by the profitable supply of services that will efficiently satisfy consumers by retailing high quality products at reasonable prices, while remunerating farmers sufficiently for their inputs and effort, then much still has to be done. This was illustrated recently when the marked depreciation of the South African Rand against other currencies described in section 15.2.2 led to sharp increases in local field crop and livestock product prices, as both import and export parity increased sharply. Prices of food immediately rose to compensate the food trade for higher input costs. However, when the local currency strengthened in the second half of 2002 and early 2003 with concomitant lower local prices for commodities, retail prices remained more sticky than could be ascribed to time lags (purchase of expensive raw materials) or fixed costs. The problem probably lies in competitive structure, as many South African food retail markets, and indeed supply chains, have oligopolistic structures. More efficient monopoly control is needed in South African food marketing.

The dualistic nature of South African agricultural production was outlined in section 15.2.1. Food consumption too is dualistic, due to the very skew distribution of income. The real challenge to agricultural marketing is to serve established commercial farmers, emerging farmers and potential emerging farmers with access to markets in ways that will ensure they all prosper. It is simultaneously an important part of the challenge to serve different consumer groups better. Those with few resources must be entitled to obtain sufficient nutritious food at affordable prices; this will necessitate food programmes such as food stamp programmes and school feeding schemes for the poorest of the poor, or otherwise possible basic income grants. At the same time, the marketing system must be able to deliver products demanded at prices that consumers are willing to pay. The nature of products demanded by the different groups differs. This all probably implies a need for greater diversification and competition in South African food and fibre supply chains; market segmentation remains a hallmark of good marketing. For example, Darroch (2002a) identifies marketing opportunities in fruit export niche markets, and food distribution in local informal settlements as urbanisation increases.

The erstwhile 'black homelands' of South Africa today have severely underdeveloped agricultural marketing systems. Small farmers in those areas have limited market access for many reasons, such as lack of storage, transport, social and telecommunication infrastructure, poor information systems, etc. This problem needs urgent attention.

Deregulation of agricultural marketing, globalisation and trade agreements have opened new opportunities for South African farmers and agro industries, but have also exposed South African farmers and agribusiness to foreign competition (some of which are subsidised) on previously protected domestic markets. New, emerging farmers are thus put under more pressure, as are established commercial producers. The South African government has a responsibility to intervene in cases of unfair, subsidised foreign competition. It is fair to expect local producers to compete with their foreign counterparts, but not if foreign governments are added to the equation. Lack of successful action – for example, with regard to improved harbour inspection in collaboration with private enterprise – will inevitably aggravate rural poverty and retard efforts to successfully create more equitability.

15.3.3 Regional progress and international co-operation

The international trade scene, notwithstanding progress made at the World Trade Organisation (WTO), has not reverted to international unfettered, *laissez faire* free trade. Since approximately the mid-nineteenth century, world trade was *competitive*, aimed at finding customers and serving them. International protectionism and the success, particularly of Japan, in using new and powerful management and technology to achieve high productivity and virtually take over markets, changed trade into *adversarial trade* (Drucker, 1990). Adversarial trade aims at taking over markets either by destroying competitors, keeping them out of a market(s),or gaining so much market power that it becomes almost impossible to challenge the market leader.

Under such circumstances, as Drucker points out, trade is not necessarily completely beneficial. The answer lies in neither protectionism nor *laissez faire*. It rather lies in *countervailing power*, obtained by forming economic regions or blocs, which can form effective trade policies that transcend both protectionism and free trade (Drucker, 1990). Free trade is encouraged within blocs, but reciprocity can also be practised within blocs. The EU and the North American Free Trade Agreement (NAFTA) are certainly the most powerful blocs. Reciprocity occurs when two blocs conclude agreements allowing each other's businesses to enjoy the same, or more entrance than outsiders, in their markets. A regional or trading bloc has to be beneficial to all its members – it must be a club worth belonging to (Sunter, 1992).

South Africa, besides being a beneficiary of the US government's African Growth and Opportunity Act (AGOA) and having concluded a free trade agreement (with some limitations) with the EU, is a member of the Southern African Customs Union (SACU), the Southern African Development Community (SADC), and a founder member of the New Partnership for Africa's Development (NEPAD). Whether or not NEPAD and the SADC are really clubs worth belonging to needs urgent and serious consideration. If these two are not presently really worth belonging to, how can they be improved or transformed? Or, what can be put in their place or used instead?

South Africa's Department of Agriculture did not participate in the Uruguay Round of the General Agreement on Tariffs and Trade (GATT) negotiations in 1994 and the expertise in modelling expected economic outcomes of trade arrangements was still not available in South Africa when negotiations took place with the EU. The models are now available and operative (for example, at the University of the Free State), and should be used when future negotiations occur with countries like the US. Such models should also be employed in efforts to improve the economic performance of the SACU, the SADC and NEPAD policies. The SADC and NEPAD should also have panels of experts to employ predictive models during negotiations with other entities. There are no other sectors for which such trade relationships are more important than for the agricultural sectors of Africa. The development of agriculture will shape the welfare and destiny of Africa for the next few generations.

South Africa also has an important duty to assist in agricultural growth elsewhere in sub-Saharan Africa. Improvements in the current South African agricultural research establishment are a critical necessity, as is intra-continental communication. Poor, famine-stricken people flock to where they hope to find jobs and enough food.

If South Africa does not contribute to food security in Africa, and does not really become part of Africa, she will be swamped by Africa.

15.3.4 Water marketing in South Africa

Because water is a scarce resource in South Africa, it must be used as efficiently as possible. South Africa's Water Act, 36 of 1998, gives priority to basic human needs and ecological sustainability above that of agriculture and other industries. The new Act respects water rights and farmers may apply for water licences that have a maximum span of 40 years and are subject to review at intervals not exceeding five years. This review period is relatively short and will create uncertainty regarding long-term investments in, for example, fruit farming.

Section 25(1) of the Act makes provision for the temporary transfer of water entitlement between users. Recent research (Conradie, 2002; Louw, 2001; Hosking *et al.*, 2002) undertaken by the South African Water Research Commission (WRC) and the Department of Water Affairs and Forestry (DWAF) indicates that the value of water differs significantly between sectors, between geographic areas and within geographic areas for urban, commercial forestry, environmental and agricultural uses. As agriculture is a large consumer of water, several studies along different rivers were undertaken, including studies on water quality. Irrigation farming is an important employer of labour, while it contributes 30% to the value of farm output in South Africa. A marginal approach and estimated water demand elasticities (municipal demand for water is price inelastic) indicate that non-agricultural uses generally place a high value on water assurance but *little* value on more than what they already use. This may indicate that although water is expected to transfer out of agriculture in the longer run, in the short run agriculture may be its best use.

Water efficiency could be significantly enhanced if water transfers within and between river reaches are promoted, as water shadow prices differ. Inputs from DWAF are important in water allocation due to the socio-economic and environmental externalities of water allocation. A partnership between the government and private initiatives is recommended. The transfer of water between major rivers such as the Orange, Vaal and Sundays/Fish could promote water-use efficiency, but external impacts on local communities and job creation will be more important. It appears, for instance, that surplus water may be available in the Sundays/Fish River. It also seems that DWAF has stabilised water supplies in the Orange River, for example, which has reduced the uncertainty of water availability.

In summary, if water is be allocated for the social benefit of all people in South Africa, then a partnership between the government and markets is required.

15.3.5 Other important challenges

Some of the many other factors that impact on agriculture and/or are impacted upon by agriculture in South Africa are described below:

(a) Health: Southern Africa is the region with the highest incidence of AIDS in the world, while some other diseases such as malaria are endemic. The effects on agricultural productivity and food security are obvious. Food security influences both productivity and health; infant mortality and a host of pathological conditions such as gastroenteritis, cholera, typhoid, dysentery, measles, tuberculosis, kwashiorkor, scurvy and marasmus are associated with poverty and undernourishment. Improved health services by both the public and private sector are essential for future development. South Africa's record to date concerning AIDS treatment is far from encouraging.

(b) Crime: Rural areas in South Africa have experienced high levels of crime, including farm attacks, stock thefts and robberies. High and improved farm productivity and successful land reform will be difficult to achieve in the absence of improved law enforcement in rural areas. For example, Darroch (2002b) reports that game (wildlife)-based tourism operators in northern Maputaland, KwaZulu-Natal, perceived crime as the major source of risk to their operations in 2000. It is in the interests of local and provincial governments, the national government, farmers' associations and trade unions to co-operate to improve this situation.

(c) Science and education: The challenges of transformation from the old apartheid system to a democratic dispensation in the South African economy, and particularly in agriculture, put a high premium on agricultural research and education at all levels. Considerable improvements in funding and focus are needed in the country's research establishment; and progress has not been satisfactory. Agricultural education – at primary and high school level, at colleges and universities – other adult education and extension should stress the skills needed for a modern agriculture. These must include accountancy, financial management, marketing, and strategy crafting and implementation, and, of course, the needed technical/biological aspects of modern, sustainable agriculture.

(d) Sustainability and resource conservation: The state of much of the resource base in South Africa and its neigbouring countries gives much reason for concern. Injudicious agricultural practices have caused deterioration of land quality; injudicious forestry practices and the spread of invader vegetation have become a threat to water resources (Backeberg *et al.*, 1996); and other ill-considered actions like some rail and road construction, some urban development and pollution are putting further pressure on resources. It is vital for future welfare that there should be a move towards sustainable, conservation practices. This is equally important for farmers, authorities and the public at large. It requires the dissemination of available knowledge and more research.

It is also important to preserve, as far as possible, land at different locations and of different qualities for its best use. This may at times cause a clash between private and public interests. It has been shown that deviation from the principle causes reduction in welfare (Groenewald, 1973).

15.4 CONCLUSION

Virtually all governments intervene in their agricultural sectors through a wide array of policies (Antle, 1999). Interest group theory and public choice theory (Pasour, 1981) suggest that the present commercial farming sector in South Africa will have weak lobbying power. The reason for this is that the political vote of commercial farmers is not translated into political capital as occurs in the EU, US and Japan, for example. It is, however, encouraging to see that AgriSA and the National African Farmers' Union (NAFU) have joined the government in drawing up a common agricultural perspective (Department of Agriculture, 2001) based upon improved access and participation, competitiveness and profitability and sustainable resource management. A successful land redistribution programme will ensure that the sector receives the government's support.

A strong commercial farming sector is vital to South Africa as the availability of food at affordable prices is essential in a country where many people live in an environment of food insecurity. A strong and healthy farming sector depends on sufficient investment in Research & Development, which is important not only because South Africa exports agricultural commodities like high quality deciduous and citrus fruit, but also because consumers benefit from lower food prices.

In conclusion, the agricultural sector has adapted well to the freer marketing conditions. The slow pace of land redistribution is, however, a major concern and innovative solutions are required. Labour numbers have declined in the commercial agricultural sector in recent years during a time when unemployment in the rest of the economy is high. More research needs to be undertaken on this important input (human capital) in agriculture.

REFERENCES

ANTLE, J.M. (1999). The new economics of agriculture. *American Journal of Agricultural Economics* 81(5): 993–1010.

BACKEBERG, G.R., BEMBRIDGE, T.J., BENNIE, A.T.P., GROENEWALD, J.A., HAMMES, P.S., PULLEN, R.A. & THOMPSON, H. (1996). Policy proposal for irrigated agriculture in South Africa. WRC Report No. KV96/96, Water Research Commission, Pretoria.

BARRY, P.J., ELLINGER, P.N., HOPKIN, J.A. & BAKER, C.B. (1995). *Financial Management in Agriculture*. Danville, Illinois: Interstate Publishers Inc.

BECKER, G.S. (1965). A theory of the allocation of time. *Economic Journal* 75: 493–517.

BINSWANGER H.P & TOWNSEND, R.F. (2000). The growth performance of agriculture in Subsaharan Africa. *American Journal of Agricultural Economics* 82(5): 1075–1086.

CHHIBBER, A. (1982). *Dynamics of Price and Non-price Response of Supply in Agriculture.* Stanford, CA: Stanford University Press.

CONNINGARTH ECONOMISTS. (2001). Regional comparative advantage of water use: the Orange River case study. Study commissioned by USAID Regional Centre for Southern Africa, Pretoria.

CONRADIE, B. (2002). The value of water in the Fish-Sundays Scheme of the Eastern Cape. WRC Report No. 987/1/02, Water Research Commission, Pretoria.

DARROCH, M.A.G. (2002a). Challenges facing agribusiness supply chains in South Africa beyond 2002. *Agrekon* 40(4): 505–521.

DARROCH, M.A.G. (2002b). Risk management and Internet use by game-based tourism operators in KwaZulu-Natal. *Agrekon* 40(4): 563–572.

DARROCH, M.A.G. (2003). Economic impact of a property tax on farmland values. Paper presented at a seminar hosted by the KwaZulu-Natal Branch of the South African Institute of Valuers on 'Placing the Property Rates Bill under the Spotlight', 7 February, Durban.

DEPARTMENT OF AGRICULTURE. (2001). *The Strategic Plan for South African Agriculture.* Pretoria: Directorate Agricultural Information Services, Department of Agriculture.

DIRECTORATE: AGRICULTURAL STATISTICS. (2002). *Abstract of Agricultural Statistics.* Pretoria: Directorate Agricultural Information Services, Department of Agriculture.

DISMUKES, R. (1999). Recent developments in crop yield and revenue insurance. Agricultural outlook. http://jan.mannlib.cornell.edu (Accessed 06/2000).

DRUCKER, P.F. (1990). *The New Realities.* London: Mandarin Paperbacks.

FAIRLAMB, C.D. & NIEUWOUDT, W.L. (1991). Economic factors affecting human fertility in the developing areas of Southern Africa. *Agricultural Economics* 6: 185–200.

FOGEL, R.W. (1998). Have the extent and the impact of chronic malnutrition been underestimated? In Mundlak, Y. (ed.). *Contemporary Economic Issues: Volume 2.* New York: MacMillan Press.

FRIEDMAN, M. (1962). *Price Theory.* Chicago: Aldine.

GARDNER, B.L. (1972). Minimum wages and the farm labor market. *American Journal of Agricultural Economics* 52: 473–476.

GARDNER, B.L. (2000). Economic growth and low income in agriculture. *American Journal of Agricultural Economics* 82(5): 1059–1074.

GROENEWALD, J.A. (1973). The allocation of land among uses: a theoretical view. *Agrekon* 12(3): 8–16.

HOSKING, S.G, DU PREEZ, M., CAMPBELL, E.E, WOOLDRIDGE, T.H & DU PLESSIS, L.L. (2002). Evaluating the environmental use of water: selected case studies in the eastern and southern Cape. WRC Report No. 1045/1/02, Water Research Commission, Pretoria.

ILORAH, R. (2000). A classification of different phases of development in the export sector of Nigerian agriculture. *South African Journal of Economic and Management Sciences* NS3: 234–272.

LATT, E.A. & NIEUWOUDT. W.L. (1985). A supply and demand analysis of regular black labour in Natal. *Agrekon* 24(2): 1–7.

LIANOS, T.P. (1972). Impact of minimum wages upon the level and composition of agricultural employment. *American Journal of Agricultural Economics* 52: 477–484.

LIN, J.Y. (1994) Chinese agriculture: institutional changes and performance. In Srinivasan,T.N. (ed.). *Agriculture and trade in China and India.* San Francisco: ICS Press.

LIPTON, M. (1993). Urban bias: of consequences, classes and causality. *Journal of Development Studies* 29: 229–258.

LOUW, DB. (2001). The development of a methodology to determine the true value of water and the impact of a potential water market on the efficient utilisation of water in the Berg River Basin. WRC Report No. 943/1/02, Water Research Commission, Pretoria.

LYNE M.C. (2001). Combining public and private resources to accelerate market-based land redistribution in South Africa. Unpublished paper, University of Natal, Pietermaritzburg.

MACHETHE, C.L. (1990). Factors contributing to poor performance of agricultural co-operatives in less developed areas. *Agrekon* 29(4): 305–309.

MASHATOLA, M.C. & DARROCH, M.A.G. (2003). Factors affecting the loan status of sugarcane farmers using a graduated mortgage loan repayment scheme in KwaZulu-Natal. *Agrekon* 42(4) forthcoming.

McGUIGAN S.M. & NIEUWOUDT, W.L. (2002). The expected consumption of protein feed in South Africa. *Agrekon* 41(1): 1–23.

MOSHER, A.T. (1971). *To Create a Modern Agriculture.* New York: Agricultural Development Council.

NIEUWOUDT, W.L. (1970). An econometric analysis of the demand for resources and the supply of output in South African agriculture. PhD thesis, University of Natal, Pietermaritzburg.

NIEUWOUDT, W.L. (1980). Value and rent of farm land. *South African Journal of Economics* 48(4): 389–397.

NIEUWOUDT, W.L. & HOWELL, J. (2000). Farming without drought relief: time to revisit an income equalisation deposit scheme. *Agrekon* 39(3): 332–356.

NIEUWOUDT, W.L., SCOTT, T.G., SIMKINS, C.E.W. & VINK, N. (1993). Financing land redistribution in South Africa: putting theory into practice. Proceedings of the Land Redistribution Options Conference, 12–15 October, Land and Agriculture Policy Centre, Johannesburg: 61–69.

PETERSON, W. (1979). International farm prices and the social cost of cheap food policies. *American Journal of Agricultural Economics* 61: 12–21.

PETERSON, W. (1987). International land quality indexes. Staff Paper P87–10, Department of Agricultural and Applied Economics, University of Minnesota.

PETERSON, W. (1988). International supply response. *Agricultural Economics* 2: 365–374.

PASOUR, E.C., Jnr. (1981). *Agriculture and the State.* New York: The Independent Institute.

SCHUH, G.E. (2002). Can lagging countries catch up? Paper presented at the 40th Annual Conference of the Agricultural Economics Society of South Africa, 18–20 September, Bloemfontein.

SEN, A. (1981). *Poverty and Famines.* Oxford: Clarendon Press.

SHINNS, L.H. & LYNE, M.C. (2003). Symptoms of poverty within a group of land reform beneficiaries in the Midlands of KwaZulu-Natal. Submitted for publication to *Agrekon.*

SRINIVASAN, T.R. (1994). Indian agriculture: policies and performance. In Srinivasan, T.R. (ed.). *Agriculture and Trade in China and India.* San Francisco: ICS Press.

SUNTER, C. (1992). *The New Century.* Cape Town: Human & Rousseau and Tafelberg.

SWANEPOEL, D.S., ORTMANN, G.F. & DARROCH, M.A.G. (1998). Causes of bankruptcy amongst maize and extensive beef farmers in South Africa: 1970–1994. *South African Journal of Economic and Management Sciences* NS1: 52–72.

VAN DEN HEEVER, R. (2002). Personal communication. Portfolio Manager, Ithala Bank, Durban.

VAN DER WATT, J.G. & GROENEWALD, J.A. (1996). Die langtermyn impak van die 1992-staatshulpskema vir die boerderysektor. *South African Journal of Agricultural Extension* 25: 12–17.

VAN SCHALKWYK, H.D. (1995). Modelling South African agricultural land prices. PhD thesis, University of Pretoria.

VAN SCHALKWYK, H.D. & GROENEWALD, J.A., (1993). Agricultural land price and quality. *Development Southern Africa* 10: 401–410.

WILSON, F. & RAMPHELE, M. (1989). *Uprooting Poverty: The Challenge.* Cape Town: David Philip Publishers.

WORLD BANK. (2002). South Africa data profile. World Bank Development Indicators. http://www.worldbank.org/data/dataquery.html (Accessed 15/01/03).

Index